W9-BOD-299

Bacterial Physiology and Metabolism

TO CAROL

Bacterial Physiology and Metabolism

J. R. SOKATCH

Medical Center, University of Oklahoma
Oklahoma City, U.S.A.

1969

ACADEMIC PRESS · LONDON AND NEW YORK

ACADEMIC PRESS INC. (LONDON) LTD.
24–28 Oval Road
London, NW1

U.S. Edition published by
ACADEMIC PRESS INC.
111 Fifth Avenue
New York, New York 10003

Third Printing 1970

Fourth Printing 1973

Fifth Printing 1974

Library of Congress Catalog Card Number: 68-24692

SBN: 12-654250-3

PRINTED OFFSET PHOTO-LITHOGRAPHY IN GREAT BRITAIN BY
THE WHITEFRIARS PRESS LIMITED, LONDON AND TONBRIDGE

PREFACE

This book was written with two purposes in mind: it was intended for use as a textbook for a graduate course in Microbial Biochemistry and as a reference source for workers in fields allied to Microbiology. The emphasis is placed on research done with bacteria with very little space devoted to principles of biochemistry which the student has covered in earlier courses. By eliminating material already covered in books probably owned by the user, the size and cost of this text becomes more reasonable.

A few words about the philosophy used in writing this book are in order. The first section, *Bacterial Physiology*, deals with non-metabolic aspects of Microbial Biochemistry (nutrition, growth and chemistry) which are important to understanding the features that set bacteria apart from other living organisms. The second section, *Energy Metabolism*, deals with ways bacteria attack foods that they find in nature, reduce them to a manageable size, ingest and metabolize them to make chemical energy. The third section, *Biosynthetic Metabolism*, deals with the formation of bacterial protoplasm beginning with simple materials such as inorganic carbon and nitrogen and leading to the formation of the complex molecules of the bacterial cell. Abbreviations used in the text are those recommended in the *Instructions to Authors* of the *Journal of Biological Chemistry*. Each enzyme is listed in the text and index by a common name followed by the name given to it by the Enzyme Commission of the International Union of Biochemistry. (The common name used is not necessarily the same one recommended by the Enzyme Commission.)

Many thanks are due to the large number of authors and publishers who graciously permitted portions of their work to be reproduced here. I feel that these figures add a flavor that would be difficult to obtain otherwise and the figures have been reproduced as they were originally published even though there may be some differences in terminology from the rules outlined in the previous paragraph. Thanks are also due to my Chairman who not only permitted but encouraged me to work on this book and provided the sort of atmosphere that allowed the writing to be done. I would also like to thank the publishers, both in London where the book was published and in New York for their expert technological assistance. Finally, but most important, I would like to thank my wife who helped with typing, proof reading and indexing and provided the expert and dedicated assistance that she did when it was needed most.

November 1968 JOHN R. SOKATCH

CONTENTS

Part Two

ENERGY METABOLISM

Part Three

BIOSYNTHETIC METABOLISM

Part One

BACTERIAL PHYSIOLOGY

I. NUTRITION

I. Requirements for Growth

Bacteria as well as all other living organisms require certain essential nutrients in the medium or diet in order to be able to grow. Growth is understood to mean balanced growth, that is, a uniform increase in protoplasm, as opposed to an increase in one or a few components. Essential nutrients fall into two classes, those required to supply energy for growth and those required to supply the chemical elements needed for biosynthesis. Of the various forms of energy available, bacteria can use chemical and light energy for growth. Most true bacteria use organic compounds for chemical energy, but many soil bacteria are able to produce useable energy by the oxidation of inorganic chemicals. Quantitatively, the most important elements required for biosynthesis are those found in protein, namely C, H, O, N and S. These elements may suffice in their inorganic forms or may be required in the form of organic growth factors. Many other elements are required for growth, such as Mg, K, PO_4, Fe, Cu, Co, Mn and Zn, and these are also used as inorganic salts.

II. Nutritional Classification of Bacteria

In contrast to the relatively uniform nutritional requirements of plants and animals, bacteria exhibit characteristic differences in their requirements for energy and carbon sources. A study of the distribution of nutritional types in nature suggests that this might be response to environment; for example, all autotrophic bacteria are soil and water species. A system of classification of nutritional types of organisms which was proposed by a group of prominent microbiologists (Lwoff *et al.*, 1946) forms the basis of the following discussion. Organisms were classified on the basis of energy requirement and on the basis of carbon source for biosyntheses.

Developments in microbiology since 1946 have resulted in changes in the original system of classification. For example, the terms photolithotroph and photoorganotroph are not used here as the organisms formerly grouped in these categories differ in their method of assimilation of carbon (a biosynthetic process) during photosynthesis, but not in the process of converting light energy into chemical energy. If any distinction is to be made among the phototrophs, it should be between those that cleave water during photosynthesis (green plants) and those that do not (bacteria).

Similarly, the requirement for growth factors is no longer considered distinctive enough to merit a separate category, since such a requirement could occur as the result of a single genetic change. This means that the original classification on the basis of minimum nutritional requirements becomes a classification on the basis of carbon source. The organisms formerly classified as mesotrophs are now grouped together with the heterotrophs under this heading. (See review by Guirard and Snell, 1962, for nutritional requirements of bacteria.)

A. Classification on the Basis of Energy Source

1. *Utilization of light energy*

Organisms which use light energy are phototrophs. This type of energy metabolism occurs only among green plants and certain pigmented bacteria. Both plants and bacteria convert light energy into chemical energy in the form of ATP (Arnon *et al.*, 1954; Frenkel, 1954). Most probably this occurs by a process similar to oxidative phosphorylation. Excited electrons, generated during the photochemical reaction pass along the cytochrome chain with the concomitant formation of ATP (Stanier, 1961; Fig. 13.4).

Bacterial phototrophs are all soil and water species, most of which are classified in the suborder *Rhodobacteriineae*, with the exception of the genus *Rhodomicrobium*, in the order *Hyphomicrobiales*. The taxonomic distinctions among the *Rhodobacteriineae* are based on the color of the photosynthetic pigments and the preferred type of carbon source. There are three families of *Rhodobacteriineae*, the sulfur purple bacteria, family *Thiorhodaceae*, the non-sulfur purple bacteria, family *Athiorhodaceae*, and the sulfur green bacteria, family *Chlorobacteriaceae*. Sulfur purple bacteria grow well with carbon dioxide as the sole carbon source, in which case they use inorganic sulfur compounds such as hydrogen sulfide and thiosulfate in order to reduce carbon dioxide. They can also assimilate organic carbon, however, in which case no separate reducing agent is required. Non-sulfur purple bacteria grow best with organic carbon, although some species can grow with carbon dioxide as the sole carbon source, using either molecular hydrogen or thiosulfate as the reducing agent. *Rhodomicrobium* is similar to the *Athiorhodaceae* because it requires an organic carbon source and has not yet been grown with carbon dioxide as the sole carbon source. It is separated from the other photosynthetic bacteria on morphological grounds, being a stalked bacterium. Some of the *Athiorhodaceae* are also able to grow aerobically in the dark on organic energy sources. This is true also of some species of algae but not of higher plants. The sulfur green bacteria reduce carbon dioxide with inorganic sulfur compounds. Unlike the other photosynthetic bacteria, *Chlorobacteriaceae* are not able to grow with organic carbon as the sole carbon source, although they can assimilate organic carbon to a certain extent (Sadler and Stanier, 1960).

2. *Utilization of chemical energy*

Organisms which use chemical energy for growth are chemotrophs. Chemotrophs are divided into chemolithotrophs, those which use inorganic energy sources, and chemoorganotrophs, those which use organic energy sources.

Bacteria which use chemolithotrophic energy metabolism are soil and water species in the order *Pseudomonadales*, suborder *Pseudomonadineae*. Some of the non-sulfur purple bacteria are also able to grow in the dark on inorganic energy sources such as hydrogen gas and thiosulfate (van Niel, 1944). This ability does not occur widely outside of bacteria, but strains of *Scenedesmus* and other blue-green algae can be obtained which will grow in the dark with hydrogen as the energy source (Gaffron, 1940).

Chemolithotrophic bacteria whose physiology has been studied with pure cultures can be classified into four groups on the basis of energy source used. It is possible that other physiological groups might be discovered since there are many species of bacteria in the *Pseudomonadineae* which have not been obtained in pure culture. Organisms which oxidize nitrogen compounds are classified in the family *Nitrobacteriaceae*, and these constitute the first physiological group. *Nitrosomonas*, which oxidizes ammonia to nitrite, and *Nitrobacter*, which oxidizes nitrite to nitrate, are the best studied examples of this family. These organisms are strict chemolithotrophs, since they will not grow with organic carbon energy sources. *Nitrosomonas* and *Nitrobacter* are important in maintaining soil fertility because they effect an oxidation of ammonia to nitrate, the preferred nitrogen source for green plants.

Hydrogenomonas (family *Methanomonadaceae*) represents the second group of chemolithotrophs, the hydrogen oxidizers. The ability to oxidize hydrogen occurs frequently among bacteria and occasionally among lower plants such as blue-green algae. *Hydrogenomonas* is also able to grow well with organic energy sources in contrast to *Nitrosomonas* and *Nitrobacter*.

Sulfur oxidizers comprise the third group of chemolithotrophs, most of which are classified in the family *Thiobacteriaceae*. They are able to grow on inorganic sulfur compounds such as hydrogen sulfide, elemental sulfur, thiosulfate and thiocyanate, and produce tetrathionate and sulfate as end products of the oxidation of these compounds. The species of *Thiobacillus* are the best known sulfur oxidizers. At least one species of *Thiobacillus* can grow on organic media (Santer *et al.*, 1959), but this appears to be an exceptional case.

The fourth group, the iron oxidizers, is represented by *Ferrobacillus*. *Ferrobacillus* (family *Siderocapsaceae*) obtains its energy by the oxidation of ferrous iron. This organism grows best in acid media, 3·5 being the optimum pH for growth (Leathen *et al.*, 1956).

Chemoorganotrophs satisfy their energy requirement by the oxidation or fermentation (anaerobic metabolism) of organic compounds. Chemoorganotrophy is the most common type of energy metabolism among bacteria and almost the only kind found in the animal kingdom. Chemoorganotrophy occurs in the plant kingdom among the non-photosynthetic groups such as yeast and fungi. Some species of algae are facultative chemoorganotrophs, being able to grow on organic carbon sources in the dark (Danforth, 1962).

Taxonomically, chemoorganotrophs are almost all those bacteria that have not been mentioned to this point. Rickettsia are able to oxidize a limited number of organic substrates (Moulder, 1962) and should possibly be considered as chemoorganotrophs, although they have not been grown on lifeless media as yet.

The compounds which chemoorganotrophs use for energy range from simple materials such as formate and oxalate to complex hydrocarbons such as camphor. It seems possible that any organic compound which can be oxidized to produce energy is subject to attack by chemoorganotrophs.

3. *Energy supplied by metabolism of the host cell*

Organisms which obtain energy for biosynthetic reactions from the metabolism of host cells are paratrophs. This category includes bacterial, plant and animal viruses. Although rickettsia are known to have some oxidative ability, it is possible that they may obtain part of their growth energy as paratrophs. Many of the reactions involved in supplying energy to paratrophs are those of the host cell and, from that point of view, will be covered in this book. The subject of viral replication as a separate topic will not, however, be treated here.

B. Classification on the Basis of Carbon Source for Growth

Organisms are divided into three classes on the basis of their carbon requirements: (a) those that are able to use inorganic carbon, (b) those that require organic carbon, and (c) those that depend on the host cell to supply their carbon. The ability to use inorganic sources of N, S and P is very common among bacteria and therefore not a distinctive characteristic. When organic N or S are required it is usually as an amino acid or vitamin, i.e. the result of a deficiency in a biosynthetic pathway.

1. *Utilization of carbon dioxide as sole carbon source*

Organisms which are able to grow with carbon dioxide as the only source of carbon are autotrophs. Taxonomically these organisms are identical with the phototrophs and chemolithotrophs, with the possible exception of the species of *Athiorhodaceae* which have not yet been grown in completely mineral media. Bacteria which require organic carbon for growth are also able to fix carbon dioxide to some extent, but fixation occurs primarily into the dicarboxylic acids related to the Krebs cycle.

Assimilation of carbon dioxide by autotrophs requires energy, which is supplied by light in the case of phototrophs and chemicals in the case of chemolithotrophs. Assimilation of carbon dioxide is also a reductive process and therefore a reducing agent is required as well. Green plants use water as the reducing agent and water is in turn oxidized to molecular oxygen.

$$H_2O + CO_2 \xrightarrow{\text{light}} (CH_2O) + O_2$$

Balance studies with photosynthesizing plants have shown that this equation is a reasonable approximation of carbon fixation (Rabinowitch, 1945), and fixed carbon is most commonly recovered as carbohydrate storage products.

Bacteria which are photosynthetic autotrophs use reducing agents other than water to fix carbon dioxide.

$$2H_2O + H_2S + 2CO_2 \xrightarrow{\text{light}} 2(CH_2O) + H_2SO_4$$

$$2H_2 + CO_2 \xrightarrow{\text{light}} (CH_2O) + H_2O$$

Chemolithotrophic autotrophs use the same inorganic compound as an energy source and as a reducing agent; they simply divert some of the electrons obtained by the oxidation of their energy source towards the reduction of carbon dioxide rather than oxygen.

2. *Utilization of organic carbon*

Organisms which must have an organic source of carbon for growth are heterotrophs. This is the most frequently encountered situation in bacteria and almost the only kind of nutrition in the animal kingdom. Taxonomically, heterotrophs are almost identical with chemoorganotrophs. Photosynthetic bacteria which assimilate organic carbon are also considered heterotrophs. Recent evidence supports the view that rickettsia can synthesize at least a part of their own cell substance (Moulder, 1962).

Carbon sources for heterotrophs vary as much as do energy sources for chemoorganotrophs. In fact, with aerobic species such as *Pseudomonas*, the same compound may serve as both energy and carbon source. On the other hand, the fermentative bacteria such as the *Lactobacillaceae* generally require the major proportion of their carbon in the form of amino acids and vitamins. The energy source is generally a fermentable carbohydrate and is converted almost quantitatively to end products, as little as 1–2 % of the energy source being assimilated into the cell (Bauchop and Elsden, 1960).

3. *Carbon supplied by biosynthetic reactions of the host cell*

Organisms which rely on the enzymic apparatus of the host cell to duplicate themselves are hypotrophs. This category includes only viruses, although the possibility that rickettsia depend to some extent on the host cell for biosynthesis cannot be excluded. Again, many of the biosynthetic processes are those of the host cell, although certain enzymes are formed during virus biosynthesis. The unique property of hypotrophs is their ability to seize control of the genetic centers of the host cell and thereby force the host to produce virus.

References

Arnon, D. I., Allen, M. B. and Whatley, F. R. (1954). *Nature* **174,** 394.

Bauchop, T. and Elsden, S. R. (1960). *J. Gen. Microbiol.* **23,** 457.

Danforth, W. F. (1962). *In* "Physiology and Biochemistry of Algae" (R. A. Lewin, ed.). Academic Press, New York, p. 99.

Frenkel, A. W. (1954). *J. Am. Chem. Soc.* **76,** 5568.

Gaffron, H. (1940). *Am. J. Botany* **27,** 273.

Guirard, B. M. and Snell, E. E. (1962). *In* "The Bacteria" (I. C. Gunsalus and R. Y. Stanier, eds.). Academic Press, New York, Vol. IV, p. 33.

Leathen, W. W., Kinsel, N. A. and Braley, S. A. (1956). *J. Bacteriol.* **72,** 700.

Lwoff, A., van Niel, C. B., Ryan, F. J. and Tatum, E. L. (1946). Nomenclature of nutritional types of microorganisms. *Cold Spring Harbor Symp. Quant. Biol.* Appendix **XI,** 302.

Moulder, J. W. (1962). "The Biochemistry of Intracellular Parasitism." The University of Chicago Press, Chicago.
Rabinowitch, E. I. (1945). "Photosynthesis." Interscience, New York, Vol. I.
Sadler, W. R. and Stanier, R. Y. (1960). *Proc. Nat. Acad. Sci. U.S.* **46,** 1328.
Santer, M., Boyer, J. and Santer, U. (1959). *J. Bacteriol.* **78,** 197.
Stanier, R. Y. (1961). *Bacteriol. Rev.* **25,** 1.
Van Niel, C. B. (1944). *Bacteriol. Rev.* **8,** 1.

2. GROWTH OF BACTERIAL CULTURES

I. Measurement of Bacterial Growth

The measurement of bacterial growth presents certain problems because of the microscopic size of the organisms. There are several methods available, however, and usually at least one satisfactory technique can be selected for a given purpose.

(1) The simplest method for measurement of bacterial growth is to follow the increase in turbidity of a culture spectrophotometrically. Turbidity must be equated to some independent measure of bacterial concentration such as cell count or dry weight per unit volume, because turbidity is frequently not directly proportional to cell concentration. In addition, cell size varies with the stage of the cell cycle, the division rate and possibly other factors as yet unknown, and these problems must be considered for each case.

(2) The method of choice for genetic or medical experiments is the direct counting of bacterial cells. This may be done by a viable count, in which case an aliquot of the culture is plated onto nutrient agar, or by estimating the total count by the use of the Petroff–Hausser counting chamber. Another method of obtaining the total count is by using the Coulter counter, an electronic particle counter (Toennies *et al.*, 1961). This apparatus is very rapid, accurate and easy to use. It has the additional advantage that particles can be scored for size as they are counted.

Other methods such as the titration of acids produced during growth (Snell, 1957) or the measurement of the volume of packed cells with a calibrated centrifuge tube have been used. Acid titration can be used only with fermentative organisms, however, and measurement of cell volume is only feasible with cells that sediment easily, because glass tubes must be used.

II. Mathematics of Growth

Most populations of living organisms increase geometrically. In the case of organisms with complex life patterns such as human beings, many factors, such as food supply, disease and economic conditions, affect the rate of increase of the community. In the case of bacteria, however, almost ideal conditions for reproduction can be obtained, resulting in population growth which is close to theoretical expectations.

Bacteria grow and reproduce by asexual binary fission, in which case each generation increases by a factor of two. The mathematical expression for this type of growth, usually referred to as exponential growth, is

$$N_2 = N_1 2^n \quad (1)$$

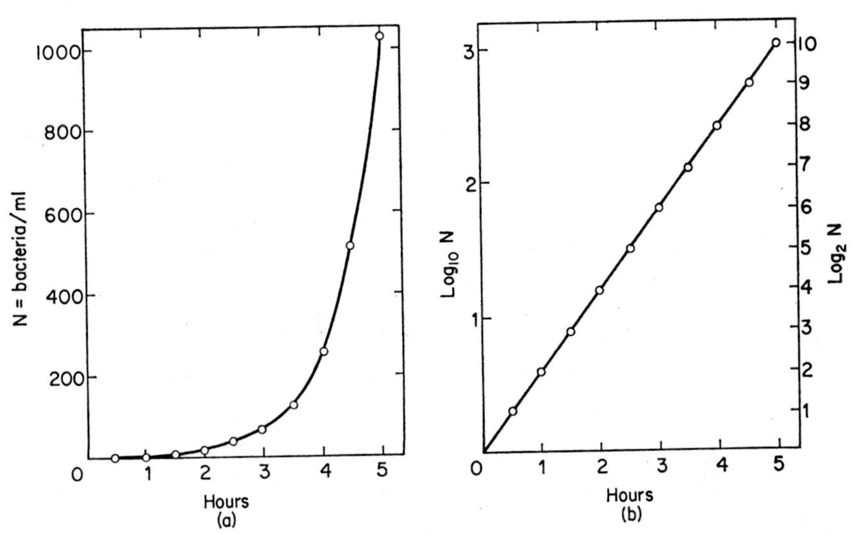

FIG. 2.1. Exponential growth of bacteria. In a the population resulting from the growth of a hypothetical bacterial culture where N_1 is one bacterium/ml and the generation time is 0·5 hours is plotted arithmetically. In b the same data are plotted logarithmically as functions of $\log_{10}$ and $\log_2$.

where N_1 is the cell concentration at time $t=1$, N_2 is the cell concentration at time $t=2$, and n is the number of generations between t_2 and t_1. Cell concentration is expressed as cells per ml or as dry weight per ml.

Equation (1) is rarely used in this form because of the enormous changes in population during exponential growth. For example, after 10 generations the cell count has increased one thousand fold (Fig. 2.1 a). Plotting growth data logarithmically overcomes this difficulty. Taking the logarithm of both sides of (1), we have

$$\log N_2 = \log N_1 + n \log 2 \quad (2)$$

and when log N is plotted as a function of time, a linear plot is obtained (Fig. 2.1 *b*). Solving (2) for n, we obtain

$$n = \frac{\log N_2 - \log N_1}{\log 2} \tag{3}$$

The number of generations per unit time is defined as the exponential growth rate, R.

$$R = \frac{n}{t_2 - t_1} = \frac{\log N_2 - \log N_1}{\log 2(t_2 - t_1)} \tag{4}$$

If logarithms to the base 2 are used as suggested by Monod (1949), equation (4) simplifies to

$$R = \frac{\log_2 N_2 - \log_2 N_1}{t_2 - t_1} \tag{5}$$

and R is simply the slope of a plot of $\log_2 N$ as a function of time (Fig. 2.1 *b*). Tables of $\log_2$ have been published (Finney *et al.*, 1955), or the more commonly available logarithms to the base 10 can be used. Since $\log_{10} 2$ is 0·301, (4) becomes

$$R = \frac{\log N_2 - \log N_1}{0{\cdot}301\ (t_2 - t_1)} \tag{6}$$

The reciprocal of R is the generation time, G

$$G = \frac{1}{R} = \frac{t_2 - t_1}{n} \tag{7}$$

In the event that not all daughter cells are viable, eq. (1) and those following cannot be applied since the population does not double with each generation. For example, if 10 % of the daughter cells are non-viable, then the cell population will increase 1·8 times per generation. This number can be substituted into eq. (2) in the place of 2, or the more general expression derived in the following paragraphs can be used.

Even though not all daughter cells are viable, growth will still be exponential and the cell concentration will increase by a constant fraction per unit time. This is stated in eq. (8), which says that the rate of increase of the population is equal to a constant times the population at any given instant:

$$\frac{dN}{dt} = \alpha N \tag{8}$$

The constant α is the specific growth rate (Herbert *et al.*, 1956) and is the fraction by which the population increases per unit time. Integration of (8) leads to[1]

$$\frac{N_2}{N_1} = e^{\alpha t}$$

Solving for α in the same way that we solved for n in eqs (2) and (3), we obtain

$$\alpha = \frac{\ln N_2 - \ln N_1}{t} = \frac{\ln N_2/N_1}{t} = \frac{2{\cdot}303\ (\log_{10} N_2/N_1)}{t} \quad (9)$$

where t is the elapsed time, actually $t_2 - t_1$. It should be noted that α is different from R, the exponential growth rate. However, α can be related to R in the special case where viability of the daughter cells is 100%. In this case, the population of bacteria doubles with each generation, and hence $\ln N_2/N_1 = \ln 2$ and t is equal to the generation time G.

$$\alpha = \frac{\ln 2}{G} = \ln 2(R) = (0{\cdot}692)\ (R) \quad (10)$$

III. Growth of Bacteria in Batch Culture

An actual growth curve for an aerobic diphtheroid in batch culture is illustrated in Fig. 2.2. The data are plotted arithmetically and logarithmically for comparison with Fig. 2.1.

A. Lag Phase

It is apparent from Fig. 2.2 *b* that exponential growth did not commence immediately after inoculation of the medium (see Fig. 2.1 *b* for comparison). This delay is termed the lag phase of the growth curve

[1] Steps in the integration

$$\frac{dN}{dt} = \alpha N$$

$$\frac{dN}{N} = \alpha d_t$$

By integration, $\ln N_2 = \alpha t + k$

At $t = 1$, $N = N_1$, $\ln N_2 = \alpha t + \ln N_1$

at $t = 2$, $N = N_2$ $\ln \frac{N_2}{N_1} = \alpha t$

$$\frac{N_2}{N_1} = e^{\alpha t}$$

and is probably the result of several factors. The more complete the medium, the sooner exponential growth takes place (Lichstein, 1959). Studies by Hinshelwood (1952) have shown that the lag time can be reduced by using young cells or a large number of cells for the inoculum. Hinshelwood and others have proposed the idea that the inoculum must build up a critical concentration of one or more essential metabolites before exponential growth can occur. Pertinent to this point, Lichstein (1959) has shown that carbon dioxide added to the medium

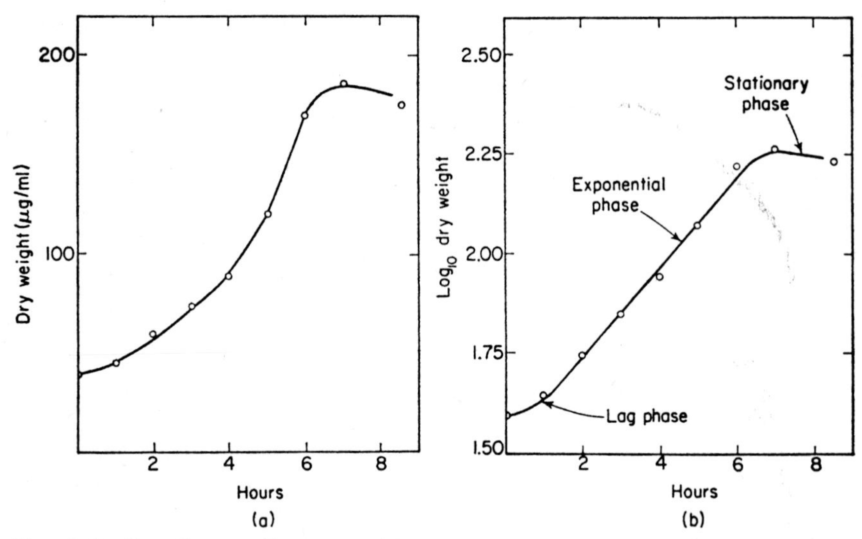

FIG. 2.2. Growth curve for an aerobic organism in batch culture. These data are for a soil diphtheroid growing in a medium with 2×10^{-3} M glucose as the sole source of carbon and energy. (L. R. Runkle and J. R. Sokatch, unpublished data).

of *Propionibacterium* greatly reduces the lag time. Physical factors such as pH, temperature and the reducing potential of the medium also affect lag time (Lichstein, 1959).

B. Exponential Growth Phase

During the next phase of growth, the exponential phase, the bacteria multiplied at a constant rate of growth (Fig. 2.2 *b*). This is the usual situation and continues until some essential nutrient is exhausted, which in this case was the carbon and energy source. In most of the cases examined carefully, virtually all the daughter cells are viable (Kelly and Rahn, 1932; Gunsalus, 1951).

R and α are affected by the composition of the medium. Nutritionally more complete media result in more rapid division rates (Senez, 1962).

Increasing the concentration of individual components of the medium, such as the energy source, results in more rapid growth rates. Monod (1942) has shown that α as a function of the substrate concentration obeys Michaelis–Menten kinetics. Therefore, the specific growth rate is related to the substrate concentration by an adaptation of the usual Michaelis–Menten equation.

$$\alpha = \alpha_m \left(\frac{S}{K_S + S} \right) \tag{11}$$

where α_m is the maximum specific growth rate, S is the substrate concentration and K_S is the substrate concentration at $\alpha = \frac{1}{2}\alpha_m$. The growth rate is also a function of the organism, although little is known of the complex factors which influence the growth rate.

The growth rate is also affected by physical factors such as pH and temperature (Gunsalus, 1951; Senez, 1962).

C. Stationary Phase of Growth

In Fig. 2.2 *b*, the exponential phase was followed by the stationary phase, a phase of no growth. Growth ceases usually for one of two reasons. Either one of the classes of essential nutrients listed on p. 3 has been exhausted, or toxic products, such as fermentation acids, have accumulated (Hinshelwood, 1952). The principle of the microbiological determination of amino acids and vitamins is based on the former case. Bacteria are used which have specific nutritional requirements for these factors, and are grown in a medium which contains an excess of all nutrients except the one to be assayed. Growth is then limited by this factor and is proportional to the amount of the factor in the specimen analyzed (see Snell, 1957, for further details).

Bauchop and Elsden (1960) have defined the yield coefficient Y as the grams of dry weight of the organism produced per mole of substrate consumed, the yield when glucose is the energy source being further designated as Y_{glucose}. Several groups of investigators have determined the yield of cells when the energy source is limiting (Monod, 1949; Bauchop and Elsden, 1960). Studies of this type have produced interesting observations on the amount of cell material which is formed per mole of substrate used during anaerobic utilization of the energy source (fermentation). The total cell crop of *S. faecalis* as a function of the energy source, in this case glucose, is illustrated in Fig. 2.3. The yield coefficient is the slope of this curve or about 20 μg/dry wt per μmole glucose (in Bauchop and Elsden's units, 20 g/mole). Y_{ATP} can be estimated from data of this type if two kinds of information are available, (a) the yield of ATP during glucose dissimilation and (b) the amount of the energy source which is diverted into cell material. In the case of *S. faecalis*, glucose is fermented by the Embden–Meyerhof pathway (Gibbs *et al.*, 1955) and less than 1 % of the energy source is converted into cell material (Bauchop and Elsden, 1960). Since two moles of ATP are produced per mole of glucose fermented by the Embden–Meyerhof pathway, it can be estimated that 10 g dry wt of *S. faecalis* are produced per mole of ATP. The small amount of glucose used for cell synthesis can be neglected in making this calculation. Yield coeffi-

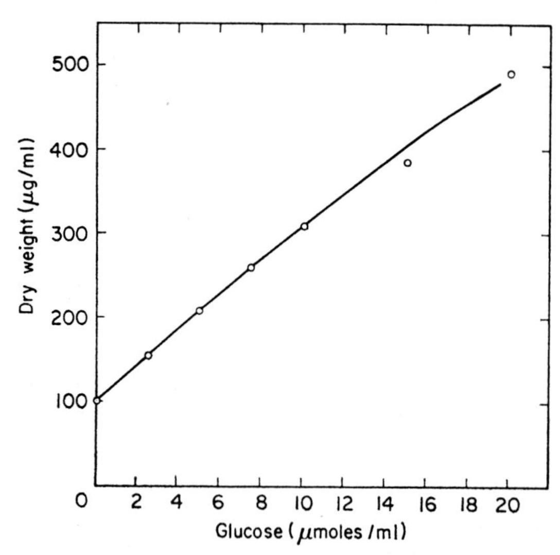

FIG. 2.3. Yield of *Streptococcus faecalis* as a function of the energy source. (Reproduced with permission from Sokatch and Gunsalus, 1957.)

cients of other fermentative organisms with other energy sources have been determined and are all in the range of 9–11 g dry wt/mole of ATP (Bauchop and Elsden, 1960; Senez, 1962).

In the case of organisms growing aerobically, Y for the energy source has a margin of uncertainty in it since about half of the carbon of the energy source is used for cellular carbon. Whitaker (1963) has attempted to overcome this difficulty by measuring the total cell crop as a function of the oxygen consumed during growth, i.e. the Y_{O_2}. It is assumed that all the oxygen consumed is used for energy production via oxidative phosphorylation. The yield of ATP during oxidative phosphorylation by bacteria is not known and therefore it is not possible to calculate a Y_{ATP} from data of this type. On the other hand, assuming that the Y_{ATP} for aerobic organisms is the same as for fermentative organisms, it is possible to estimate the yield of ATP per atom of oxygen consumed from Y_{O_2}. Since Y_{O_2} has been found to be in the range of 15–25 g dry wt/atom of oxygen used, it would appear that about 2–3 moles of ATP are formed per atom of oxygen consumed (Whitaker, 1963).

Calculations of the theoretical amount of cell material which should be possible per mole of ATP used have been made by Gunsalus and Schuster (1961). These calculations are based on known biosynthetic pathways and indicate that about 30 g of cell dry wt should be produced per mole of ATP used. The reason for the difference between this figure and the observed value of about 10 g/mole of ATP is not known. Certain energy requiring processes such as permeation, motility and cell maintenance could account for part of the difference, but it seems unlikely that it would account for such a large portion. The requirement for maintenance has been measured (Marr *et al.*, 1963) for *E. coli* and was found to be a very small portion of the total energy requirement.

IV. Growth of Bacteria in Continuous Culture

A. Continuous Culture Techniques

The development of continuous culture techniques has provided methods to study problems of bacterial physiology and genetics which would be difficult or impossible to investigate with the use of batch cultures. Continuous culture methods have also provided solutions to such practical problems as growing large amounts of microorganisms for industrial processes and biochemical studies.

Two main types of apparatus are in use for continuous culture studies, the externally controlled and the internally controlled systems. In the former case, growth is limited by the concentration of an essential nutrient in the medium (Monod, 1950; Novick and Szilard, 1950). This type of apparatus is illustrated in Fig. 2.4 *a*. Sterile medium is contained in a reservoir and is composed so that one nutrient is present in a limiting concentration. Medium is dripped into the growth tube at a constant rate. As we have already seen, R and α are functions of the concentration of nutrients in the medium and therefore the growth rate will be determined by the concentration of the limiting nutrient and the rate of flow of medium into the growth tube. This system is referred to as externally controlled because the growth rate depends on the composition of the medium, i.e. on a factor external to the cell.

Internally controlled systems utilize regulation of medium input by a photoelectric cell which measures the density of the bacterial population (Fig. 2.4 *b*) (Novick, 1955). In this case, the sterile medium contains an excess of all nutrients. As the turbidity in the growth tube

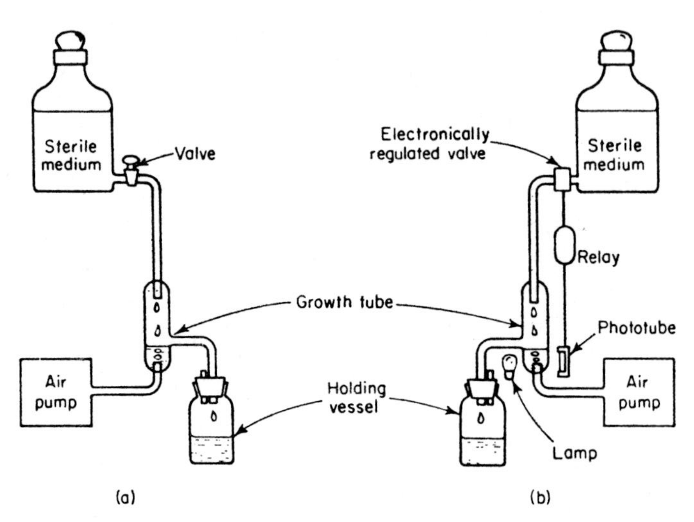

FIG. 2.4. Schematic representation of the two main types of apparatus in use for continuous growth of bacterial cultures. The externally controlled system is illustrated in *a* and the internally controlled system is illustrated in *b*.

changes, the flow of medium into the growth tube is regulated by a valve which is under the control of the photoelectric cell. This type of apparatus can be pre-set to maintain a given bacterial density in the growth tube. This apparatus is referred to as internally controlled because the growth rate is determined by the nature of the organism used, rather than the composition of the medium.

B. MATHEMATICS OF GROWTH IN CONTINUOUS CULTURES

The following discussion applies to both externally controlled and internally controlled systems. The rate of increase of the cell concentration in the growth tube is equal to the specific growth rate α, times the cell population N at any given instant minus the population, P, lost through the overflow port (Novick, 1955).

$$\frac{dN}{dt} = \alpha N - P \tag{12}$$

P is equal to N times the volume of culture fluid leaving per unit time. If the rate of flow of medium is W ml/hour and the volume of culture in the growth tube is V, then the fraction of culture changed per unit time is W/V. Therefore, the population lost per unit time is

$$P = \frac{W}{V} N \tag{13}$$

Substituting this term in eq. (12), we have

$$\frac{dN}{dt} = \alpha N - \frac{W}{V} N = N\left(\alpha - \frac{W}{V}\right) \tag{14}$$

Note that eq. (14) applies to both internally and externally controlled systems.

It is apparent from (14) that when αN is larger than the washout rate, $(W/V)N$, dN/dt will be positive, and the cell concentration will increase. Conversely, when $(W/V)N$ exceeds αN, the population will decrease. When the steady state has been achieved, $dN/dt = 0$ and it follows from (14) that

$$\alpha = \frac{W}{V} \tag{15}$$

These theoretical expectations have been verified for continuous culture systems (Herbert *et al.*, 1956). Cell populations inoculated at either higher or lower concentrations than the expected steady state concentrations do change to the expected density. When equilibrium is reached, the specific growth rate α is equal to the washout rate (eq. (15)). Since α is equal to $\ln 2/G$ when viability is 100% (eq. (10)), generation times can be determined quite easily from the washout rate.

The specific growth rate in continuous culture as a function of the substrate concentration is also described by eq. (11). For studies of continuous culture systems S is taken as the concentration of substrate in the growth tube.

In the externally controlled system, the steady state population in the growth tube will depend on the amount of the limiting nutrient which is actually used for growth. If S_r is the concentration of the controlling nutrient in the reservoir and S is the concentration in the growth tube, then $S_r - S$ is the amount used for growth. Q is defined as the amount of limiting nutrient required to form one cell. Therefore, the steady state population attained in the growth tube will be

$$N = \frac{S_r - S}{Q} \tag{16}$$

Practically speaking, most of the nutrient is used up during growth so that S becomes very small and can be neglected (Herbert *et al.*, 1956). Equation (16) then becomes

$$N = \frac{S_r}{Q} \tag{17}$$

It is apparent from (17) that the cell population in the growth tube will be directly proportional to the amount of the controlling nutrient in the sterile medium. If N is expressed in terms of cell dry wt/unit volume, N_m, rather than cell population, then from (17) we have

$$\frac{1}{Q} = \frac{N_m}{S_r} = Y \tag{18}$$

In this special case, Q becomes the reciprocal of the yield coefficient Y, and Y can be determined by a study of the total cell crop as a function of the concentration of the limiting nutrient.

C. Applications of Continuous Flow Systems

Continuous flow systems such as the chemostat and turbidostat (Novick, 1955) have been used to study a wide variety of problems in

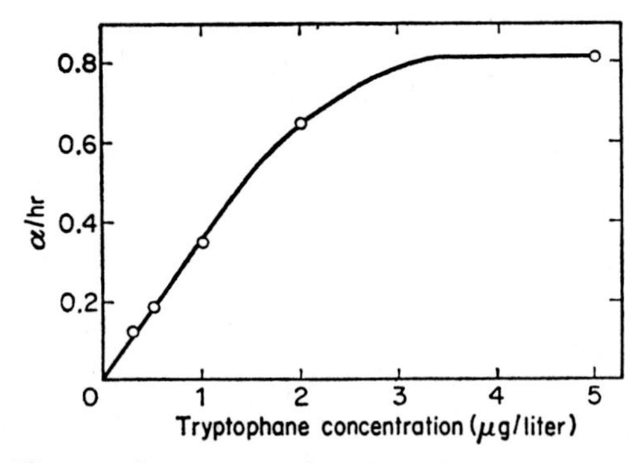

Fig. 2.5. Specific growth rate as a function of the tryptophane concentration. (Reproduced with permission from Novick and Szilard, 1950.)

the field of bacterial physiology. The dependence of the growth rate constant on the concentration of the limiting nutrient (Fig. 2.5), the kinetics of enzyme induction, the yield coefficient for energy sources, and changes in cell properties such as mass per cell as a function of the specific growth rate, have all been studied with continuous flow apparatus (James, 1961). Genetic problems such as the rate of spontaneous mutation in a stable population have also been studied (Novick, 1955).

V. Synchronous Growth of Bacterial Cultures

A. Methods of Obtaining Synchrony of Cell Division

Synchronous cultures are cultures in which most of the organisms are in the same stage of the cell division cycle (Fig. 2.6). Various methods of producing this result have been used (Maaløe, 1962), but they can be categorized as either forced or mechanical selection methods.

Forced methods of producing synchrony utilize some artificial means of inducing the organisms either to divide or to retard cell division. Temperature shocks, both hot and cold, and combinations of heat and cold have been used to induce synchrony. Another class of forced methods employs nutritional techniques such as allowing the growth medium to become depleted with respect to one of the nutrients and then transferring the organisms to fresh complete medium. Addition of thymine to organisms which have a nutritional requirement for thymine has also been used to induce synchrony. A problem with forced methods

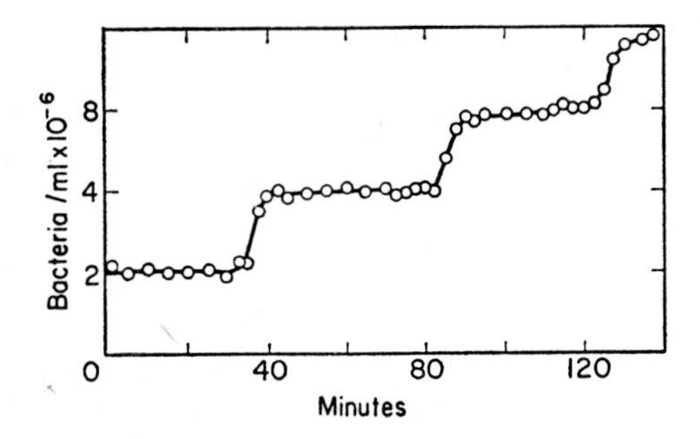

FIG. 2.6. Synchronous growth of *Escherichia coli*. In this case, synchronous cell division was obtained by filtration through a filter paper disc. (Reproduced with permission from Helmstetter and Cummings, 1963.)

is that they interfere with normal metabolism of the cell and data obtained from this type of study must be interpreted with this reservation in mind.

Mechanical selection methods depend on some means of selecting organisms of a given size. Transfer of these organisms to fresh medium frequently results in synchronous cell division. The most commonly used selection method is filtration through fine filter paper, although differential centrifugation has also been used. The probable reason for success of the mechanical methods is that cell size changes during the cell cycle and therefore cells of the same size are in the same part of the cell cycle.

Synchronous division never lasts for more than 4–5 generations and may occur for only one generation. The reason for this is that the generation times of the dividing cells vary considerably with respect to the individual cells (Powell, 1958).

B. The Bacterial Cell Cycle

The changes which occur during the cell duplication cycle are summarized in Fig. 2.7. In synchronously dividing cultures, RNA and protein are synthesized continually at exponential rates (Kuempel and Pardee, 1963). Cell length also increases exponentially during synchronous division (Schaechter *et al.*, 1962). Therefore, changes in all the parameters of cell mass which have been measured support the conclusion that growth occurs exponentially during the division cycle. The

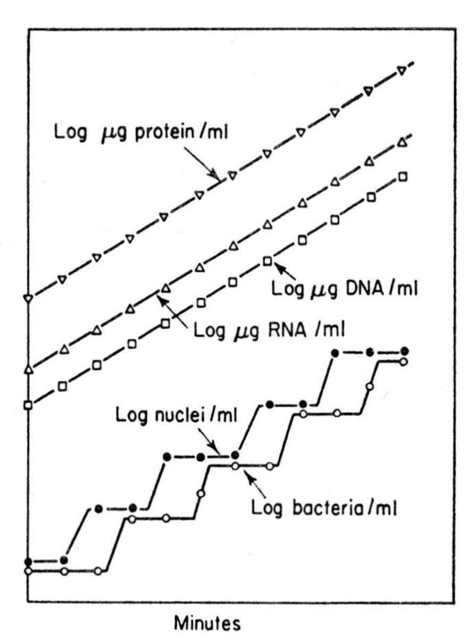

Fig. 2.7. Schematic representation of changes in cellular components during the cell cycle. Protein, DNA and RNA are synthesized continuously at logarithmic rates during synchronous growth. The nuclei (generally more than one per cell) divide about the middle of the cell cycle.

new cytoplasm is divided equally between the parent and daughter cells after cell division (van Tubergen and Setlow, 1961).

Nuclear division occurs about the middle of the cell cycle (Schaechter *et al.*, 1962). DNA synthesis, however, is continuous during all stages of synchronous growth except possibly when the generation times are very long (Maaløe, 1963).

Duplication of the nucleus occurs by a process whereby the existing chromosome serves as a template for the production of a replicate (Meselson and Stahl, 1958). Duplication starts at one end of the chromosome and continues linearly to the other end. This has been determined by following the production of genetic markers with time on the chromosome of *E. coli* where the linkage groups are known (Nagata, 1963; Yoshikawa and Sueoka, 1963). The bacterial nucleus exists as a circular chromosome, but this does not affect the concept of nuclear reproduction in any manner. Cairns (1963) exposed *E. coli* to a pulse of labeled thymidine and prepared radioautographs of isolated nuclei, and obtained pictures which provide strong support for the concept of linear reproduction of the nucleus.

After the increase in protoplasm and division of the nucleus, the new cell is pinched off between the new nucleus and the parent nucleus (Schaechter *et al.*, 1962) by the growth of the cell membrane and the cell wall. The cell wall of *E. coli* is synthesized at all points along the cell wall with the result that both new and old cell wall substance is equally divided between the parent and daughter cells (van Tubergen and Setlow, 1961).

The elucidation of the events of the bacterial duplication cycle is very recent and has been possible largely through the development of techniques for producing synchronously dividing cultures. Very little is known at this time of the control mechanisms that initiate the various events of the cell cycle.

References

Bauchop, T. and Elsden, S. R. (1960). *J. Gen. Microbiol.* **23**, 457

Cairns, J. (1963). *J. Mol. Biol.* **6**, 208.

Finney, D. J., Hazlewood, T. and Smith, M. J. (1955). *J. Gen. Microbiol.* **12**, 222.

Gibbs, M., Sokatch, J. R. and Gunsalus, I. C. (1955). *J. Bacteriol.* **70**, 572.

Gunsalus, I. C. (1951). *In* "Bacterial Physiology" (C. H. Werkman and P. W. Wilson, eds.). Academic Press, New York, p. 101.

Gunsalus, I. C. and Schuster, C. W. (1961). *In* "The Bacteria" (I. C. Gunsalus and R. Y. Stanier, eds.). Academic Press, New York, Vol. II, p. 1.

Helmstetter, C. E. and Cummings, D. J. (1963). *Proc. Nat. Acad. Sci. U.S.* **50**, 767.

Herbert, D., Elsworth, R. and Telling, R. C. (1956). *J. Gen. Microbiol.* **14**, 601.

Hinshelwood, C. (1952). "The Chemical Kinetics of the Bacterial Cell." Clarendon Press, Oxford.

James, T. W. (1961). *Ann. Rev. Microbiol.* **15**, 27.

Kelly, C. D. and Rahn, O. (1932). *J. Bacteriol.* **23**, 147.

Kuempel, P. L. and Pardee, A. B. (1963). *J. Cellular Comp. Physiol.* **62**, 15.

Lichstein, H. C. (1959). *Bacteriol. Rev.* **23**, 261.

Maaløe, O. (1962). *In* "The Bacteria" (I. C. Gunsalus and R. Y. Stanier, eds.). Academic Press, New York, Vol. IV, p. 1.

Maaløe, O. (1963). *J. Cellular Comp. Physiol.* **62**, suppl. 1, p. 31.

Marr, A. G., Nelson, E. H. and Clark, D. J. (1963). *Ann. N.Y. Acad. Sci.* **102**, 536.

Meselson, M. and Stahl, F. W., (1958). *Proc. Natl. Acad. Sci. U.S.* **44**, 671.

Monod, J. (1942). "Recherches sur la Croissance des Cultures bactériennes." Hermann et Cie, Paris.
Monod, J. (1949). *Ann. Rev. Microbiol.* **3**, 371.
Monod, J. (1950). *Ann. Inst. Pasteur.* **79**, 390.
Nagata, T. (1963). *Proc. Nat. Acad. Sci. U.S.* **49**, 551.
Novick, A. (1955). *Ann. Rev. Microbiol.* **9**, 97.
Novick, A. and Szilard, L., (1950). *Proc Natl. Acad. Sci. U.S.* **36**, 708.
Powell, E. O. (1958). *J. Gen. Microbiol.* **18**, 382.
Schaechter, M., Williamson, J. P., Hood, J. R. and Koch, A. L. (1962). *J. Gen. Microbiol.* **29**, 421.
Senez, J. C. (1962). *Bacterial Rev.* **26**, 95.
Snell, E. E. (1957). *In* "Methods in Enzymology" (S. P. Colowick and N. O. Kaplan, eds.). Academic Press, New York, Vol. III, p. 477.
Sokatch, J. R. and Gunsalus, I. C. (1957). *J. Bacteriol.* **73**, 452.
Toennies, G., Iszard, L., Rogers, N. B. and Shockman, G. D. (1961). *J. Bacteriol.* **82**, 857.
van Tubergen, R. P. and Setlow, R. B. (1961). *Biophys. J.* **1**, 589.
Whitaker, A. M. (1963). Thesis, University of Sheffield.
Yoshikawa, H. and Sueoka, N. (1963). *Proc. Nat. Acad. Sci. U.S.* **49**, 559.

3. CHEMICAL COMPOSITION OF BACTERIA

I. Proteins

Under any conditions of growth, protein is a major component of the bacterial cell. Several bacterial proteins have been isolated in pure or crystalline form and the amino acid compositions of a selected few of these are given in Table 3.1. Epsilon-N-Methyl lysine is an amino acid which has so far been found only in bacterial flagella (Ambler and Rees, 1959). All other amino acids identified in hydrolysates of bacterial proteins are conventional amino acids of the L series. Hydroxylysine and hydroxyproline, which are limited to collagen type proteins, have not been detected in bacterial proteins. D-Amino acids have also not been

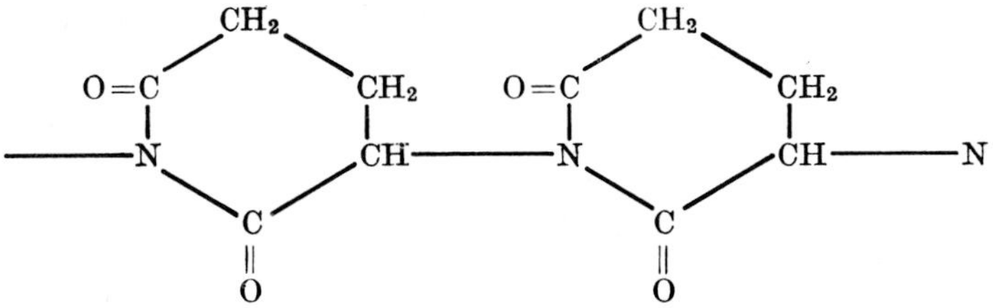

CH_2 CH_2
O=C CH_2 O=C CH_2
—N CH—N CH—N
C C
O O

FIG. 3.1. Polyglutamyl capsule of *Bacillus* species.

found in bacterial proteins, but have been found in the cell wall mucopeptide of bacteria, in capsules of *Bacillus* species, and in antibiotics produced by bacteria (Abraham, 1957; Miller, 1961) and other organisms. D-Glutamic acid is the sole amino acid in the capsule of *Bacillus anthracis* polypeptide capsule, although other species of *Bacillus* have up to 50 % of L-glutamic acid (Thorne, 1956). The probable structure of this polyglutamyl peptide is shown in Fig. 3.1. (Housewright 1962).

The existence of certain classes of conjugated proteins such as phosphoproteins and glycoproteins has not been established for bacteria. If lipoprotein exists in bacteria, it probably exists in the cell membrane, since this structure is known to contain large amounts of both lipid and

TABLE 3.1. Amino Acid Composition of Crystalline Bacterial Proteins

	Tryptophane synthetase A protein (a)	Botulinum toxin Type A (b)	Flagellin, *S. typhimurium* (c)	α-Amylase *B. subtilis* (d)	α-Amylase *B. stearothermophilus* (e)	Ferredoxin, *C. pasteurianum* (f)
Molecular weight	29 000	15 000 (i)	—	48 700	15 600	6000
Composition	Moles/ mole pr.	Moles/ mole pr.	g/10⁵ g pr.	Moles/ mole pr.	Moles/ mole pr.	Moles/ mole pr.
Amide N	24	1370 (g)	156	49·7	3	— (h)
Alanine	40·9	394	147	29	8	8
Arginine	11·7	239	24	17	3	0·0
Aspartic acid	23·6	1370 (g)	155	53	11	8
Cysteine	—	20	—	8·0	0	—
Half Cystine	3·1	40	0	0	4	7
Glycine	20·2	166	79	39	9	4
Glutamic acid	31·4	953	96	43	22	4
Histidine	4·3	60	4	12	4	0
Isoleucine	20·2	820	49	17	7	5
Leucine	27·9	708	74	23	9	0
Methionine	4·8	64	3	5	3	0
ε-N-Methyl lysine	—	—	28	—	—	—
Phenylalanine	12·0	64	13	18	6	1
Proline	19·7	203	13	14	22	3
Serine	11·6	374	59	24	6	5
Threonine	9·7	642	101	23	8	1
Tryptophane	0·0	82	0	15	0	0
Tryosine	7·2	672	20	24	3	1
Valine	18·3	406	60	25	11	6
Lysine	14·2	477	33	25	6	1

References and footnotes: (a) Henning *et al.*, 1962; (b) Buehler *et al.*, 1947; (c) Ambler and Rees, 1959; (d) Junge *et al.*, 1959; (e) Campbell *et al.*, 1961; Lovenberg *et al.*, 1963; (g) assumed that aspartic acid = amide N; (h) also contains 7 moles ferrous iron and 7 moles sulfide per mole ferredoxin; (i) this is the molecular weight of the monomer (Van Alstyne *et al.*, 1966); earlier estimations of a molecular weight of 900 000 appear to have been made with an aggregated form of the toxin.

protein (McQuillen, 1960). Conjugated proteins which have been studied in bacteria are mainly enzymes which participate in electron transport pathways (Table 3.2). Difference spectra of oxidized and reduced bacterial cytochrome enzymes studied in whole cell or cell-free preparations of many species of bacteria generally fall into well recognized patterns (Smith, 1961). On the other hand, the chromophores of certain bacterial cytochromes such as the cytochrome c_3 of the anaerobic *Desulfovibrio desulfuricans* (Postgate, 1961) and the haemoprotein of purple photosynthetic bacteria (Kamen and Bartsch, 1961) appear to be unique to bacteria (see also Morton, 1958). Another interesting enzyme is *Pseudomonas aeruginosa* blue protein, which was found to be a copper enzyme (Yamanaka *et al.*, 1963). Reduced blue protein is able to react with cytochrome oxidase from this same organism in the presence of oxygen and may therefore have a respiratory function in *P. aeruginosa*.

TABLE 3.2. Prosthetic Groups of Conjugated Bacterial Proteins

Organism	Enzyme	Prosthetic group	Reference
S. faecalis	NADH Peroxidase	FAD	Dolin, 1957
S. faecalis	Diaphorase	FMN	Dolin and Wood, 1960
E. coli	Pyruvate oxidase	FAD	Hager, 1959
E. coli	Cytochrome b_1	Iron protoporphyrin	Deeb and Hager, 1964
P. aeruginosa	Cytochrome oxidase	Haem a_2	Yamanaka and Okunuki, 1963
P. aeruginosa	Blue protein	Cu^{2+}	Yamanka *et al.*, 1963
B. anitratum	Glucose oxidase	Quinone?	Hauge, 1964
M. lysodeikticus	Catalase	Iron protoporphyrin	Herbert and Pinsent, 1948
D. desulfuricans	Cytochrome c_3	Haematohaemin	Postgate, 1961

Isoelectric points of purified bacterial proteins are usually in the region of 5, which is the result of a "typical" assortment of amino acids. The range of pI's extends from 3·7 for ferrodoxin, which has a high proportion of acidic amino acids, to 10·5 for cytochrome c_3 of *Desulfovibrio desulfuricans* (Postgate, 1961) which would be expected to have a high proportion of basic amino acids. The majority of proteins, bacterial or otherwise, are in the 20 000 to 100 000 molecular weight range.

Not many detailed studies of the fine structure of bacterial proteins have been made, but they exhibit the usual sensitivity to heat and other denaturing agents, and it is almost certain that they possess the same types of internal structure known to occur in other proteins. *Bacillus stearothermophilus*, a thermophilic spore former, produces an α-amylase which is an interesting exception to this generalization. This enzyme loses only 29% of its enzyme activity after 20 hours at 85°C (Manning *et al.*, 1961). Campbell and Manning (1961) studied the optical rotation of this enzyme and reached the conclusion that the enzyme is randomly coiled, i.e. "denatured", in its biologically active form. This is probably a consequence of the high proportion of proline which prevents folding in the relatively short protein molecule.

One class of bacterial proteins which has been studied by X-ray diffraction techniques comprises the bacterial flagellar proteins (Astbury *et al.*, 1955). Astbury and his associates found that these proteins, which they term flagellins, belong to the keratin-myosin-epidermin-fibrinogen group of elastic proteins, and reached the conclusion that bacterial flagella are the equivalent of monomolecular muscles.

II. Lipids

A. Simple Lipids

Simple lipids are unsubstituted lipids which have been extracted from bacteria. In the main, these are fatty acids, although a few neutral compounds have been obtained as well. Fatty acids occur as free acids and as bound acids which are released by saponification.

Sterols have been detected in only one species of true bacteria, *Azotobacter chroococcum* (Sifferd and Anderson, 1936). Either all other reports have been negative or the amounts detected were so small that the possibility of artifacts cannot be ruled out. Cholesterol and cholesterol esters, or similar compounds, have been detected in several species of PPLO, however (Lynn and Smith, 1960; Tourtellotte *et al.*, 1963).

1. *Gram Negative Bacteria*

Most of the fatty acids of Gram negative bacteria are conventional even numbered acids, but there are some unique acids as well (Table 3.3). Marr and Ingraham (1962) found that palmitic acid formed the largest share of the even numbered saturated acids of *E. coli* lipids. In another study by Kaneshiro and Marr (1961), the positions of the double bonds of the unsaturated acids were determined. The C_{16} unsaturated acid was palmitoleic acid and the C_{18} acids were a mixture of about 70% *cis*-vaccenic acid and 30% oleic acid. *Cis*-vaccenic acid was first isolated from *Lactobacillus arabinosus* by Hofmann *et al.* (1952) and identified as *cis*-octadec-11-enoic acid (Fig. 3.2). Two other unique acids produced by *E. coli* are the C_{17} and C_{19} cyclopropane acids. Lactobacillic acid, the C_{19} acid, was the first to be discovered, again by Hofmann *et al.* (1952), in the lipids of *L. arabinosus*. The cyclopropane bridge of lactobacillic acid occurs across the same two carbons which have the double bond in *cis*-vaccenic acid (Fig. 3.2), which is actually the precursor to lactobacillic acid. Kaneshiro and Marr (1961) were the first to establish the structure of the C_{17} cyclopropane acid as *cis*-9,10 methylenehexadecanoic acid. This means that the C_{17} cyclopropane acid of *E. coli* bears the same structural relationship to palmitoleic acid that lactobacillic acid bears to *cis*-vaccenic acid. Another fatty acid

TABLE 3.3. Fatty Acids of Gram-negative Bacteria

Acid	*E. coli* Total acids (a) %	*E. coli* Phospholipid acids (b) %	*S. marcescens* (c) %	*A. tumefaciens* (d) %
Lauric	Trace	—	Trace	—
Myristic	5·6	5·7	6·0	—
β-Hydroxymyristic	10·9	—	—	—
Palmitic	33·1	39·8	43·8	10·2
Palmitoleic	28·4 (e)	6·7	2·3 (g)	—
Cis-9,10-methylene-hexadecanoic	1·8	21·8	32·4	—
Stearic	—	—	0·5	—
Oleic / *Cis*-vaccenic	20·2 (e)	18·6 (f)	10·5 (h)	68·6 (h)
Lactobacillic	—	7·4	3·1	13·1

References and footnotes: (a) Marr and Ingraham, 1962; (b) Kaneshiro and Marr, 1961; (c) Bishop and Still, 1963; (d) Hofmann and Tausig, 1955; (e) position of double bond not determined in this study; (f) mono-unsaturated acids in this study consisted of about 70% *cis*-vaccenic and 30% oleic acids; (g) palmitoleic acid was 93% of C_{16} monounsaturated acids; (h) *cis*-vaccenic acid was predominant C_{18} monounsaturated acid.

peculiar to bacteria is β-hydroxy myristic acid, (-)-3-(D)-hydroxytetradecanoic acid, which Ikawa *et al.* (1953) isolated from the endotoxin of *E. coli*. Another β-hydroxy acid was isolated from *Pseudomonas aeruginosa* as part of a rhamnoside produced by the organism (Bergstrom *et al.*, 1946). This acid was identified as (-)-3-(D)-hydroxydecanoic acid (Jarvis and Johnson, 1949). Hofmann and Tausig (1955) found that *cis*-vaccenic acid was the principal fatty acid of *Agrobacterium tumefaciens* (Table 3.3). Bishop and Still (1963) found that the principal fatty acids of *Serratia marcescens* were palmitic and 9,10-methylenehexadecanoic acid.

Many species of bacteria produce poly-β-hydroxy butyrate which serves as a lipid storage material. This substance, which was first isolated by Lemoigne (1927), was rediscovered recently as a storage material formed by gram positive sporeformers and photosynthetic bacteria, and it is now known to be produced by many species of both gram negative and gram positive bacteria. Poly-β-hydroxybutyrate occurs in crystalline form in bacteria with a molecular weight estimated as high as 250 000 (Lundgren *et al.*, 1965). The polymer seems to be identical in all species studied.

2. *Mycoplasma*

Tourtellotte *et al.* (1963) separated the lipids of the PPLO organism *Mycoplasma galisepticum* on silicic acid columns and characterized the fractions. The fatty acid spectrum of this organism more closely resembled that of animals than that of bacteria. The principal acids were the conventional even numbered saturated and unsaturated acids in the C_{12} to C_{18} range, although lesser amounts of caproic, caprylic and capric acids were found. The fatty acids occurred mainly as free fatty acids and as di- and triglycerides and as phospholipids.

3. *Gram Positive Bacteria*

In their study of the fatty acids of *L. arabinosus*, Hofmann *et al.* (1952) first isolated and characterized *cis*-vaccenic and lactobacillic acids (Fig. 3.2). Both of these acids were present in other species of *Lactobacillus*, but lactobacillic acid was not present in a group C *Streptococcus* (Table 3.4). It is probable that palmitoleic acid is the C_{16} monounsaturated fatty acid and *cis*-vaccenic is the principal C_{18} unsaturated fatty acid in the lactobacilli, although the positions of the double bonds were not established in all cases. Palmitic acid is the

TABLE 3.4. Fatty Acids of *Lactobacillus* and *Streptococcus*. (Hofmann *et al.*, 1955)

Acid	*L. arabinosus* %	*L. casei* %	*L. delbruekii* %	Group C. streptococcus %
Decanoic	1·1	2·1	0·5	0·5
Lauric	2·3	2·8	1·1	5·2
Myristic	1·2	2·1	2·5	4·4
Palmitic	18·7	24·3	27·5	26·6
Stearic	—	7·0	10·8	18·0
Lactobacillic	30·1	12·6	6·4	—
$C_{16}+C_{18}$ mono-unsaturated	35·6	37·6	45·5	38·0

principal saturated acid with lesser amounts of stearic acid and the lower fatty acids.

Saito (1960) and Kaneda (1963) studied the lipids of *Bacillus subtilis* and discovered the presence of unique branched-chain fatty acids. These acids were of either the *iso* configuration, where methyl branching occurred at the second from the end (penultimate) carbon, or the *anteiso* configuration, where branching occurred at the third from the end (antepenultimate) carbon. Isomers of C_{14}, C_{15}, C_{16}, and C_{17} acids were found with the C_{15} and C_{17} anteiso acids predominating.

4. *Actinomycetales*

By far the most interesting and unique collection of fatty acids described in nature belong to species of the *Actinomycetales*. There has always been an active interest in the chemistry of the tubercle bacillus with efforts to relate the lipids of this organism to the pathogenesis of tuberculosis and the acid-fast staining properties of this organism (Asselineau and Lederer, 1955). A considerable body of information has resulted concerning the chemistry of *Mycobacterium* species, but the original problems remain unsolved. The earliest fruitful chemical studies were by Anderson and Chargaff (1929), when they isolated tuberculostearic acid from the tubercle bacillus (Fig. 3.2). This acid was later identified as (-)-(D)-10-methyloctadecanoic acid (Spielman, 1934).

Phthioic acid, which was also first isolated from the tubercle bacillus by Anderson (1932), has since proved to be a mixture of acids. One component was named mycolipenic acid by Polgar (1954a) and is an unsaturated acid with the probable structure indicated in Fig. 3.2. There are known to be other members of α-β-unsaturated acids in this series, however (Asselineau and Lederer, 1960).

Mycocerosic acid is another acid with methyl side chains which has been isolated from the tubercle bacillus. Like mycolipenic acid, there is now known to be a family of these compounds, the main component of which appears to have the structure indicated in Fig. 3.2. (Polgar, 1954b; Asselineau *et al.*, 1959). Mycocerosic acid has been found in the tubercle bacillus as the diester of the alcohol phthiocerol (Philpot and Wells, 1952). Phthiocerol was discovered in the tubercle bacillus by Stodola and Anderson (1936) and is now known to be a mixture of the two alcohols illustrated in Fig. 3.2 (Asselineau and Lederer, 1960).

The mycolic acid series of acids are among the most interesting compounds isolated from the tubercle bacillus. There is a family of these acids with the general structure shown in Fig. 3.2. Mycolic acids were first isolated by Stodola *et al.* (1938) and have since been the object of

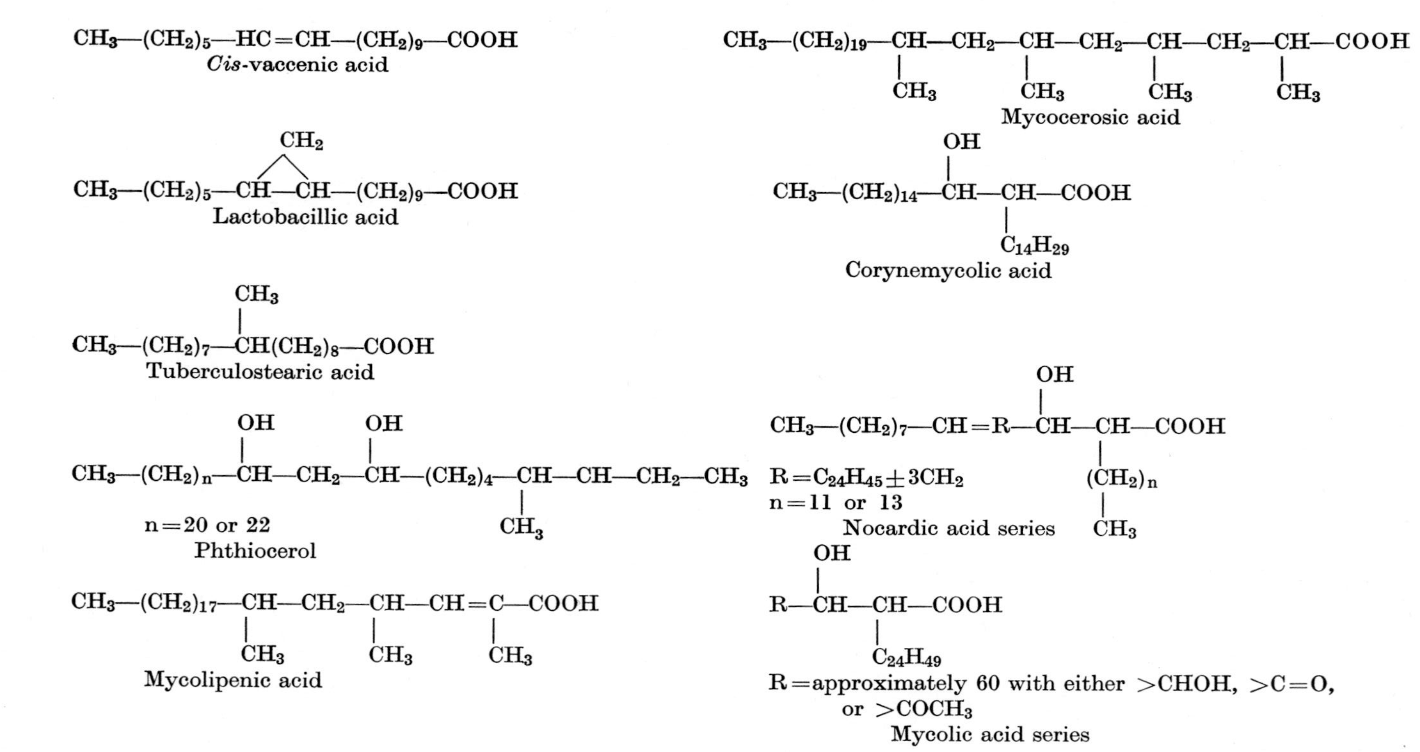

FIG. 3.2. Structure of some fatty acids and alcohols isolated from acid fast bacteria and related species.

considerable study. The properties of the known mycolic acids were summarized by Asselineau and Lederer (1960). In addition to the β-hydroxyl group, there may be another oxygen functional group in the R chain, either a carbonyl group, a methoxyl group or another hydroxyl group.

After the discovery of mycolic acids, Lederer and Pudles (1951) obtained a similar compound, but shorter chain length, from *Corynebacterium diphtheriae*, which they named corynemycolic acid in view of its structural similarity to the mycolic acids. Michel *et al.* (1960) isolated another similar series of acids, in this case unsaturated, from *Nocardia asteroides*. These acids were named nocardic acids and a partial structure is shown in Fig. 3.2.

B. Complex Lipids

Very few chemical studies of the structures of conventional complex lipids from bacteria have been made. Glycerol esters have been demonstrated in many species of gram positive and gram negative bacteria (Asselineau and Lederer, 1960) and are presumed to be similar to glycerides of higher organisms. The situation is similar in the case of phospholipids. Hydrolysis products have been reported for phospholipid preparations from several species of bacteria (Asselineau and Lederer, 1960). Choline and ethanolamine have been detected but sphingosine has not. Serine and several other amino acids have also been reported, but in many cases it is not certain whether these amino acids are actually part of the phospholipid structure or form a contaminating component. Inositol and several other sugars have been shown to be present in phospholipid preparations of bacteria. Fatty acids were usually C_{16} or C_{18} acids, even in the case of the tubercle bacillus, although tuberculostearic and phthienoic acid have been identified as well.

An unusual situation exists in the case of *Thiobacillus thiooxidans*, which excretes a phospholipid into the medium during growth on sulfur (Schaeffer and Umbreit, 1963). This phospholipid is composed of phosphatidyl glycerol with lesser amounts of phosphatidyl choline and ethanolamine (Jones and Benson, 1965). The latter authors suggest the possibility that phospholipid may aid the organism in its digestion of sulfur. The principle phospholipid isolated from human and BCG strains of the tubercle bacillus is phosphatidyl inositodimannoside (Vilkas and Lederer, 1956) (Fig. 3.3).

More attention has been paid to the unusual complex lipids which occur in bacteria. In the rhamnolipid of *P. aeruginosa*, β-hydroxydecanoic acid is attached to L-rhamnose by a glycosidic linkage rather than the conventional ester link (Fig. 3.4). This lipid is composed of two

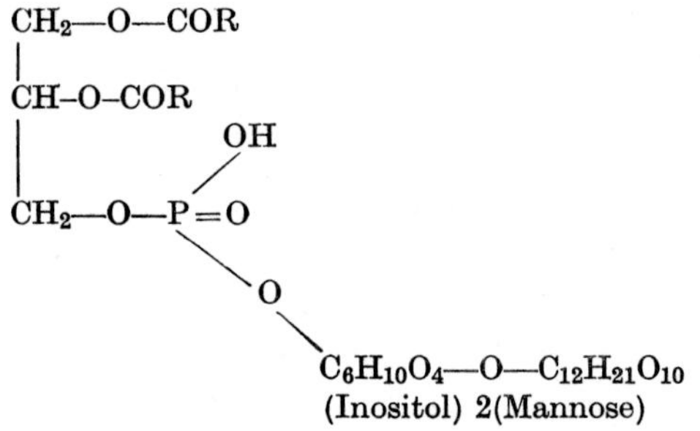

CH_2—O—COR
|
CH–O–COR
|
CH_2—O—P(OH)(=O)—O—$C_6H_{10}O_4$—O—$C_{12}H_{21}O_{10}$
(Inositol) 2(Mannose)

FIG. 3.3. Phosphatidyl inositodimannoside from *Mycobacterium tuberculosis*. R is a mixture of palmitic and stearic acids.

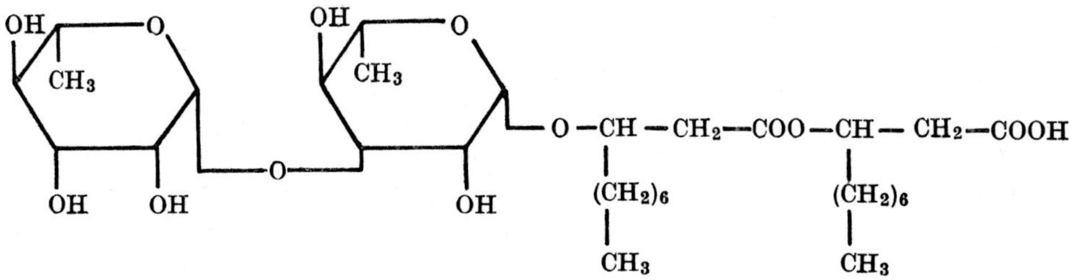

OH, O, CH_3, OH, OH, O, OH, O, CH_3, OH
—O—CH($(CH_2)_6CH_3$)—CH_2—COO—CH($(CH_2)_6CH_3$)—CH_2—COOH

FIG. 3.4. Rhamnolipid of *Pseudomonas aeruginosa*. (Jarvis and Johnson, 1949.)

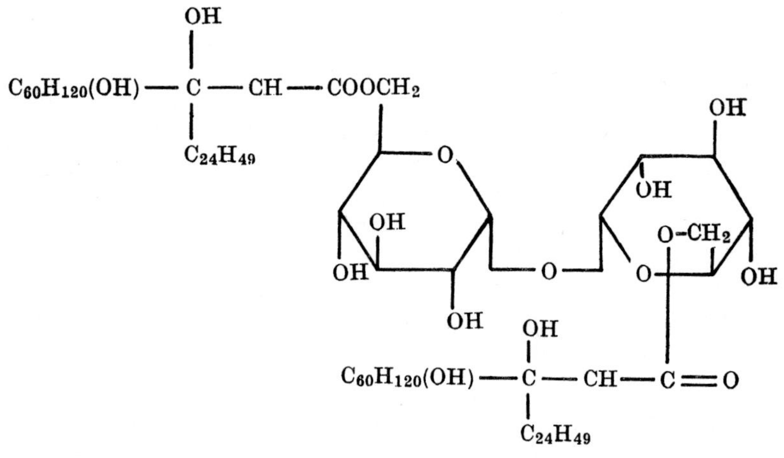

$C_{60}H_{120}(OH)$—C(OH)($C_{24}H_{49}$)—CH—$COOCH_2$
OH, O, OH, OH, OH, O, OH, OH, O–CH_2, O, OH
$C_{60}H_{120}(OH)$—C(OH)($C_{24}H_{49}$)—CH—C=O

FIG. 3.5. Cord factor from *Mycobacterium tuberculosis*.

molecules of L-rhamnose and two of β-hydroxydecanoic acid (Jarvis and Johnson, 1949).

Cord factor is a lipid formed by some strains of tubercle bacilli which causes them to string together in chains or ropes of organisms. This lipid was discovered by Bloch (1950), who is responsible for the name and who also discovered the toxicity of this lipid. Cord factor was later shown to be 6,6′-dimycolyltrehalose (Noll *et al.*, 1956) (Fig. 3.5).

In their studies of the endotoxin of Gram negative bacteria, Westphal and Lüderitz (1954) separated a lipid component from the lipopolysaccharide by mild acid hydrolysis which they designated lipid A.

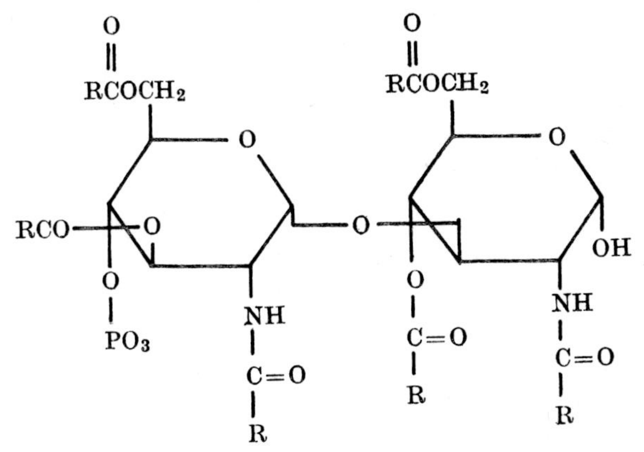

FIG. 3.6. Structure of lipid A according to Burton and Carter (1964). R Groups are fatty acid residues, about half of which are β-hydroxy acids esterified with acetyl residues. On the average, there are five R groups per mole of lipid A and possible locations for these are indicated in this figure.

Burton and Carter (1964) prepared highly purified lipid A from *E. coli*. Their best preparations contained 2 moles glucosamine, one mole of phosphate, three or four acetyl residues and five long chain fatty acids per mole of lipid A. The fatty acids found were lauric, myristic and β-hydroxymyristic, and another acid, possibly β-hydroxydecanoic acid. Their studies indicated a molecular weight of about 1700. About half the fatty acid residues are hydroxyl acids and they proposed that the acetyl residues were located on these hydroxyl groups rather than on the glucosamine molecule. On the basis of these studies they proposed the structure shown in Fig. 3.6.

III. Carbohydrates

Bacterial carbohydrates occur as intermediates in biochemical reactions, as intracellular storage materials, as a portion of the nucleic acid molecule, as a part of the cell wall structure, and as extracellular capsules. From a chemical point of view, polysaccharides may be divided into homopolysaccharides, which are composed of one monosaccharide, heteropolysaccharides, which are composed of more than one monosaccharide, and complex polysaccharides, which contain an appreciable quantity of non-carbohydrate substance, such as lipids or amino acids.

A. Homopolysaccharides

Several glucans have been isolated from bacteria (Table 3.5). Materials resembling starch and glycogen have been detected in bacteria in a number of instances. Not many structural studies have been

Table 3.5. Homopolysaccharides Produced by Bacteria

	Linkage	Reference
Glucans		
Starch	α-(1-4) with 1-6 branching	Hehre *et al.*, 1947
Glycogen	α-(1-4) with 1-6 branching	Hehre and Hamilton, 1946; Barker *et al.*, 1950
Cellulose	β-(1-4) linear	Barclay *et al.*, 1954
Crown gall polysaccharide	β-(1-2)	Putman *et al.*, 1950
Dextran	α-(1-6) with (1-4) and (1-3) links	Jeanes *et al.*, 1954
Levans		
A. levanicum levan	β-(2-6) with (2-1) branching	Feingold and Gehatia, 1957
Galactans		
M. mycoides galactan	β-(1-6)	Plackett *et al.*, 1962
Mannans		
B. polymyxa	Probably α-(1-2) and (1-3) with (1-2-6) branches	Ball and Adams, 1959

done, but Barker *et al.* (1950) studied the structure of a storage polysaccharide from *Neisseria perflava* and established that it was a glycogen-like amylopectin with an average chain length of 11–12 residues.

The production of true cellulose by several species of *Acetobacter* was recognized many years ago (Kaushal and Walker, 1951; Hibbert and Barsha, 1931). Barclay *et al.* (1954) studied the structure of the extracellular polysaccharide produced by *Acetobacter acetigenum* and confirmed the β-(1–4) linkage of the glucose residues. They also established that the average chain length was about 600 residues.

Other glucans formed by bacteria are the crown gall polysaccharide, a β-(1-2) glucan (Putman *et al.*, 1950) and the dextrans. Dextrans are β-(1-6) glucans with 1-4 and 1-3 branching. Dextrans are produced by a number of genera of bacteria, but have been most frequently found and studied in the *Leuconostoc* and *Streptococcus* species. Jeanes *et al.* (1954) studied the dextrans of several strains of organisms and found that there were three groups of dextrans which they designated as A, with up to 2% 1-3 linkages, B with 3–6% 1-3 linkages, and C with over 6% 1-3 linkages.

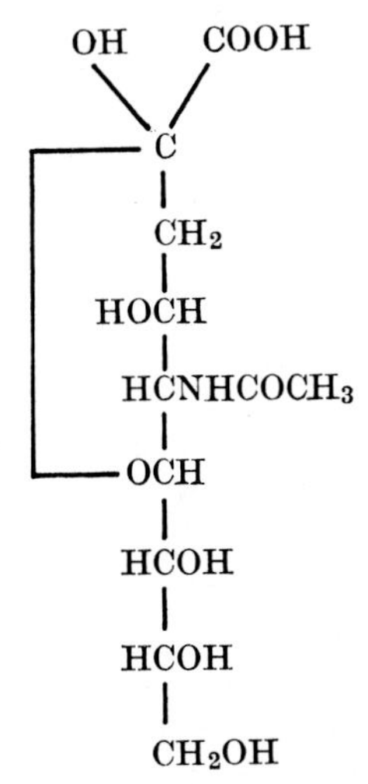

FIG. 3.7. N-Acetylneuraminic acid.

Other homopolysaccharides produced by bacteria are levans, mannans and galactans and the structures of these polymers insofar as they are known are indicated in Table 3.5.

The Vi antigen which is found in *Escherichia coli* and several enteric pathogens has been purified and found to be composed of D-galactose-aminuronic acid (Heyns *et al.*, 1959), but the linkage groups of this acidic polymer are not known. Colominic acid is another acid polysaccharide formed by *E. coli*, and N-acetylneuraminic acid (Fig. 3.7) has been identified as the sole sugar in this polymer (Barry, 1958) connected by 2,8-ketoglycoside linkages (McQuire and Binkley, 1964).

B. HETEROPOLYSACCHARIDES

Because of the difficult chemical problems involved in discerning the fine structure of heteropolymers, it will probably be a number of years before the linkages are known of a number of unusual heteropolysaccharides produced by bacteria. Unfortunately, there is not enough space to be able to review all of these interesting polymers; instead, this discussion will consider only those polymers where some details of the structure are known.

Hyaluronic acid was first isolated from group A streptococci by Kendall *et al.* (1937), and has since been recognized as the capsular material formed by several species of *Streptococcus*. Bacterial hyaluronic acid is composed of N-acetyl glucosamine and glucuronic acid in equimolar amounts, as is the animal polymer, and the glycoside linkages are probably also the same.

C-Substance was isolated from a group A streptococcus by Heyman *et al.* (1963). This polymer is composed of L-rhamnose and N-acetyl glucosamine. On the basis of methylation studies, the main linkage appears to be α-(1-3) from rhamnose to rhamnose. Branching occurs by α-(2-1) links from rhamnose to rhamnose. N-Acetyl glucosamine seems to be situated mainly at the end of the chains with β-(3-1) links from rhamnose to N-acetyl glucosamine.

The most thoroughly studied bacterial heteropolysaccharides are the capsular substances of *Diplococcus pneumoniae* which are responsible for the serological specificity of the types of this organism. The structures of several of these polysaccharides are shown in Fig. 3.8 insofar as they are known. The structure shown for type II polysaccharide is one of several possible structures which fit the chemical data. The serological specificity of this polysaccharide is due mainly to the non-reducing end groups of glucopyranosyluronic acid (Heidelberger, 1962). The structure of type III capsule was the first structure of a capsule of *D. pneumoniae* to be determined (Reeves and Goebel, 1941). Type III capsule consists of a linear polymer with a repeating unit of cellobiuronic acid (4-0-β-D-glucuronosyl-D-glucose). Type VI capsule has the repeating unit shown in Fig. 3.8. Antisera against type VI polysaccharide cross-react with type II capsular substance and vice versa. The reason for the serological cross-reactivity is the presence in both polymers of (1-3)-rhamnopyranose linkages. Type VIII capsule is also a linear polymer with the repeating unit shown in Fig. 3.8. Antisera against type VIII capsule will also react with type III polysaccharide. In this case, the common unit is the 4-0-β-D-glucuronosyl-D-glucose link. Type XIV capsule is a highly branched polysaccharide with the probable structure shown in Fig. 3.8. The serological specificity of antisera against this capsular substance is apparently due mainly to the presence of the non-reducing end groups of galactose.

Gram positive bacteria contain glycerol phosphate or ribitol phosphate compounds, named teichoic acids, located at or near the cell surface. The structure of glycerol teichoic acid from *Lactobacillus casei* is shown in Fig. 3.9. This material is a polymer of 1,3 linked glycerol phosphate units with carbon 2 of glycerol esterified with D-alanine. Other glycerol teichoic acids are variations of this formula. For example, glycerol teichoic acid from *Streptococcus faecalis* contains the disaccharide kojibiose (2-0-α-D-glucopyranosyl-D-glucose) attached glycosidically to

```
--|-3)-L-Rhap-(1-3)-L-Rhap-(1-3)-L-Rhap-(1-4)-D-Gpu-(1-4)-D-Gp-(1-|--
  |                                                       6)      | n
                                                          |
                                                  D-Gpu-(1
```

Type II. Heidelberger, 1962

```
-|--3)-β-D-Gpu-(1-4)-β-D-Gp-(1-|--
 |                             | n
```

Type III. Reeves and Goebel, 1941

```
--|-2)-α-D-Galp-(1-3)-α-D-Gp-(1-3)-α-L-Rhap-(1-3)-D or L-R1-1 or 2-PO3-|--
  |                                                                   | n
```

Type VI. Rebers and Heidelberger, 1961

```
-|--4)-β-D-Gpu-(1-4)-β-D-Gp-(1-4)-α-D-Gp-(1-4)-α-D-Galp-(1-|--
 |                                                         | n
```

Type VIII. Jones and Perry, 1957

```
        β-D-Galp-(1
                 |
                 6)
β-D-Galp-(1-4)-β-D-GpNAc-(1-3)-β-D-Galp-(1
                                        |
                                        6)
                  β-D-Gp-(1-4)-β-D-GpNAc-(1-4)-β-Gp-(1-4)-β-D-GpNAc-(1-|--
                                                                   6)  |
                                                                   |
                                                           β-D-Galp-(1
```

Type XIV. Barker *et al.*, 1958

FIG. 3.8. Heteropolysaccharides of *Diplococcus pneumoniae* serological types. Abbreviations used: Gp = glucopyranose; Galp = galactopyranose; GpNAc = N-acetyl-glucosaminopyranose; Rhap = rhamnopyranose; Gpu = glucopyranosyl uronic acid; and R1 = ribitol.

carbon 2 of glycerol (Wicken and Baddiley, 1963). D-Alanine is esterified to either carbon 3 or 4 of kojibiose. Teichoic acid from *S. faecalis* has been identified as the D antigen (Wicken *et al.*, 1963). Ribitol teichoic acid from *Bacillus subtilis* is a polyribitol phosphate with 1,5-phosphodiester links (Fig. 3.9). Glucose is linked to carbon 4 of ribitol and D-alanine is esterified to carbon 2 or 3 of ribitol. *Bacillus subtilis* teichoic acid is nine ribitol residues in length. Teichoic acid from *Staphylococcus aureus* is similar to that from *B. subtilis* except that N-acetyl glucosamine is attached to carbon 4 of ribitol in place of

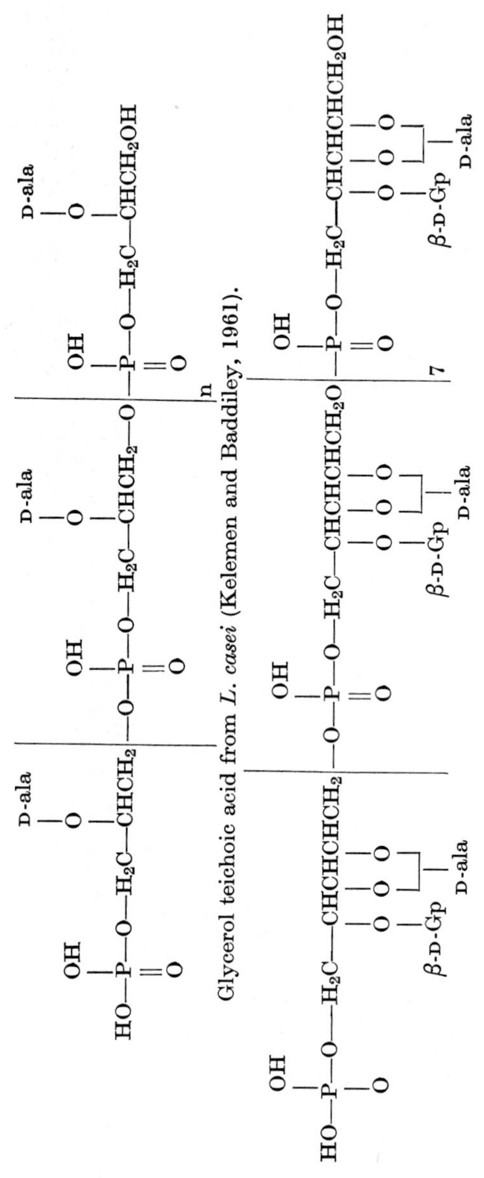

Glycerol teichoic acid from *L. casei* (Kelemen and Baddiley, 1961).

Ribitol teichoic acid from *B. subtilis* (Armstrong *et al.*, 1961).

Fig. 3.9. Structure of teichoic acids from bacteria. Gp = glucopyranose; ala = alanine.

glucose and the polymer is eight ribitol residues in length (Baddiley *et al.*, 1962).

C. Complex Polysaccharides

Cell walls of gram positive bacteria are complex polysaccharides composed of carbohydrates and amino acids. Methods used for the preparation of cell walls generally involve mechanical breakage of the cell, followed by differential centrifugation and washing of the cell walls (Salton, 1960). The backbone of the cell wall of gram positive bacteria appears to be a mucopeptide composed of N-acetyl glucosamine, muramic acid and amino acids. The amino acids identified in cell wall hydrolysates are glutamic acid, alanine, glycine, aspartic acid, lysine or diaminopimelic acid, but not both, and occasionally serine. Much of the alanine and aspartic acid and all of the glutamic acid is present as the

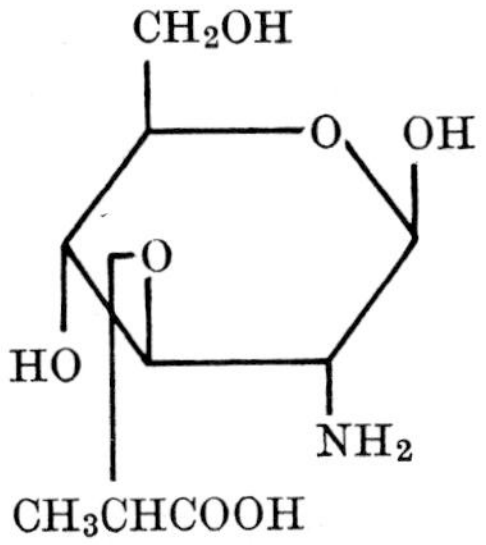

CH_2OH
O OH
O
HO
NH_2
$CH_3CHCOOH$

Fig. 3.10. Structure of muramic acid.

D-isomer (Ikawa and Snell, 1960). Diaminopimelic acid is usually present as the *meso* or LL isomer (Hoare and Work, 1957). Muramic acid was discovered in hydrolysates of cell walls and spores by Strange and Dark (1956) and identified as 3-0-carboxyethyl-D-glucosamine (Strange and Kent, 1959) (Fig. 3.10). Muramic acid or mucopeptide-like materials have been isolated from gram negative bacteria, rickettsia, and even the large viruses (see review by Perkins, 1963).

Knowledge of the arrangement of the components of the mucopeptide has been obtained by isolation and identification of fragments obtained as a result of the action of hydrolytic enzymes such as lysozyme. One such product has been identified as O-(N-acetyl-β-D-glucosaminyl)-(1-4)-N-acetylmuramic acid (Jeanloz *et al.*, 1964), and this compound presumably identifies the linkage between glucosamine and muramic acid of the peptide. The peptide is joined to muramic acid through the carboxyl carbon of the lactyl and the nitrogen of L-alanine (Fig. 3.11; Park, 1964). Mucopeptide backbone chains are cross-linked by bonds involving the epsilon nitrogen of lysine or diaminopimelic acid and the carboxyl carbon of the terminal D-alanine (Fig. 3.11). *Staphylococcus aureus* contains a pentaglycine bridge between alanine and lysine, while other organisms are known to contain bridges of alanine or alanine plus threonine (Petit *et al.*, 1966).

Besides the mucopeptide, many cell walls contain polysaccharide. The C substance of group A streptococci which has already been mentioned appears to be located in the cell wall. Teichuronic acid, a polymer of glucuronic acid and N-acetyl galactosamine which was isolated from *Bacillus subtilis* by Janczura *et al.* (1961), is another polysaccharide located in the cell wall fraction.

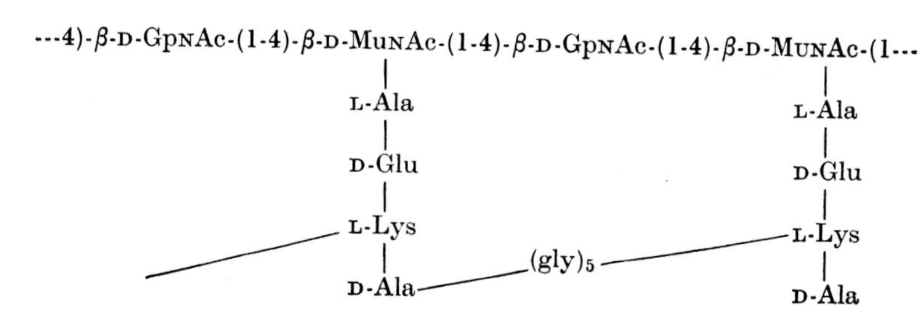

FIG. 3.11. Tentative structure for *Staphylococcus aureus* mucopeptide. Abbreviations used: MuNAc = N-acetylmuramic acid; GpNAc = N-acetylglucosaminopyranose.

Cell walls of gram negative bacteria are composed of protein, lipid and carbohydrate plus mucopeptide similar to that of gram positive bacteria (Salton, 1960). A lipopolysaccharide complex can be isolated from cell walls of gram negative bacteria and this complex seems to be identical with the endotoxin of this group of organisms. At least part of the lipid is lipid A, although another cephalin-like lipid, lipid B, has also been separated from endotoxin preparations (Westphal and Lüderitz, 1954).

The polysaccharide portion of the lipopolysaccharide is the O-antigen of gram negative bacteria. The sugars listed in Fig. 3.12 are the ones which occur in *Salmonella* polysaccharides, which are the best characterized polysaccharides. All *Salmonella* serotypes contain glucosamine, glucose, galactose and a heptose. The only heptose isolated from the enteric bacteria so far is L-glycero-D-mannoheptose, but other heptoses are known to be produced by the gram negative bacteria (Davies, 1960). The remainder of the polysaccharide is composed of up to three of the sugars listed under "side chains" in Fig. 3.12, in addition to those already mentioned. Fucoseamine (2-amino-2,6-dideoxy-D-galactose (Crumpton and Davies, 1958)) and viosamine (4-amino-4,6-dideoxy-D-glucose (Stevens *et al.*, 1963)) occur in lipopolysaccharides of *Chromobacterium violaceum.*

The dideoxy hexoses are an unusual group of sugars which were discovered in the chemical studies of lipolysaccharides. The trivial names given to these sugars denote the species from which the sugar was first isolated: tyvelose from *S. typhosa*, abequose from *S. abortus equi*, paratose from *S. paratyphi* A and colitose from *E. coli*. Ascarylose

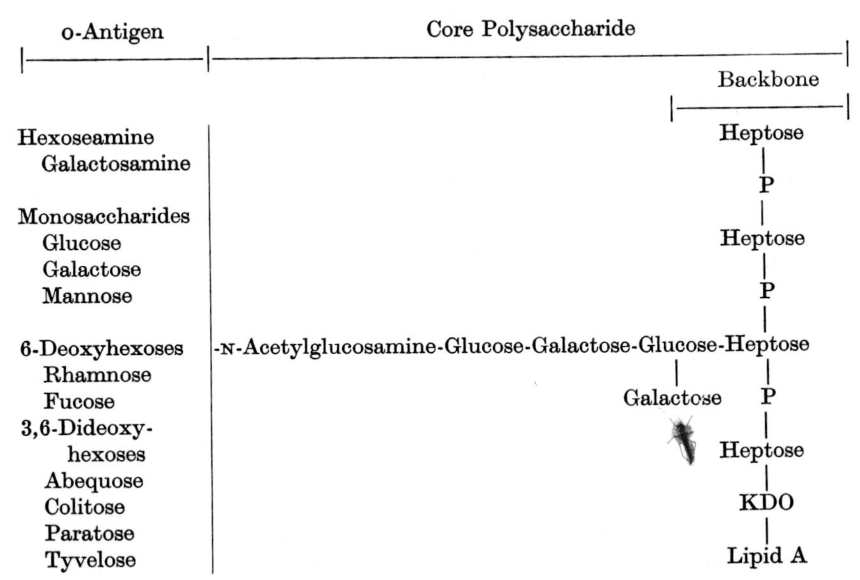

FIG. 3.12. Structure of lipopolysaccharide of *Salmonella* species. All *Salmonella* serotypes contain at least the four sugars of the core polysaccharide plus heptose and KDO and in certain rough strains these may be the only sugars present (Horecker, 1966). The smooth organisms contain up to three additional sugars in the O-antigen which is attached to the core. In *S. typhimurium* and probably other *Salmonella* species, ethanolamine occurs in the backbone structure (Grollman and Osborn, 1964). P = phosphate, KDO = 2-keto-3-deoxyoctulosonic acid.

was first isolated from egg cell membranes of *Ascaris* worms, but has been detected in *P. pseudotuberculosis* (Davies, 1960). The configurations of these sugars were established by synthesis (Fouquey *et al.*, 1958); tyvelose is 3,6-dideoxy-D-mannose; abequose is 3,6-dideoxy-D-galactose; paratose is 3,6-dideoxy-D-glucose; colitose is 3,6-dideoxy-L-galactose; and ascarylose is 3,6-dideoxy-L-mannose.

A tentative partial structure of the lipopolysaccharide of salmonellae which has been deduced from both chemical and biosynthetic studies is shown in Fig. 3.12 (Nikaido, 1964; Horecker, 1966; see also Chapter

15). The core polysaccharide consists of a heptose phosphate backbone with side chains composed of glucose, galactose and N-acetylglucosamine. In addition to these sugars, 2-keto-3-deoxyoctulosonic acid (KDO) has been identified in lipopolysaccharides (Heath and Ghalambor, 1963). The absolute configurations of the hydroxylated carbons have not been established, but carbons 1 to 3 are derived from phosphoenolpyruvate and carbons 4 to 8 from D-arabinose-5-phosphate (Levin and Racker, 1959). Some of the KDO occurs as end groups of the polysaccharide and is thought to link the polysaccharide with lipid A (Osborn, 1963; Horecker, 1966). The core polysaccharide appears to be common to all *Salmonella* serotypes and possibly to other gram negative bacteria as well. The O-antigen specificity of lipopolysaccharide resides in the highly branched side chains (Fig. 3.12), and certain aspects of the fine structure of the antigenic side chains have been determined as a consequence of studies of the biosynthesis of lipopolysaccharide (Chapter 15).

IV. Nucleic Acids

A. Composition

Deoxyribonucleic acid (DNA) of bacteria is similar in composition to DNA from higher plants and animals. Bacterial DNA contains no sugars other than deoxyribose (Belozersky and Spirin, 1960), but DNA of the T-even phages of *E. coli* contains glucose in addition to deoxyribose. Adenine, guanine, cytosine and thymine comprise the major bases of DNA, (Table 3.6), but small amounts of 6-methylaminopurine are present in DNA of *E. coli* (Dunn and Smith, 1955) and other bacteria.

TABLE 3.6. Bases of DNA and RNA of Bacteria and Bacteriophage

Deoxyribonucleic acid	Ribonucleic acid
Major Bases	Major Bases
Adenine	Adenine
Thymine	Uracil
Guanine	Guanine
Cytosine	Cytosine
Minor Bases	Minor Bases
6-Methylaminopurine	Thymine
5-Hydroxymethylcytosine	5-Methylcytosine
(In T-even phages only)	6-Methylaminopurine
	6-Dimethylaminopurine
	2-Methyladenine

6-Methylaminopurine does not occur in DNA outside of bacteria (Dunn and Smith, 1958). On the other hand, 5-methylcytosine, a minor base of plant and animal DNA, has not been detected in bacterial DNA (Brawerman and Shapiro, 1962). 5-Hydroxymethylcytosine entirely replaces cytosine in DNA of T-even phages, but does not occur in DNA of T-odd phages (Wyatt and Cohen, 1953; Table 3.7).

TABLE 3.7. Base Ratios of DNA from Bacteria and Bacteriophage

Bacteria	Base ratios, mole % A	T	G	C	$\frac{A+T}{G+C}$	Reference
E. coli	23	27	25	26	0·99	(a)
P. aeruginosa	18	18	34	31	0·55	(b)
A. tumefaciens	22	20	30	28	0·72	(b)
V. cholerae	29	28	20	23	1·31	(b)
S. marcescens	21	21	29	29	0·72	(b)
S. enteritidis	25	25	25	25	1·00	(b)
C. diphtheriae	24	24	25	27	0·92	(b)
L. bifidus	22	21	28	29	0·74	(b)
B. subtilis	29	29	21	21	1·36	(b)
C. perfringens	37	36	14	13	2·70	(b)

Bacteriophage	A	T	G	C	HMC	$\frac{A+T}{G+C}$	Reference
T2	33	33	18	—	17	1·89	(c)
T4	32	33	18	—	16	1·91	(c)
T6	32	33	18	—	17	1·85	(c)
T1	27	25	23	25	—	1·08	(d)
T5	30	31	20	20	—	1·53	(c)
ΦX174	25	33	24	18	—	—	(e)

Mole $\% = \frac{\text{moles base}}{\text{moles A+T+G+C}}$ (100). Abbreviations: A=adenine, T=thymine, G=guanine, C=cytosine, HMC=5-hydroxymethylcytosine. Literature data are rounded off to the nearest whole number. References: (a) Smith and Wyatt, 1951; (b) Lee *et al.*, 1956; (c) Wyatt and Cohen, 1953; (d) Creaser and Taussig, 1957; (e) Sinsheimer, 1959.

Similarly, ribonucleic acid (RNA) of bacteria contains the conventional components ribose, adenine, guanine, cytosine and uracil, together with minor amounts of unusual bases. Thymine (Littlefield and Dunn, 1958) and 5-methylcytosine (Amos and Korn, 1958) are present in RNA of *E. coli* and other bacteria. A series of methylated adenines has been found in RNA from bacteria (Littlefield and Dunn, 1958), and these, as well as methylated guanines, have been found in plant and animal RNA (Brawerman and Shapiro, 1962.)

B. Structure

Native DNA is a polymer of deoxyribonucleotides with a molecular weight in the millions and is either linear or has a very small amount of branching (Ulbricht, 1963; Brown, 1963). The deoxyribonucleosides are attached to each other by 5,3-phosphodiester bridges. Purines are attached to C-1 of deoxyribose by a β-glycosidic linkage with position 9, and pyrimidines are similarly linked through position 3. Glucose in DNA

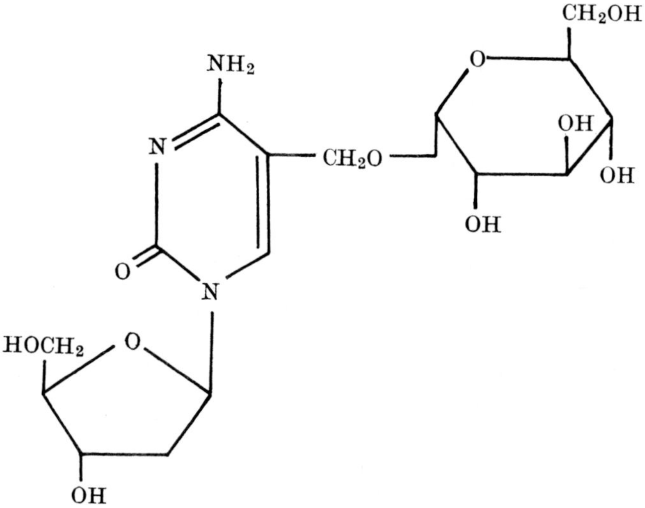

Fig. 3.13. Glucosylated hydroxymethylcytidine from T2 and T4 phage DNA. (Lehman, 1960.)

of the T-even phages is linked to 5-hydroxymethylcytosine by an α-glycosidic linkage to the hydroxymethyl substituent (Lehman, 1960; Fig. 3.13). Phages T2 and T4 contain only one glucose residue per base, but T6 contains two glucose units linked to 5-hydroxymethylcytosine. The glucose units are linked to each other by a β-(1-6) linkage.

The molar proportions of bases in DNA obey certain rules discovered by Chargaff (1955); adenine equals thymine and guanine equals cytosine (Table 3.7). Moreover, the base ratios are characteristic for each species studied. Discovery of these rules was important in the elucidation of the fine structure of DNA, for they substantiated Watson's and Crick's proposal (1953) that DNA was a two-stranded helical chain, the two strands being held together by hydrogen bonds between the 6-amino and 6-keto bases (Fig. 3.14). This structure explained the adenine-

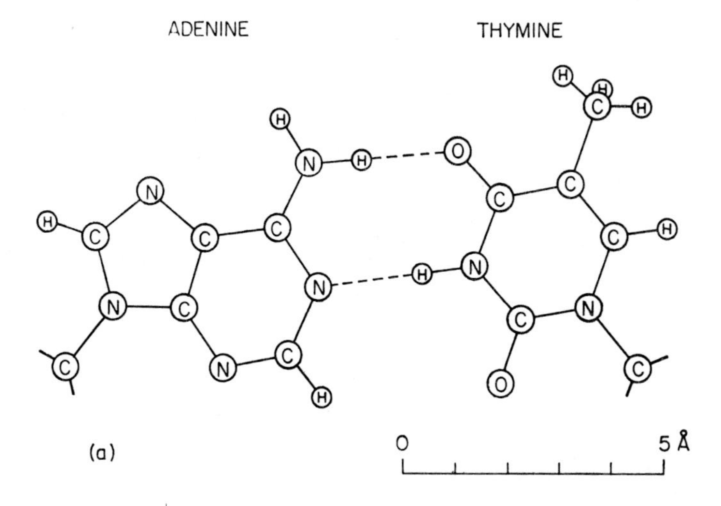

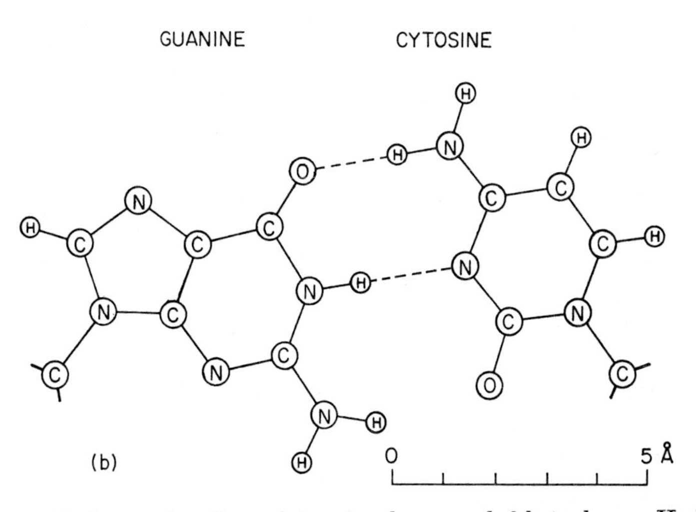

FIG. 3.14. Hydrogen bonding of 6-amino bases and 6-keto bases. Hydrogen bonds are indicated by dotted lines. Carbon-1 of deoxyribose is shown in each case. Reproduced with permission from Watson and Crick, 1953.

thymine and guanosine-cytosine equivalences, and subsequent studies have borne out this relationship. An interesting exception to Chargaff's rule is the bacteriophage ΦX174 DNA (Sinsheimer, 1959; Table 3.7), where the proportions of the bases bear no relationship to each other and it is known that this species of DNA is single-stranded.

TABLE 3.8. Base Ratios of Bacterial RNA

Organism	Base ratios, mole % A	U	G	C	Reference
E. coli	24	19	31	24	(a)
P. aeruginosa	25	19	32	24	(a)
S. marcescens	20	24	31	24	(b)
S. typhosa	26	19	31	24	(a)
C. diphtheriae	23	22	32	24	(a)
B. cereus	25	23	31	21	(c)
C. perfringens	28	20	30	22	(a)

Base ratios are expressed in the same manner as in Table 3.7; abbreviations used are also the same with the additional abbreviation U=uracil. References: (a) Belozersky and Spirin, 1960; (b) Elson and Chargaff, 1955; (c) Stuy, 1958.

The base ratios of total cell RNA do not exhibit the species specificity that DNA does, and do not follow Chargaff's rules (Table 3.8). On the other hand, there are quantitatively minor fractions of RNA which do show specificity. This type of RNA is designated messenger RNA because this is the RNA which carries the code message for the synthesis of a specific protein from the chromosome to the protein-synthesizing enzyme system. Messenger RNA was discovered by Volkin and Astrachan (1956) in a study of RNA from phage-infected *E. coli*. They found that the base ratios of this type of RNA reflected the base ratio of the infecting phage.

References

Abraham, E. P. (1957). "Biochemistry of Some Peptide and Steroid Antibiotics". John Wiley and Sons, New York.
Ambler, R. P. and Rees, M. W. (1959). *Nature* **84**, 56.
Amos, H. and Korn, M. (1958). *Biochim. Biophys. Acta.* **29**, 444.
Anderson, R. J. (1932). *J. Biol. Chem.* **97**, 639.
Anderson, R. J. and Chargaff, E. (1929–30). *J. Biol. Chem.* **85**, 77.
Armstrong, J. J., Baddiley, J. and Buchanan, J. (1961). *Biochem. J.* **80**, 254.
Asselineau, C., Asselineau, J., Ryhage, R., Stallberg-Stenhagen, S. and Stenhagen, E. (1959). *Acta Chem. Scand.* **13**, 822.
Asselineau, J. and Lederer, E. (1955). "Ciba Found. Symp. Exp. Tuberc. Bacillus Host." Little, Brown and Co., Boston. p. 14.
Asselineau, J. and Lederer, E. (1960). *In* "Lipid Metabolism" (K. Bloch, ed.). John Wiley and Sons, New York, p. 337.
Astbury, W. T., Beighton, E. and Weibull, C. (1955). *Symp. Soc. Exptl Biol.* **9**, 282.
Baddiley, J., Buchanan, J. G., Martin, R. O., and Rajbhandary, U. L. (1962). *Biochem. J.* **85**, 49.
Ball, D. H. and Adams, G. A. (1959). *Can. J. Chem.* **37**, 1012.

Barclay, K. S., Bowne, E. J., Stacey, M. and Webb, M. (1954). *J. Chem. Soc.* 1501.

Barker, S. A., Bourne, E. J. and Stacey, M. (1950). *J. Chem. Soc.* 2884.

Barker, S. A., Heidelberger, M., Stacey, M. and Tipper, D. J. (1958). *J. Chem. Soc.* 3468.

Barry, G. T. (1958). *J. Exptl Med.* **107**, 507.

Belozersky, A. N. and Spirin, A. S. (1960). *In* "The Nucleic Acids" (E. Chargaff and J. N. Davidson, eds.). Academic Press, New York, Vol. III, p. 147.

Bergstrom, S., Theorill, H. and Davide, H. (1946). *Arch. Biochem.* **10**, 165.

Bishop, D. G. and Still, J. L. (1963). *J. Lipid Res.* **4**, 81.

Bloch, H. (1950). *J. Exptl Med.* **91**, 197.

Brawerman, G. and Shapiro, H. S. (1962). *In* "Comparative Biochemistry" (M. Florkin and H. S. Mason, eds.). Academic Press, New York, Vol. IV, p. 107.

Brown, D. M. (1963). *In* "Comprehensive Biochemistry" (M. Florkin and E. H. Stotz, eds.). Elsevier Publishing Company, Amsterdam, Vol. 8, p. 208.

Buehler, H. J., Schantz, E. J. and Lamanna, C. (1947). *J. Biol. Chem.* **169**, 295.

Burton, A. J. and Carter, H. E. (1964). *Biochemistry* **3**, 411.

Campbell, L. L. and Manning, G. B. (1961). *J. Biol. Chem.* **236**, 2962.

Chargaff, E. (1955). *In* "The Nucleic Acids" (E. Chargaff and J. N. Davidson, eds.). Academic Press, New York, Vol. I, p. 307.

Creaser, E. H. and Taussig, A. (1957). *Virology* **4**, 200.

Crumpton, M. J. and Davies, D. A. L. (1958). *Biochem. J.* **70**, 729.

Davies, D. A. L. (1960). *Advan. Carbohydrate Chem.* **15**, 271.

Deeb, S. S. and Hager, L. P. (1964). *J. Biol. Chem.* **239**, 1024.

Dolin, M. (1957). *J. Biol. Chem.* **225**, 557.

Dolin, M. I. and Wood, N. P. (1960). *J. Biol. Chem.* **235**, 1809.

Dunn, D. B. and Smith, J. O. (1955). *Nature* **175**, 336.

Dunn, D. B. and Smith, J. O. (1958). *Biochem. J.* **68**, 627.

Elson, D. and Chargaff, E. (1955). *Biochim. Biophys. Acta* **17**, 367.

Feingold, D. S. and Gehatia, M. (1957). *J. Polymer Sci.* **23**, 783.

Fouquey, C., Lederer, E., Luderitz, O., Polonsky, J., Staub, A. M., Stirm, M. S., Tinelli, R. and Westphal, O. (1958). *Compt. Rend.* **246**, 2417.

Ghuysen, J. M. (1961). *Biochim. Biophys. Acta* **47**, 561.

Grollman, A. P. and Osborn, M. J. (1964). *Biochemistry* **3**, 1571.

Hager, L. P. (1959). *Bacteriol Proc.* p. 109.

Hauge, J. G. (1964). *J. Biol. Chem.* **239**, 3630.

Heath, E. C. and Ghalambor, M. A. (1963). *Biochem. Biophys. Res. Commun.* **10**, 340.

Hehre, E. J., Carlson, A. S. and Neill, J. M. (1947). *Science* **106**, 523.

Hehre, E. J. and Hamilton, D. M. (1946). *J. Biol. Chem.* **166**, 777.

Heidelberger, M. (1962). *Arch. Biochem.* Suppl. **1**, 169.

Henning, U., Helinski, D. R., Chao, F. C. and Yanofsky, C. (1962). *J. Biol. Chem.* **237**, 1523.

Herbert, D. and Pinsent, J. (1948). *Biochem. J.* **43**, 193.

Heyman, H., Manniello, J. M. and Barkulis, S. S. (1963). *J. Biol. Chem.* **238**, 502.

Heyns, K., Kiessling, G., Lindenberg, W., Paulsen, H. and Webster, M. E. (1959). *Chem. Ber.* **92**, 2435.

Hibbert, H. and Barsha, J. (1931). *J. Am. Chem. Soc.* **53**, 3907.

Hoare, D. S. and Work, E. (1957). *Biochem. J.* **64**, 441.

Hofmann, K., Hsigo, C. Y. Y., Henis, D. and Panos, C. (1955). *J. Biol. Chem.* **217**, 49.

Hofmann, K., Lucas, R. N. and Sax, S. M. (1952). *J. Biol. Chem.* **195**, 473.
Hofmann, K. and Tausig, F. (1955). *J. Biol. Chem.* **213**, 425.
Horecker, B. L. (1966). *Ann. Rev. Microbiol.* **20**, 253.
Housewright, R. D. (1962). *In* "The Bacteria" (I. C. Gunsalus and R. Y. Stanier, eds.). Academic Press, New York, Vol. III, p. 389.
Ikawa, M., Koepfli, J. B., Mudd, S. G. and Niemann, C. (1953). *J. Am. Chem. Soc.* **75**, 1035.
Ikawa, M. and Snell, E. E. (1960). *J. Biol. Chem.* **235**, 1367.
Janczura, E., Perkins, H. R. and Rogers, H. J. (1961). *Biochem. J.* **80**, 82.
Jarvis, F. G. and Johnson, M. J. (1949). *J. Am. Chem. Soc.* **71**, 4124.
Jeanes, A., Haynes, W. C., Wilham, C. A., Rankin, J. C., Melvin, E. H., Austin, M. J., Cluskey, J. E., Fisher, B. E., Tsuchiya, H. W. and Rist, C. E. (1954). *J. Am. Chem. Soc.* **76**, 5041.
Jeanloz, R. W., Sharon, N. and Flowers, H. M. (1964). *Bacterial Endotoxins*, Proc. Symp., Rutgers State Univ., New Brunswick, N.J., 49.
Jones, G. E. and Benson, A. A. (1965). *J. Bacteriol.* **89**, 260.
Jones, J. K. N. and Perry, M. B. (1957). *J. Am. Chem. Soc.* **79**, 2787.
Junge, J. M., Stein, E. A., Neurath, H. and Fischer, E. H. (1959). *J. Biol. Chem.* **234**, 556.
Kamen, M. D. and Bartsch, R. G. (1961). *Haematin Enzymes*, Symp. Inter. Union Biochem. Canberra **2**, 417.
Kaneda, T. (1963). *J. Biol. Chem.* **238**, 1222.
Kaneshiro, T. and Marr, A. G. (1961). *J. Biol. Chem.* **236**, 2615.
Kaushal, R. and Walker, J. K. (1951). *Biochem. J.* **48**, 618.
Kelemen, M. V. and Baddiley, J. (1961), *Biochem. J.* **80**, 246.
Kendall, F. E., Heidelberger, M. and Dawson, M. H. (1937). *J. Biol. Chem.* **118**, 61.
Landy, M. and Braun, W. (eds.) (1964). *Bacterial Endotoxins*, Proc. Symp., Rutgers State Univ., New Brunswick, N.J.
Lederer, E. and Pudles, J. (1951). *Bull. Soc. Chim. Biol.* **33**, 1003.
Lee, K. Y., Wahl, R. and Barbo, E. (1956). *Ann. Inst. Pasteur* **91**, 212.
Lehman, I. R. (1960). *J. Biol. Chem.* **235**, 3254.
Lemoigne, M. (1927). *Ann. Inst. Pasteur* **41**, 148.
Levin, D. H. and Racker, E. (1959). *J. Biol. Chem.* **234**, 2532.
Littlefield, J. N. and Dunn, D. B. (1958). *Biochem. J.* **70**, 642.
Lovenberg, W., Buchana, B. B. and Rabinowitz, J. C. (1963). *J. Biol. Chem.* **238**, 3899.
Lundgren, D. G., Alper, R., Schnaitman, C. and Marchessault, R. H. (1965). *J. Bacteriol.* **89**, 245.
Lynn, R. J. and Smith, P. F. (1960). *Ann. N.Y. Acad. Sci.* **79**, 493.
McQuillen, K. (1960). *In* "The Bacteria" (I. C. Gunsalus and R. Y. Stanier, eds.). Academic Press, New York, Vol. I, p. 249.
McQuire, E. J. and Binkley, S. B. (1964). *Biochemistry* **3**, 247.
Manning, G. B., Campbell, L. L. and Foster, R. J. (1961). *J. Biol. Chem.* **236**, 2959.
Marr, A. G. and Ingraham, J. L. (1962). *J. Bacteriol.* **84**, 1260.
Michel, G., Bordet, C. and Lederer, E. (1960). *Compt. Rend.* **250**, 3518.
Miller, M. W. (1961). "The Pfizer Handbook of Microbial Metabolites." McGraw-Hill Book Co., New York.
Morton, R. K. (1958). *Rev. Pure Appl. Chem.* **8**, 161.
Nikaido, H. (1964). *Bacterial Endotoxins*, Proc. Symp., Rutgers State Univ., New Brunswick, N.J., **67**.

Noll, H., Bloch, H., Asselineau, J. and Lederer, E. (1956). *Biochim. Biophys. Acta* **20**, 299.
Osborn, M. J. (1963). *Proc. Natl Acad. Sci. U.S.* **50**, 499.
Park, J. T. (1964). Bacterial Endotoxins, Proc. Symp., Rutgers State Univ., New Brunswick, N.J., 63
Perkins, H. R. (1963). *Bacteriol. Rev.* **27**, 18.
Petit, J., Munoz, E. and Ghuysen, J. (1966). *Biochemistry* **5**, 2764.
Philpot, F. J. and Wells, A. Q. (1952). *Am. Rev. Tuberc.* **66**, 28.
Plackett, P., Buttery, S. H. and Cottew, G. S. (1962). *Recent Progr. Microbiol.* Symp. Intern. Congr. Microbiol., 8th, Montreal, Can., p. 535.
Polgar, N. (1954a). *J. Chem. Soc.* 1003.
Polgar, N. (1954b). *J. Chem. Soc.* 1011.
Postgate, J. (1961). *Haematin Enzymes*, Symp. Intern. Union Biochem. Canberra **2**, 407.
Putman, E. W., Potter, A. L., Hodgson, R. and Hassid, W. Z. (1950). *J. Am. Chem. Soc.* **72**, 5024.
Rebers, P. A. and Heidelberger, M. (1961). *J. Am. Chem. Soc.* **83**, 3056.
Reeves, R. E. and Goebel, W. F. (1941). *J. Biol. Chem.* **139**, 511.
Saito, K. (1960). *J. Biochem.* **47**, 710.
Salton, M. R. J. (1960). "Microbial Cell Walls." John Wiley and Sons, Inc., New York.
Schaeffer, W. I. and Umbreit, W. W. (1963). *J. Bacteriol.* **85**, 492.
Sifferd, R. H. and Anderson, R. J. (1936). *Z. Physiol. Chem.* **239**, 270.
Sinsheimer, R. L. (1959). *J. Mol. Biol.* **1**, 43.
Smith, J. D. and Wyatt, G. R. (1951). *Biochem. J.* **49**, 144.
Smith, L. (1961). *In* "The Bacteria" (I. C. Gunsalus and R. Y. Stanier eds.). Academic Press, New York, Vol. II, p. 365.
Spielman, M. A. (1934). *J. Biol. Chem.* **106**, 87.
Stevens, C. L., Blumbergs, P., Daniher, F. A., Wheat, R. W., Kujomoto, A. and Rollins, E. L. (1963). *J. Am. Chem. Soc.* **85**, 3061.
Stodola, F. H. and Anderson, R. J. (1936). *J. Biol. Chem.* **114**, 467.
Stodola, F. H., Lesuk, A. and Anderson, R. J. (1938). *J. Biol. Chem.* **126**, 505.
Strange, R. E. and Dark, F. A. (1956). *Nature* **177**, 186.
Strange, R. E. and Kent, L. H. (1959). *Biochem. J.* **71**, 333.
Stuy, J. H. (1958). *J. Bacteriol.* **76**, 179.
Thorne, C. B. (1956). *Sixth Symp. Soc. Gen. Microbiol.* **68**.
Tourtellottee, M. E., Jensen, R. G., Gander, G. W. and Morowitz, H. J. (1963). *J. Bacteriol.* **86**, 370.
Ulbricht, T. L. V. (1963). *In* "Comprehensive Biochemistry" (M. Florkin and E. H. Stotz, eds.). Elsevier Publishing Co., Amsterdam, **8**, p. 58.
Van Alstyne, D., Gerwing, J. and Tremaine, J. H. (1966). *J. Bacteriol.* **92**, 796.
Vilkas, E. and Lederer, E. (1956). *Bull. Soc. Chim. Biol.* **38**, 111.
Volkin, E. and Astrachan, L. (1956). *Virology* **2**, 149.
Watson, J. D. and Crick, F. H. C. (1953). *Nature* **171**, 964.
Westphal, O. and Lüderitz, O. (1954). *Angew. Chem.* **66**, 407.
Wicken, A. J. and Baddiley, J. (1963). *Biochem. J.* **87**, 54.
Wicken, A. J., Elliot, S. D. and Baddiley, J. (1963). *J. Gen. Microbiol.* **31**, 231.
Wyatt, G. R. and Cohen, S. S. (1953). *Biochem. J.* **55**, 774.
Yamanaka, T., Kijimoto, S. and Okunuki, K. (1963). *J. Biochem.* (*Tokyo*) **53**, 256.
Yamanaka, T. and Okunuki, K. (1963). *Biochem. Z.* **338**, 62.

Part Two

ENERGY METABOLISM

4. OLIGOSACCHARIDE CATABOLISM

Life on our planet continues because the elements of protoplasm become available for re-use by another generation of organisms by the action of saprophytic microorganisms. The carbon cycle begins with the production of sugars by green plants using energy from sunlight; plants use the energy stored in sugars for growth and animals use it by feeding on the plants; plants and animals die and microorganisms convert their complex structures partly to cell material and partly to carbon dioxide thereby obtaining energy.

I. Digestion of Starch and Glycogen

A. Amylases

Starch and glycogen, which are the storage carbohydrates of plants and animals, are composed of varying proportions of amylose, a linear α-(1-4) glucoside, and amlopectin, an α-(1-4) glucoside with (1-6) branching. The two best known hydrolytic enzymes active against starch and

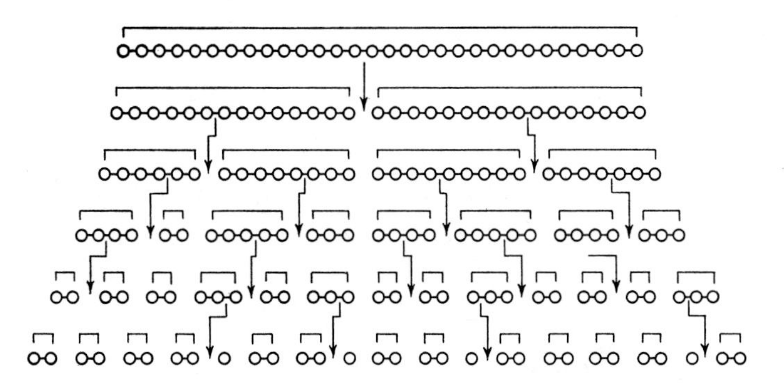

Fig. 4.1. Action of α-amylase on amylose. (-O-O-) represents glucose units connected by (1-4) linkages, ↓ are points of attack by α-amylase. The bottom row shows the ultimate products of α-amylase action on amylose, although not necessarily in the proportions actually obtained. (Reproduced with permission from Bernfeld, 1951.)

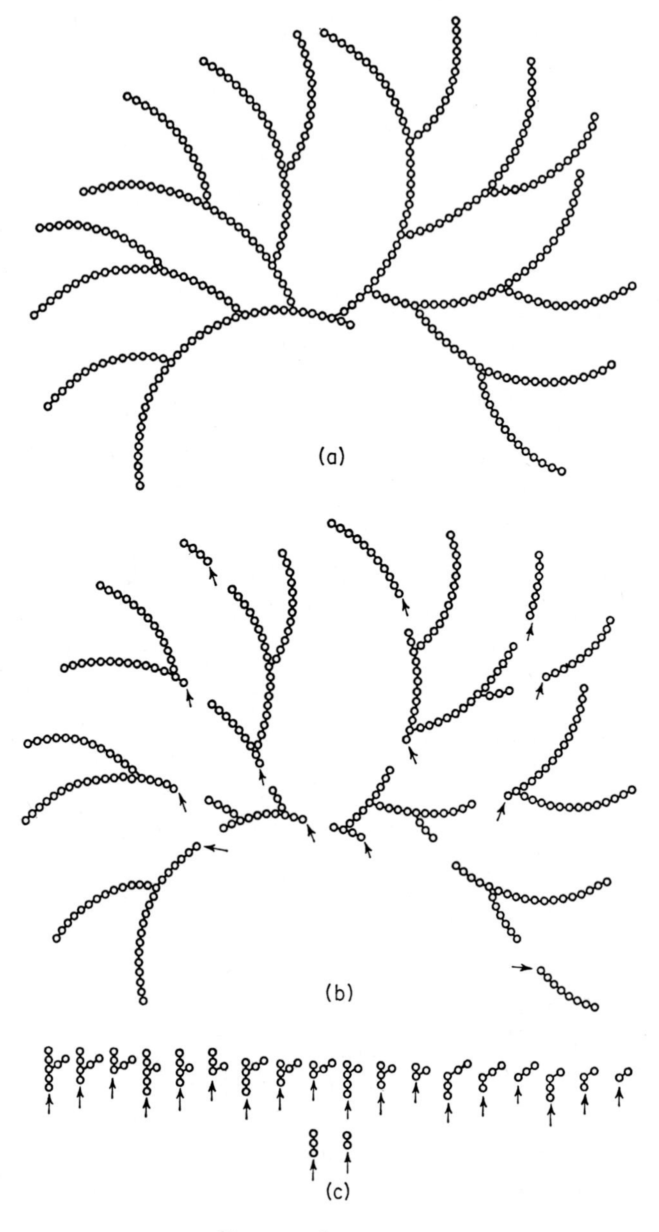
(a)
(b)
(c)

∘ Glucose unit
↑ Reducing end group
α-1,4-glucosidic linkage
α-1,6-glucosidic linkage

glycogen are α- and β-amylases. α-Amylases (α-1,4-glucan 4-glucano-hydrolase) cleave amylose to produce dextrins in the early stages of the reaction, but ultimately a mixture of approximately 6 parts maltose and 1 part glucose (Fig. 4.1; Bernfeld, 1962). With amylopectins, the products depend on the degree of branching, since α-amylases will bypass but not hydrolyze (1-6) linkages. Therefore the products consist of a mixture of maltose, glucose and dextrins with the (1-6) linkages intact (Fig. 4.2). These facts, plus kinetic data, indicate that α-amylases attack at random points along the amylose chain.

In contrast, β-amylases (α-1,4-glucan maltohydrolase) attack amylose at the non-reducing end of the chain and hydrolytically remove successive maltose residues, resulting in the production of only maltose (Fig. 4.3). It is for this reason that α- and β-amylases are also referred

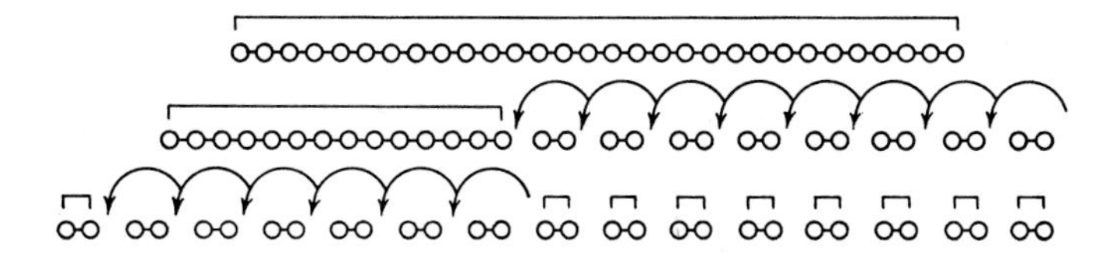

FIG. 4.3. Action of β-amylase on amylose. (-O-O-) represents glucose units connected by (1-4) linkages, ↓ are points of β-amylase attack. The non-reducing end of the amylose chain is to the right. The only products are maltose. (Reproduced with permission from Bernfeld, 1951.)

to as dextrinogenic and saccharogenic amylases, respectively. The linkage broken by β-amylase is inverted, resulting in the production of β-maltose, that is, maltose with the reducing carbon in the β-configuration. This is in contrast to the action of α-amylases where the α-configuration is retained. This difference was first noticed by Kuhn (1924), who proposed the terminology for α- and β-amylases based on the type of maltose produced. β-Amylases are stopped at the (1-6) branch points of amylopectin, resulting in the formation of maltose and β-limit dextrin (Fig. 4.4).

α-Amylases are widely distributed in nature amongst plants, animals and microorganisms (Bernfeld, 1962), including bacteria of the genera *Bacillus*, *Clostridium* and *Pseudomonas* (Rogers, 1961), but very little

---

FIG. 4.2. Action of α-amylase on amylopectin. The native, branched amylopectin is shown at the top, next is the partially digested amylopectin showing the random attack of the enzyme, and in the bottom row are shown the ultimate products of α-amylase action on amylopectin with the (1-6) linkages intact. (Reproduced with permission from Bernfeld, 1951.)

fundamental work has been done on the nature of the reaction catalyzed by the bacterial enzymes.

β-Amylases are common in plants, unknown in animals, and rare in bacteria. There are reports of bacterial amylases with more saccharifying than dextrinizing activity, but no detailed studies on the nature of the reaction. The amylase of *Bacillus polymyxa*, which is discussed later, has properties of both α- and β-amylases.

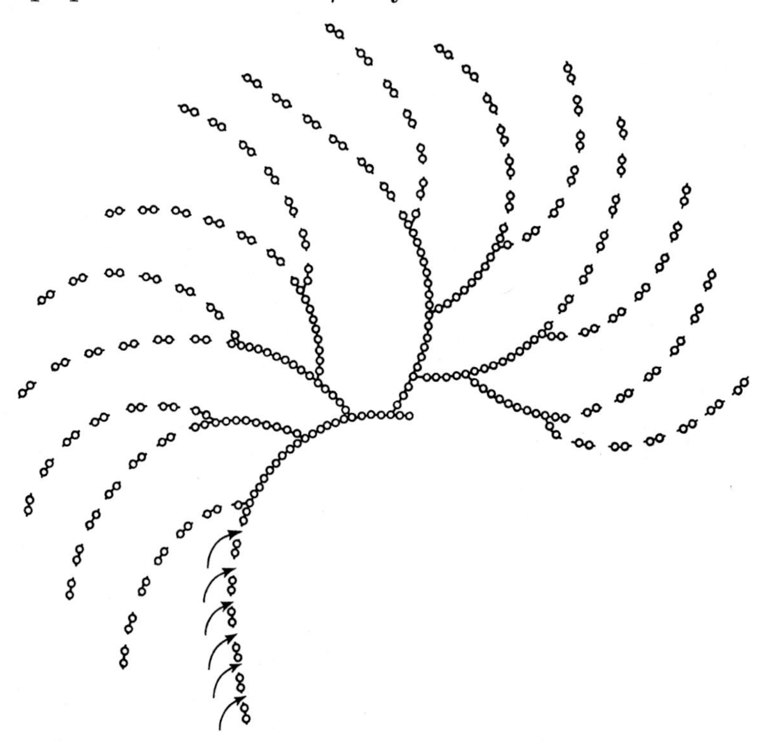

FIG. 4.4. Action of β-amylase on amylopectin. The structure of the β-limit dextrin is shown in the center of the figure. (Reproduced with permission from Bernfeld, 1951.)

Some bacteria form polysaccharidases different in action from α- and β-amylases. One such enzyme is the "amylase" of *Bacillus macerans* which is actually a transglycosylase with little or no hydrolytic activity (French, 1957). This remarkable enzyme produces cyclic dextrins from starch and glycogen which were first discovered by Schardinger in 1903, and are still known as Schardinger dextrins. The predominant cyclodextrins produced contain six, seven and eight glucose residues and are known as α-, β-, and γ-dextrins, respectively.

French *et al.* (1954) found that *B. macerans* amylase (α-1,4-glucan 4-glycosyltransferase, cyclizing) converted Schardinger dextrins to linear oligosaccharides if a co-substrate containing glucose at the non-reducing end was supplied, such as glucose itself, maltose, maltobionic acid or cellobiose. The products contained glucose at the reducing end (Fig. 4.5) and were themselves substrates for the enzyme, producing a mixture of linear oligosaccharides. For example, α-dextrin and glucose yielded glucosides from 2 to 10 glucose units in length. These studies can be extrapolated to explain the formation of Schardinger dextrins from starch where the non-reducing end of the amylose chain acts as the co-substrate (Fig. 4.5).

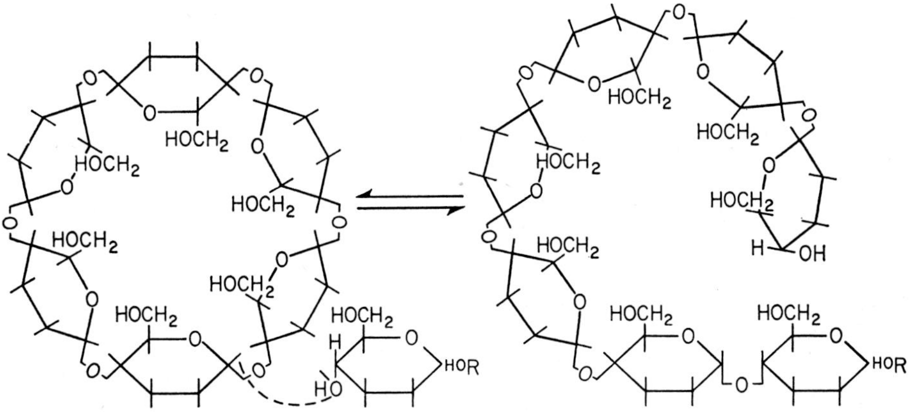

FIG. 4.5. Action of *Bacillus macerans* amylase. The reaction from left to right is the cleavage of α-dextrin by a co-substrate to produce a linear oligosaccharide. (Reproduced with permission from Hehre, 1951).

Another unusual bacterial polysaccharidase is the amylase of *B. polymyxa* which has properties of both α- and β-amylases (Robyt and French, 1964). *Bacillus polymyxa* amylase will attack amylose, amylopectin and β-limit dextrin, indicating that the enzyme can bypass the (1-6) branch points. On the other hand, β-maltose is the main product and therefore an inversion occurs at the reducing carbon during hydrolysis. A unique feature of the enzyme is that it will hydrolyze Schardinger dextrins and is thus the first purified bacterial enzyme to have this ability.

B. PHOSPHORYLASES

Polysaccharide phosphorylase (α-1,4-glucan:orthophosphate glucosyltransferase) catalyzes the phosphorylytic cleavage of α-D-glucose-1-

phosphate from the non-reducing end of the amylose chain (Brown and Cori, 1961). It is noteworthy that the product of the reaction is in the α-configuration and hence no net inversion occurs during phosphorolysis. Arsenate will replace phosphate (arsenolysis), but in this case the product is free glucose, presumably because glucose-1-arsenate is unstable. The reaction is reversible; in fact, the initial interest in phosphorylase was for its role in glycogen biosynthesis. All polysaccharide phosphorylases require an oligosaccharide primer, although the length of the primer varies with the enzyme source. All polysaccharide phosphorylases contain pyridoxal phosphate as the prosthetic group, but the means by which this co-factor functions is unknown.

Phosphorylases have been reported in animals, plants and bacteria, but the enzymes from animal and plant sources are the best characterized. Phosphorylase has been reported in species of *Clostridium* (Volchok *et al.*, 1955; Shemanova and Blagoveshchenskii, 1957), *Streptococcus* (Ushijima and McBee, 1957), *Neisseria* (Hehre, 1949), *Corynebacterium* (Carlson and Hehre, 1949) and *Agrobacterium* (Madsen, 1961). In many cases, the observed activity appeared only when the organism was grown in the presence of starch. There are no studies with purified bacterial enzymes and it is not known if the action of the bacterial enzymes is similar to animal and plant phosphorylases.

II. Digestion of Cellulose

Cellulose is one of the world's most abundant polysaccharides, and the ability to digest this substance is wide-spread among a number of diverse microorganisms. Cellulolytic bacteria include a number of gram negative aerobic saprophytes (Siu, 1951), anaerobic rumen bacteria of several genera (Hungate, 1950), and anaerobic thermophilic sporeformers (McBee, 1950). One very unusual group of organisms engaged in the digestion of cellulose is the genus *Cytophaga* the slime bacteria, (Stanier, 1942).

Whitaker (1953, 1957) purified a cellulase from the mold *Myrothecium verrucaria* which, in some respects, has an action analogous to α-amylase. With the aid of kinetic studies, he determined that *Myrothecium* cellulase (β-1,4-glucan 4-glucanohydrolase) attacked the substrate at random points along the cellulose chain to produce both cellobiose and glucose as the products. Several pieces of evidence indicated that the observed enzymic activity was due to the action of a single enzyme.

On the other hand, *Cellvibrio gilvus*, an organism isolated from cow manure, produces a cellulase with a mode of action analogous to β-amylase. Storvick *et al.* (1963) studied the action of this enzyme with β-(1-4) glucosides up to six hexose units in length. They found that this enzyme began digestion of cellulose at the non-reducing end of the molecule. With up to five glucose units (cellopentaose), the attack was almost exclusively at the second glucoside linkage from the non-reducing end, resulting in the formation of cellobiose; with six glucose units in the chain, there was appreciable hydrolysis at the third glucose linkage from the non-reducing end, resulting in the release of cellotriose as well as cellobiose. Storvick and his co-workers found that an inversion occurred during hydrolysis resulting in the formation of α-cellobiose.

III. Digestion of Other Polysaccharides by Bacteria

Hyaluronic acid is formed by several species of streptococci as well as being a common constituent of connective tissue. Hyaluronidase (hyaluronate lyase) is also formed by several species of bacteria, but the

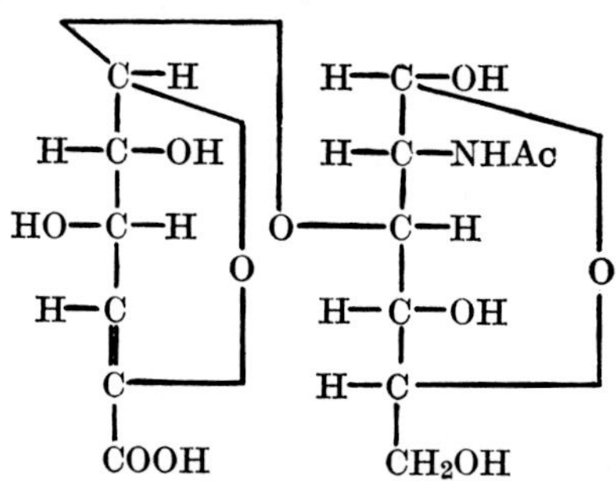

FIG. 4.6. Structure of unsaturated disaccharide produced by bacterial hyaluronidase. (Reproduced with permission from Linker *et al.*, 1956.)

product of the bacterial enzyme is different from that of the other known hyaluronidases. Bacterial hyaluronidase produces an unsaturated disaccharide 3-0-(β-D-4,5-glucoseenpyranosyluronic acid)-2-deoxy-2-acetamido-D-glucose (Fig. 4.6), which suggests that the mechanism of enzyme action is an elimination reaction (Meyer *et al.*, 1960). Hyaluronidase has been reported in species of *Diplococcus*, *Streptococcus*, *Staphylococcus*, *Clostridium*, *Proteus* and *Flavobacterium*, and apparently all bacterial hyaluronidases from the same product. Testicular and snake venom hyaluronidase form tetrasaccharides as the main product of their action on hyaluronic acid by a mechanism which involves transglycosylation (Meyer *et al.*, 1960). Bacteria are able to attack several other important polysaccharides such as chitin, agar, blood group substances and pneumococcal polysaccharides, but in most cases very little is known about either the products or the mechanism (Rogers, 1961; Pollock, 1962).

IV. Cleavage of Galactosides

The most common natural galactoside is milk lactose, and the best characterized enzyme for the cleavage of galactose is β-galactosidase

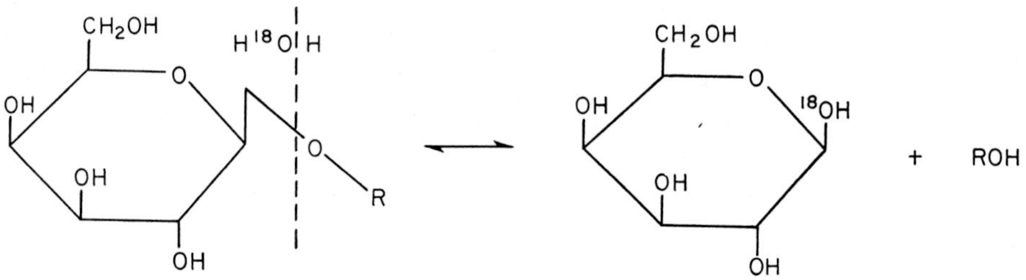

FIG. 4.7. Action of β-galactosidase. The bond between C-1 of galactose and oxygen is the bond which is split since the ^{18}O is found in galactose and not the aglycone, ROH.

(β-D-galactoside galactohydrolase) of *E. coli* (Wallenfels and Malhotra, 1961). β-Galactosidase catalyzes the cleavage of β-galactosides to β-D-galactose and the corresponding aglycone, which is glucose when lactose is the substrate (Fig. 4.7). The bond between C-1 of galactose and oxygen is the bond which is split (Wallenfels and Malhotra, 1961). This type of cleavage was predictable on the basis of the rule proposed by

TABLE 4.1. Relative Rates of Substrate Hydrolysis by β-Galactosidase of *E. coli*. (Wallenfels and Malhotra, 1961.)

Substrate	Relative rate of hydrolysis
Ethyl-β-D-galactoside	1
β-D-Galactosyl-(1-4)-β-D-glucose (β-lactose)	33
β-D-Galactosyl-(1-4)-α-D-glucose (α-lactose)	90
Equilibrium mixture of α- and β-lactose	70
β-D-Galactosyl-(1-3)-D-glucose	2
β-D-Galactosyl-(1-6)-D-glucose	7
o-Nitrophenyl-β-D-galactoside	2125
p-Nitrophenyl-β-D-galactoside	300
α-L-Arabinosyl-(1-3)-D-glucose	56
α-L-Arabinosyl-(1-6)-D-glucose	9
o-Nitrophenyl-α-L-arabinoside	196
p-Nitrophenyl-α-L-arabinoside	136

Koshland (1954), that the cleavage of a glycoside occurs between the carbon and oxygen of the sugar for which the enzyme has the most specific requirement. β-Galactosidase of *E. coli* is specific for the glycone (galactose); α-L-arabinosides are also hydrolyzed by the enzyme but L-arabinose differs from D-galactose only by the substitution of H at position 5 for —CH_2OH. The enzyme requires the β-linkage, but (1-3), (1-4) and (1-6) linkages are all attacked (Table 4.1). On the other hand, the aglycone can vary within rather wide limits; for example, *o*-nitrophenyl-β-D-galactoside is a much better substrate than lactose.

In addition to hydrolysis, β-galactosidase also catalyzes transfer reactions which can be regarded as a special case of the general reaction;

$$R—OR' + R''OH \rightarrow R—OR'' + R'OH$$

when R″OH is water, the result is hydrolysis, and when R″OH is another sugar, the result is transglycosylation.

V. Cleavage of Glucosides

The principal source of maltose and cellobiose is the enzymic degradation of starch, glycogen, and cellulose. Maltase (α-D-glucoside glucohydrolase) has been reported in several species of bacteria, and some studies have been done with the enzyme from *Clostridium acetobutylicum* (French and Knapp, 1950). An indirect utilization of maltose occurs in *E. coli* where maltose is converted to glucose and a polymer of glucose by amylomaltase (α-1,4-glucan:D-glucose 4-glucosyltransferase) (Monod and Torriani, 1950; Wiesmeyer and Cohen, 1960). The glucose poly-

$$\text{Maltose} + H_2O \rightarrow 2 \text{ glucose}$$

$$n \text{ Maltose} \rightarrow n \text{ glucose} + (\text{glucose})_n$$

mer is cleaved by phosphorylase to glucose-1-phosphate (Doudoroff *et al.*, 1949).

The phosphorylytic cleavage of maltose and cellobiose occurs in bacteria which decompose starch and cellolose, but apparently not in higher organisms. Maltose and cellobiose phosphorylase have properties in common which set them apart from sucrose phosphorylase; both cause an inversion during phosphorolysis, and in both cases arsenate will replace phosphate in the overall reaction. Inorganic phosphate

$$\text{Maltose} + H_3PO_4 \rightarrow \beta\text{-D-glucose-1-phosphate} + \text{glucose}$$

$$\text{Cellobiose} + H_3PO_4 \rightarrow \alpha\text{-D-glucose-1-phosphate} + \text{glucose}$$

does not exchange with the phosphate of glucose-1-phosphate and aresenate will not cause the decomposition of glucose-1-phosphate to free glucose. Maltose phosphorylase (maltose: orthophosphate glucosyltransferase) has been studied in extracts of *Neisseria meningitidis* (Fitting and Doudoroff, 1952), and cellobiose phosphorylase in extracts of *Ruminococcus flavifaciens* (Ayers, 1959), *Clostridium thermocellum* (Alexander, 1961) and *Cellvibrio gilvus* (Hulcher and King, 1958).

VI. Cleavage of Fructosides

The most abundant natural fructoside is sucrose. Invertase occurs in many organisms, including bacteria (Myrback, 1960), but very little is known of the bacterial enzymes.

$$\text{Sucrose} + H_2O \rightarrow \text{glucose} + \text{fructose}$$

Amylosucrase of *Neisseria perflava* catalyzes a reaction similar to

$$n \text{ Sucrose} \rightarrow n \text{ fructose} + (\text{glucose})_n$$

amylomaltase (Hehre, 1949).

Sucrose phosphorylase (disaccharide glucosyltransferase) of *P. saccharophila* catalyzes the phosphorylytic cleavage of sucrose with retention of the configuration of the glycosidic bond (Hassid and Doudoroff, 1950). Arsenate will replace phosphate in the overall reaction resulting in the formation of free glucose. However, inorganic phosphate will exchange with the phosphate of glucose-1-phosphate and arsenate

$$\text{Sucrose} + H_3PO_4 \rightarrow \alpha\text{-D-glucose-1-phosphate} + \text{fructose}$$

will cause the decomposition of glucose-1-phosphate with the formation of free glucose. The reaction catalyzed by sucrose phosphorylase is reversible and, in fact, sucrose formed by the reverse reaction was used as evidence to substantiate the 1-1 linkage of sucrose. Sucrose phosphorylase is specific with respect to the glucosyl portion, but several other sugars can substitute for fructose. The cleavage occurs between C-1 of glucose and oxygen (Cohn, 1949), which agrees with Koshland's rule.

VII. Other Oligosaccharides

The unsaturated disaccharide formed by the action of bacterial hyaluronidase on hyaluronic acid is hydrolyzed by an enzyme from a *Flavobacterium* species to the corresponding 4-deoxy-5-ketoglucuronic acid and N-acetylglucosamine (Linker *et al.*, 1960; Fig. 4.8).

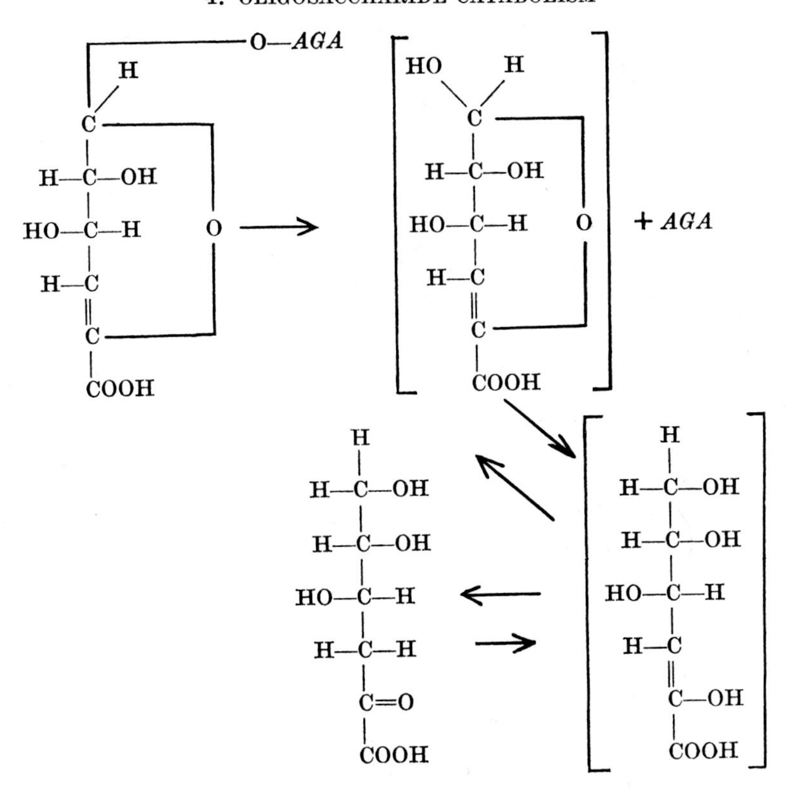

FIG. 4.8. Hydrolysis of unsaturated disaccharide by an enzyme from *Flavobacterium* species. (Reproduced with permission from Linker *et al.*, 1960.) AGA = N-acetylglucosamine.

References

Alexander, J. K. (1961). *J. Bacteriol.* **81**, 903.

Ayers, W. A. (1959). *J. Biol. Chem.* **234**, 2819.

Bernfeld, P. (1951). *Advan. Enzymol.* **12**, 379.

Bernfeld, P. (1962). *In* "Comparative Biochemistry" (M. Florkin and H. W. Mason, eds.). Academic Press, New York, Vol. III, p. 355.

Brown, D. H. and Cori, C. F. (1961). *In* "The Enzymes" (P. D. Boyer, H. Lardy and K. Myrback, eds.). Academic Press, New York, Vol. 5, p. 207.

Carlson, A. S. and Hehre, E. J. (1949). *J. Biol. Chem.* **177**, 281.

Cohn, M. (1949). *J. Biol. Chem.* **180**, 771.

Doudoroff, M., Hassid, W. Z., Putman, E. W., Potter, A. L. and Lederberg, J. (1949). *J. Biol. Chem.* **179**, 921.

Fitting, C. and Doudoroff, M. (1952). *J. Biol. Chem.* **199**, 153.

French, D. (1957). *Advan. Carbohydrate Chem.* **12**, 189.

French, D. and Knapp, D. W. (1950). *J. Biol. Chem.* **187**, 463.

French, D., Levine, M. L., Norberg, E., Nordin, P., Pazur, J. H. and Wild, G. M. (1954). *J. Am. Chem. Soc.* **76**, 2387.
Hassid, W. Z. and Doudoroff, M. (1950). *Advan. Enzymol.* **10**, 123.
Hehre, E. J. (1949). *J. Biol. Chem.* **177**, 267.
Hehre, E. J. (1951). *Advan. Enzymol.* **11**, 297.
Hulcher, F. H. and King, K. W. (1958). *J. Bacteriol.* **76**, 571.
Hungate, R. E. (1950). *Bacteriol. Rev.* **14**, 1.
Koshland, D. (1954). *In* "The Mechanism of Enzyme Action" (W. D. McElroy and B. Glass, eds.). The Johns Hopkins Press, Baltimore, Maryland, p. 608.
Kuhn, R. (1924). *Chem. Ber.* **57**, 1965.
Linker, A., Hoffman, P., Meyer, K., Sampson, P. and Korn, E. D. (1960). *J. Biol. Chem.* **235**, 3061.
Linker, A., Meyer, K. and Hoffman, P. (1956). *J. Biol. Chem.* **219**, 13.
McBee, R. H. (1950). *Bacteriol. Rev.* **14**, 51.
Madsen, N. B. (1961). *Biochim. Biophys. Acta.* **50**, 194.
Meyer, K., Hoffman, P. and Linker, A. (1960). *In* "The Enzymes" (P. Boyer, H. Lardy and K. Myrback, eds.). Academic Press, New York, Vol. 4, p. 447.
Monod, J. and Torriani, A.-M. (1950). *Ann. Inst. Pasteur* **78**, 65.
Myrback, K. (1960). *In* "The Enzymes" (P. D. Boyer, H. Lardy and K. Myrback, eds.). Academic Press, New York, Vol. 4, p. 379.
Pollock, M. R. (1962). *In* "The Bacteria" (I. C. Gunsalus and R. Y. Stanier, eds.). Academic Press, New York, Vol. IV, p. 121.
Robyt, J. and French, D. (1964). *Arch. Biochem. Biophys.* **104**, 338.
Rogers, H. J. (1961). *In* "The Bacteria" (I. C. Gunsalus and R. Y. Stanier eds.). Academic Press, New York, Vol. II, p. 257.
Shemanova, G. F. and Blagoveshchenskii, V. A. (1957). *Biokhimiya* **22**, 799.
Siu, R. G. H. (1951). "Microbial Decomposition of Cellulose". Reinhold, New York.
Stanier, R. Y. (1942). *Bacteriol. Rev.* **6**, 143.
Storvick, W. O., Cole, F. E. and King, K. W. (1963). *Biochemistry* **2**, 1106.
Ushijima, R. N. and McBee, R. H. (1957). *Proc. Montana Acad. Sci.* **17**, 33.
Volchok, A. K., Ivanov, V. J. and Lobanova, A. V. (1955). *Biokhimiya* **20**, 522.
Wallenfels, K. and Malhotra, O. P. (1961). *Advan. Carbohydrate Chem.* **16**, 239.
Whitaker, D. R. (1953). *Arch. Biochem. Biophys.* **43**, 253.
Whitaker, D. R. (1957). *Can. J. Biochem. Physiol.* **35**, 733.
Wiesmeyer, H. and Cohn, M. (1960). *Biochim. Biophys. Acta* **39**, 417.

5. TRANSPORT OF SUGARS

At some point in the metabolism of an energy source, the substrate must enter the cell to be acted upon by the enzymes of the organism. Although no transport enzymes have been isolated, the bacterial cell must assist actively in the entry process because (a) substrates often accumulate in concentrations higher than the concentration of the medium, (b) entry processes are selective for stereoisomers, and (c) transport systems are under genetic control. Transport systems of bacteria are referred to as "permease systems" and the specific transport protein as the "permease" (Kepes and Cohen, 1962; Cohen and Monod, 1957).

I. Transport Systems

A. Transport of Galactosides

The most thoroughly studied bacterial transport system is that responsible for the transport of β-galactosides by *E. coli*. Although lactose does not accumulate internally in wild type *E. coli* because of the high activity of β-galactosidase, it has been possible to study the transport of β-galactosidase by the use of mutants which are unable to grow on lactose because of a lack of β-galactosidase, and by use of galactosides which are substrates for the transport system but are not hydrolyzed by β-galactosidase (Kepes and Cohen, 1962).

Because of the nature of transport systems, there is no direct assay for a permease and, therefore, no permease has ever been isolated. The usual assay for permeases is accomplished by incubating whole cells with radioactive substrate, filtering quickly on a membrane filter and then determining the amount of radioactive substrate associated with the cells. Uptake of β-galactosides measured in this way shows many characteristics of enzymic reactions. The transport system is stereospecific for β-galactosides as opposed to β-glucosides, although a wide variety of compounds will substitute for the glucose part of lactose.

The initial velocity of galactoside uptake obeys Michaelis-Menten kinetics and can be expressed as

$$V_{\text{in}} = V_{\text{in}}^{\max} \frac{G_{\text{ex}}}{K_m + G_{\text{ex}}}$$

Where V_{in} is the velocity of the entry reaction, $V_{\text{in}}^{\max}$ is the maximum velocity, K_m is the Michaelis constant, and G_{ex} is the concentration of the substrate in the medium. The internal concentration of galactoside, G_{in}, depends on the external concentration according to the expression

$$G_{\text{in}} = C \frac{G_{\text{ex}}}{K_m + G_{\text{ex}}}$$

where C is the capacity, the amount of substrate accumulated when G_{ex} is at a saturating concentration (Rickenberg *et al.*, 1956). (In the

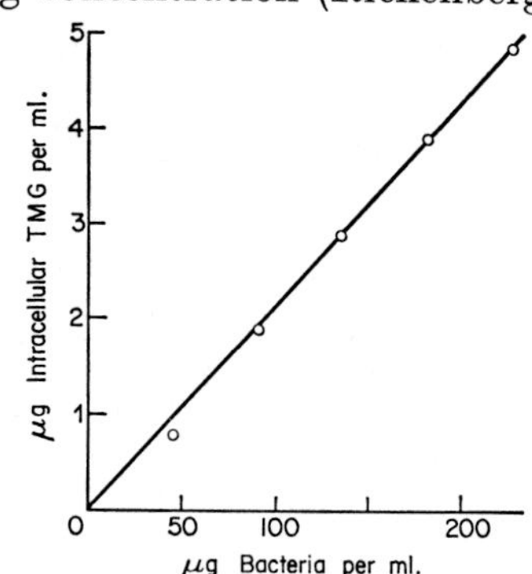

FIG. 5.1. Kinetics of induction of galactoside transport system. (Reproduced with permission from Rickenberg *et al.*, 1956.) At zero time, thiomethyl-β-D-galactoside was added to a succinate-salts medium inoculated with *E. coli* and the cells were harvested by filtration when growth reached the indicated cell concentrations. Galactoside permease appeared after the addition of the inducer and was proportional to the concentration of the cell dry weight which is characteristic of inducible enzyme systems (Cohen and Monod, 1957).

original publication, capacity was represented by the symbol Y; C is used here in order to avoid confusion with the yield coefficient.) C is not an inherent characteristic of an organism, but depends on the experimental conditions.

The proposal that a specific protein participates in the transport process is based on the finding that galactoside permease is inducible and that the kinetics of induction are typical of induced enzyme formation (Fig. 5.1). Furthermore, chloramphenicol, which is known to inhibit protein synthesis, inhibits the induction of galactoside permease (Rickenberg *et al.*, 1956).

Almost all of the galactoside can be extracted from the *E. coli* in an unchanged form, which argues against a chemical modification of the substrate during transport. On the contrary, intracellular lactose contributes to the internal osmotic pressure of the cell, which indicates that the substrate exists in simple aqueous solution rather than bound to an internal receptor (Sistrom, 1958). Energy enters into the transport of galactosides in some manner, since the accumulation of galactosides is inhibited by sodium azide and 2,4-dinitrophenol (Rickenberg *et al.*, 1956). An external energy source is usually not required, presumably because the reserve material of the cell can supply sufficient energy for transport of substrates.

Exit of galactosides from *E. coli* appears to be the result of a mechanism different from the entry reaction. Firstly, the kinetics of the two processes are different since exit is a zero order reaction and rate of exit is proportional to the internal concentration of the galactoside; secondly, the exit reaction is much more temperature-dependent than the entry reaction (Kepes and Cohen, 1962), which accounts for the marked change in C with change in temperature.

B. Transport of Galactose

Galactose transport by *E. coli* was studied in a mutant of this organism which was devoid of galactokinase and therefore unable to grow on galactose as the energy and carbon source (Horecker *et al.*, 1960a). In many respects, this system is similar to, although distinct from, the β-galactoside transport system of *E. coli*. Galactose permease was specific for galactose since none of the other sugars tested, including β-galactosides, were concentrated. The sugar was extracted from the cells and accounted for entirely as free galactose. It was possible to demonstrate the accumulation of galactose as much as several thousand-fold over the concentration in the medium. At about 2×10^{-5} M galactose, the system was saturated with an internal concentration of about 6×10^{-2} M galactose. Horecker and his co-workers calculated that about 5500 calories were required per mole galactose transported. As with β-galactoside permease, 2,4-dinitrophenol inhibited the uptake of galactose suggesting an energy requirement for uptake.

Horecker *et al.* (1960b) were able to demonstrate a clear-cut distinction between the entry and exit reactions of galactose. Cells grown on succinate plus galactose accumulated about one-fifth the total amount of galactose accumulated by cells grown on succinate alone. These workers suspected that galactose was acting as an inducer of the exit reaction, and by kinetic analysis of the induced and uninduced cells they found that the rate of uptake of galactose was unchanged but that the exit rate of galactose was accelerated approximately five-fold in induced cells. They were able to demonstrate the induction of the exit reaction by galactose and found that the kinetics of induction were similar to those found for galactoside transport illustrated in Fig. 5.1, except that there was an appreciable rate of exit of galactose in uninduced cells. 2,4-Dinitrophenol added to cells preloaded with galactose resulted in a rapid loss of sugar, and the rate of exit was close to that of untreated cells; hence it appeared that the energy requirement was for the entry and not for the exit reaction.

C. Transport of Other Sugars by Bacteria

Maltose permease is another transport system which has been studied in *E. coli* (Wiesmeyer and Cohn, 1960). Most, but not all, of the radioactive substrate was extracted as free maltose, as apparently some maltose was metabolized. No exit could be demonstrated for maltose, but in other respects the permease was similar to those already described. Other bacterial permeases are the β-D-glucuronide permease of *E. coli* (Stoeber, 1957; Kepes and Cohen, 1962) and the α-glucoside permease of the same organism. α-Glucoside permease may be responsible for glucose transport since glucose strongly inhibits the uptake of α-methyl glucoside (Cohen and Monod, 1957).

II. Mechanism of Sugar Transport by Bacteria

It is not yet possible to reconstruct a bacterial transport system with purified components, but some conclusions can be drawn from the kinetic and genetic experiments. Several models have been proposed, but the one which accounts for most of the observations is the carrier mediated transport system (Cohen and Monod, 1957; Cirillo, 1961; Kepes and Cohen, 1962) shown in Fig. 5.2. In this model, the bacterial cell is surrounded by an impermeable membrane which prevents the entrance of polar compounds except by way of the specific transport system. This membrane accounts for the occurrence of "cryptic" mutants which are known to have the internal enzymic apparatus for the metabolism of a substrate but are unable to grow on the compound, presumably because of an inability to form the specific permease. The most logical candidate for the impermeable barrier is the lipoprotein cell membrane (McQuillen, 1960) which would be a poor solute for sugars and other polar compounds. Transport of sugars by bacteria and small intestine (Hokin and Hokin, 1963) is generally an energy-requiring process evidenced by the accumulation of sugars in excess of the concentration in the medium. In yeast the internal sugar concentration is close to that of the medium, and this type of stereo-specific transport is viewed as a facilitated diffusion process (Cirillo, 1961). The

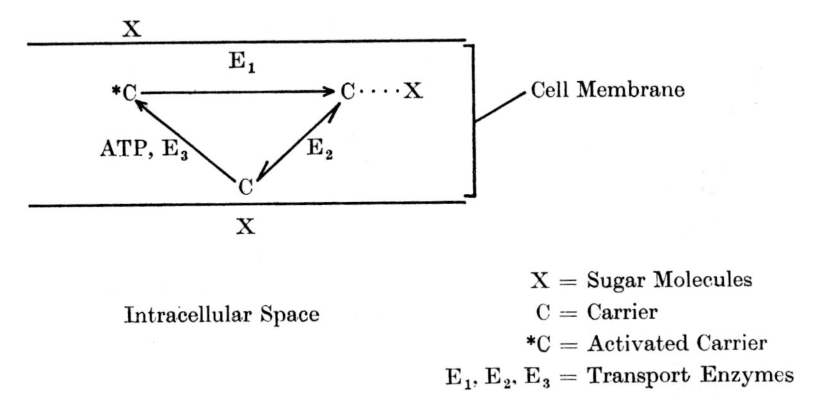

FIG. 5.2. Representation of the carrier theory of active transport.

sugar first reacts in a specific manner with an activated carrier which may be catalyzed by an enzyme, E_1 (Fig. 5.2). The carrier-substrate complex diffuses to the internal side of the cell membrane where the reversible dissociation of the complex is catalyzed by enzyme E_2. The carrier is activated again by enzyme E_3 which requires energy as represented by ATP; in the case of facilitated diffusion, activation of the carrier is not necessary. Fox and Kennedy (1965) have partially purified a protein from the cytoplasmic membrane of *E. coli*, which they suggest functions as a carrier in the β-galactoside transport system of this organism.

References

Cirillo, V. P. (1961). *Ann. Rev. Microbiol.* **15**, 197.
Cohen, G. N. and Monod, J. (1957). *Bacteriol. Rev.* **21**, 169.
Fox, C. F. and Kennedy, E. P. (1965). *Proc. Nat. Acad. Sci. U.S.* **54**, 891.
Hokin, L. E. and Hokin, M. R. (1963). *Ann. Rev. Biochem.* **32**, 553.
Horecker, B. L., Thomas, J. and Monod, J. (1960a). *J. Biol. Chem.* **235**, 1580.
Horecker, B. L., Thomas, J. and Monod, J. (1960b). *J. Biol. Chem.* **235**, 1586.
Kepes, A. and Cohen, G. N. (1962). *In* "The Bacteria" (I. C. Gunsalus and R. Y. Stanier, eds.). Academic Press, New York, Vol. IV, p. 179.
McQuillen, K. (1960). *In* "The Bacteria" (I. C. Gunsalus and R. Y. Stanier, eds.). Academic Press, New York, Vol. I, p. 245.
Rickenberg, H. V., Cohen, G. V., Buttin, G. and Monod, J. (1956). *Ann. Inst. Pasteur* **91**, 829.
Sistrom, W. R. (1958). *Biochim. Biophys. Acta* **29**, 579.
Stoeber, F. (1957). *Compt. rend. acad. sci.* **244**, 1091.
Wiesmeyer, H. and Cohn, M. (1960). *Biochim. Biophys. Acta* **39**, 440.

6. FERMENTATION OF SUGARS

I. Methods of Study

Pasteur defined fermentation as "la vie sans air" and this definition is still applied in academic circles, although many aerobic industrial processes are also referred to by those in the trade as fermentations. The products of fermentation are organic chemicals and the products themselves can be a clue to the mechanism of the fermentation. Since fermentation is a relatively poor energy-yielding process compared to aerobic oxidation, large amounts of sugar are utilized, usually with negligible assimilation of the carbon of the energy source. This latter fact makes it feasible to prepare fermentation balances which are chemical balances of products per mole of sugar utilized. Fermentation balance studies are usually made with organisms grown in sealed flasks under an inert atmosphere, or with a washed cell suspension in a medium which contains the energy source but which will not allow proliferation of the organism.

In the construction of a fermentation balance, the data are tabulated as in Table 6.1, with the sugar consumed (in this case glucose), and the products formed listed in column (1). In column (2), the products formed per mole of glucose consumed are recorded. In column (3), the oxidation states of the energy source and products are listed. Usually neither oxygen nor any other oxidant enters into the reaction; thus the oxidation state of the products should equal the oxidation state of the energy source. In calculating the oxidation state of a compound, hydrogen is arbitrarily assigned a value of $-0{\cdot}5$ and oxygen a value of $+1{\cdot}0$. Amino nitrogen is frequently encountered in fermentation studies and has a value of $+1{\cdot}5$ in the system used; for example, glucose, $C_6H_{12}O_6$, and glucosamine, $C_6H_{13}O_5N$, both have an oxidation state of 0·0. In columns (4) and (5) are listed the values obtained by multiplying the moles of product produced by their oxidation states, oxidized products in column (4) and reduced products in column (5). If all the products have been recovered, the algebraic sum of columns (4) and (5) should equal the oxidation state of the substrate. If this is not the case then

TABLE 6.1. Calculation of a Fermentation Balance. Data taken from Friedemann (1939)

	(1)	(2)	(3)	(4)	(5)	(6)
Energy source used						
Glucose	36·5 mmoles	—	0·0	—	—	6·0
Products						
Lactic acid	54·3 mmoles	1·49	0·0	—	—	4·47
Formic acid	10·0 ,,	0·27	+1·0	+0·27	—	0·27
Acetic acid	4·6 ,,	0·13	0·0	—	—	0·26
Ethyl alcohol	5·5 ,,	0·15	−2·0	—	−0·30	0·30
Sum:				+0·27	−0·30	5·30

Sum of columns (4)+(5) = −0·03; O/R index = 0·27/0·30 = 0·90; carbon recovery = 5·30/6·00 × 100 = 88%.

Column (1) = Energy source used and products produced.
Column (2) = Moles product per mole energy source.
Column (3) = Oxidation state of energy source or product.
Column (4) = Value of oxidized products, column (2) × column (3).
Column (5) = Value of reduced products, column (2) × column (3).
Column (6) = Carbon recovery.

either there are unknown products, or some unknown oxidant or reductant is participating in the fermentation. In the case of the example cited in Table 6.1, the sum of the oxidized and reduced products is −0·03, which is in very good agreement with the theoretical value of 0·0. Frequently the sum of the oxidized products is divided by the sum of the reduced products to provide a figure designated as the O/R index. The O/R index is 1·0 for sugars with an oxidation state of 0·0, such as glucose or galactose, where the oxidized products must equal the reduced products. However, with sugars more oxidized or more reduced than glucose, the O/R index can have many values and hence is not useful in these cases.

Another check on the accuracy of the fermentation data is the carbon recovery, column (6), Table 6.1. The figures for column (6) are obtained by multiplying the moles of product obtained by the number of carbons in the product. These figures are added and compared with the number of carbons per mole substrate, in this case, 6 for glucose. The carbon recovery in the example cited in Table 6.1 was about 88%, which is acceptable.

The value of a fermentation balance is that it provides clues to the type of cleavage which occurs in the catabolism of the sugar. For example, in the balance presented in Table 6.1, three-quarters of the glucose is accounted for as lactic acid, which suggests a cleavage of glucose to two moles of a C_3 intermediate, which is, of course, pyruvate. The formation of ethyl alcohol, acetic acid, and formic acid can be explained by assuming that the C_3 intermediate undergoes further metabolism to C_2 plus C_1.

The details of a catabolic pathway can be discovered only by purifying the enzymes and identifying the intermediates. This takes a great deal of effort as there may be a dozen or more reactions involved, but there is no alternative to this step.

There is no assurance, however, that an enzyme which is active on a given substrate actually participates in the pathway in question, since a substrate can be an intermediate in more than one pathway. If the energy source can be synthesized with ^{14}C in a specific carbon, then the ^{14}C distribution in the products can be determined to see if the isotope distributions of the products agree with the proposed pathway. Isotope studies can also be used to supplement fermentation balance data and to provide guide-lines for enzymic experiments for the study of an unknown pathway. An excellent review of bacterial fermentations is that of Wood (1961), which should be consulted for further details.

II. Embden–Meyerhof Fermentations

A. Balances and Isotope Studies

1. *The homolactic fermentation*

The elucidation of the Embden–Meyerhof or glycolytic pathway for the anaerobic catabolism of glucose was largely a result of the studies with muscle and yeast, but it was apparent that the fermentation of hexoses by some species of bacteria was similar to the Embden–Meyerhof pathway. The products of glucose fermentation by all species of *Streptococcus*, many species of *Lactobacillus*, and several other species of bacteria were mainly lactic acid with minor amounts of acetic acid, formic acid and ethanol (Table 6.1). Many species of *Streptococcus* produce upwards of 90% of lactic acid based on the sugar used (Sherman, 1937), and consequently this type of fermentation is known as the homolactic fermentation. Alkaline pH favors the formation of acetic acid, ethanol and formic acid in the ratio of 1 : 1 : 2 (Gunsalus and Niven, 1942), usually assumed to be the result of the cleavage of pyruvate to $C_2 + C_1$.

Knowledge of the enzymic reactions of the Embden–Meyerhof pathway made it possible to predict that carbons 1 and 6 of glucose would

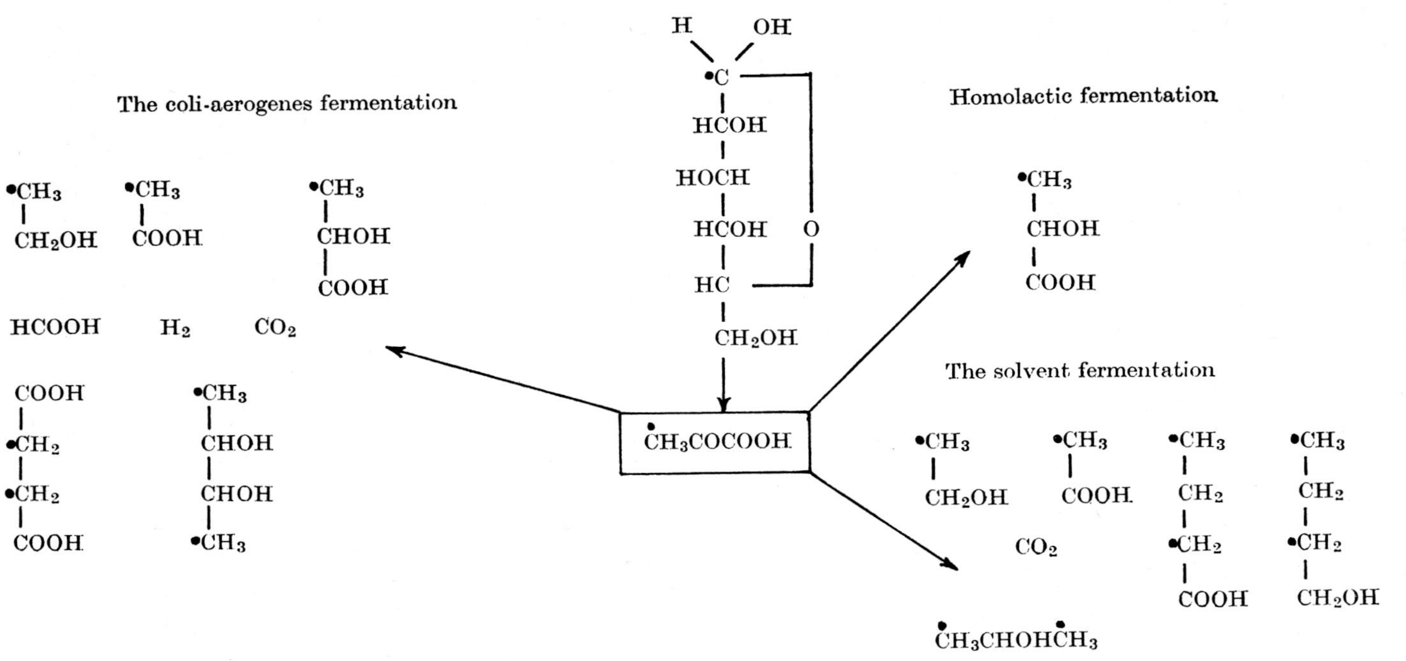

FIG. 6.1. Isotope distribution of products from the bacterial fermentations of glucose-1-^{14}C by the Embden–Meyerhof pathway.

become the methyl carbon of lactic acid, carbons 2 and 5 of glucose would become the carbinol carbon of lactic acid, and carbons 3 and 4 of glucose would become the carboxyl carbon of lactic acid (Fig. 6.1). Fermentation of glucose-1-^{14}C by several species of homolactic acid bacteria resulted in lactic acid labeled in the methyl carbon, which is the result expected from the operation of the Embden–Meyerhof pathway (Table 6.2).

TABLE 6.2. Isotope distribution of lactic acid formed by the fermentation of glucose-1-^{14}C by homofermentative lactic acid bacteria.

	L. casei (a)	*L. pentosus* (b)	*S. faecalis* (b)
	%	%	%
COOH	3	3	1
|			
CHOH	0	6	9
|			
CH_3	97	91	91

The figures given are the percent of the total activity of lactic acid found in the respective carbons.
References: (a) Gibbs *et al.* (1950), (b) Gibbs *et al.* (1955).

It is assumed that the fermentation of other hexoses by homolactic acid bacteria is also accomplished by the Embden–Meyerhof pathway. Steele *et al.* (1954) discovered that *S. pyogenes*, *S. faecalis* and *L. casei* produced less lactic acid in the fermentation of galactose than they did from glucose, and produced more acetic acid, formic acid and ethanol. Nevertheless, it is clear that galactose fermentation is an Embden–Meyerhof pathway because fermentation of galactose-1-^{14}C by *S. faecalis* resulted in the production of methyl-labeled lactic acid, acetic acid and ethanol, which would be expected from the cleavage of pyruvate formed from galactose by the operation of the Embden–Meyerhof pathway (Fukuyama and O'Kane, 1962).

2. *The solvent fermentation*

The solvent fermentation is characteristic of anaerobic spore-formers and aerobic spore-formers capable of growing anaerobically. The basic fermentation is illustrated by *Clostridium lactoacetophilum* (*tyrobutyricum*) (Table 6.3, Bhat and Barker, 1947), which ferments glucose to a mixture of acetic acid, butyric acid, carbon dioxide and hydrogen. The balance is short in reduced products, which could be accounted for by a bit of butyl or ethyl alcohol. Almost two moles of carbon dioxide

TABLE 6.3. Solvent Fermentation of Clostridia

Products	*C. lactoacetophilum* (a) moles/mole glucose	*C. butylicum* (b) moles/mole glucose
Acetic acid	0·28	0·203
Butyric acid	0·73	0·145
Carbon dioxide	1·90	2·07
Hydrogen	1·82	1·11
Butyl alcohol	—	0·502
Isopropyl alcohol	—	0·180
Oxidized products	+3·80	+4·14
Reduced products	−3·28	−3·94
Net:	+0·52	+0·20
O/R Index	1·16	1·05
Carbon recovery	90%	93%

References: (a) Bhat and Barker (1947); (b) Osburn *et al.* (1937).

are produced per mole of glucose, which suggests that glucose is catabolized to pyruvate which is then cleaved to $C_2 + C_1$. Summing up the C_2 fragments adds strength to this argument, since acetic acid represents 0·28 moles of C_2 and butyric acid represents 1·46 moles C_2 for a total of 1·74 moles of C_2. This calculation assumes that butyric acid arises by the head to tail condensation of two moles of C_2. The proportion of acetic and butyric acids formed in the fermentation is not fixed, but depends on the organisms and the conditions.

Several other clostridial fermentations such as the butyl alcohol type of *C. butylicum* (Table 6.3) are variations of this basic fermentation. Butyl alcohol is formed by the reduction of butyric acid and is necessarily accompanied by a reduction in the amount of hydrogen evolved. Isopropyl alcohol can be explained as a by-product of the formation of butyrate by assuming that acetoacetate is an intermediate in butyric acid formation and that acetoacetate is decarboxylated to acetone and carbon dioxide. Acetone is reduced to isopropyl alcohol and the overall process results in a slight increase in the yield of carbon dioxide and a further reduction in hydrogen evolved. Acetone itself is a major product in the fermentation of sugars by *C. acetobutylicum*, and this fermentation is used for the commercial production of acetone (Prescott and Dunn, 1959). Other products such as lactic acid, ethyl alcohol and formic acid have also been reported in clostridial fermentations.

Isotope studies bear out the Embden-Meyerhof nature of the clostridial fermentations. Paege *et al.* (1956) and Cynkin and Gibbs (1958) studied the fermentation of labeled glucose by *C. perfringens*, which produces a butyric type fermentation with the addition of small amounts of lactic acid and ethanol. Fermentation of glucose 3,4-^{14}C resulted in labeled carbon dioxide; glucose-1-^{14}C resulted in methyl-labeled ethanol and acetic acid; glucose-2-^{14}C resulted in carbinol-labeled ethanol and carboxyl-labeled acetic acid.

Clostridium thermoaceticum ferments glucose to three moles of acetic acid. Wood (1952) found that the fermentation of glucose-1-^{14}C by this organism resulted in methyl-labeled acetate, but glucose 3,4-^{14}C resulted in acetic acid labeled in both carbons, although at a low specific activity. The latter data agreed with the earlier observation of Barker and Kamen (1945) that *C. thermoaceticum* was able to fix carbon dioxide into both carbons of acetic acid. Thus the fermentation of glucose by *C. thermoaceticum* is regarded as an Embden–Meyerhof pathway producing two moles of C_3 in the usual manner; C_3 then undergoes a cleavage to $C_2 + C_1$. The identity of C_1 is apparently carbon dioxide which is reduced to acetate by an unknown pathway.

3. *The coli-aerogenes fermentation*

The coli-aerogenes fermentation of glucose is characteristic of the *Enterobacteriaceae* and two variations of this fermentation are illustrated in Table 6.4. Fermentation of glucose by *E. coli* is typified by a high yield of acids consisting of acetic, formic, lactic and succinic acids, while the aerogenes type of fermentation, here illustrated by *Aerobacter indologenes*, is typified by a high yield of neutral products, especially 2,3-butanediol and ethyl alcohol. These features are the basis for the methyl red and Voges–Proskauer tests used for the differentiation of *E. coli* and *A. aerogenes*. The more acid fermentation of *E. coli* changes the methyl red indicator, and 2,3-butanediol and acetoin are oxidized to diacetyl under the conditions of the Voges–Proskauer test which then reacts with the Voges–Proskauer reagents to produce the characteristic red color.

The coli fermentation fits the picture of an Embden–Meyerhof pathway with a major portion of the C_3 intermediate diverted to produce $C_2 + C_1$ products. Summation of the C_2 and C_1 products reveals that the balance is a bit short in C_1 products if succinate is considered to represent two moles of C_2. However, if succinate is formed by a $C_3 + C_1$ condensation, then the sums of C_2 and C_1 products are more nearly equal. In either case, succinate will be labeled in the methylene carbons from the fermentation of glucose-1-^{14}C, and therefore isotope studies cannot distinguish between these two possibilities. Very little is known about the origin of succinate in the coli-aerogenes fermentation, but the role of this compound in the propionate fermentation has been thoroughly examined. Another characteristic of the coli-aerogenes fermentation is the production of hydrogen, which originates mainly from the decomposition of formic acid to hydrogen and carbon dioxide by the formic hydrogenlyase enzyme system (Stephenson and Stickland, 1932).

$$HCOOH \rightarrow H_2 + CO_2$$

TABLE 6.4. *Coli-aerogenes* Fermentation of Glucose

	E. coli (a)	*A. indologenes* (b)
Products	moles/mole glucose	moles/mole glucose
Ethyl alcohol	0·498	0·704
Glycerol	0·014	—
Acetic acid	0·365	0·082
Succinic acid	0·107	—
Lactic acid	0·795	0·064
2,3-Butanediol	(trace)	0·581
Acetoin	(trace)	0·020
Formic acid	0·024	0·543
Carbon dioxide	0·880	1·400
Hydrogen	0·750	0·115
Oxidized products	+1·89	+3·34
Reduced products	−1·76	−3·30
Net:	+0·13	0·04
O/R Index	1·07	1·01
Carbon recovery	91·5%	102%

References: (a) Blackwood *et al.* (1956), (b) Reynolds and Werkman (1937).

The most plausible explanation for the origin of 2,3-butanediol formed in the aerogenes fermentation is the head to head condensation of two moles of C_2. In support of this postulate, an increased yield of carbon dioxide over that of the coli fermentation is observed and the sums of C_2 and C_1 products are nearly equal, assuming that 2,3-butanediol is equivalent to two moles of C_2.

Fermentation of glucose-1-^{14}C and glucose-2-^{14}C by *E. coli* and *A. aerogenes* resulted in products with the expected labeling patterns for the Embden–Meyerhof pathway (Table 6.5). Glucose-1-^{14}C fermentation resulted in methyl-labeled 2,3-butanediol, ethyl alcohol, acetic and

TABLE 6.5. Isotope distribution of products of glucose-1-^{14}C and glucose-2-^{14}C fermentation by *E. coli* and *A. aerogenes*.

	E. coli (a)		*A. aerogenes* (b)	
	G-1-^{14}C	G-2-^{14}C	G-1-^{14}C	G-2-^{14}C
Products	R.S.A.	R.S.A.	R.S.A.	R.S.A.
Carbon dioxide	1·6	0·5	2·9	0·6
2,3-Butanediol				
CH_3	—	—	38·7	0·9
| CHOH	—	—	0·0	37·2
|				
Ethyl alcohol				
CH_3	37·6	0·0	37·1	1·0
| CH_2OH	0·0	37·6	0·2	39·6
Acetic acid				
CH_3	38·7	0·5	44·8	0·9
| COOH	0·0	37·6	0·4	41·4
Lactic acid				
CH_3	46·2	0·5	37·1	0·9
| CHOH	0·5	41·6	0·1	33·5
| COOH	1·1	0·0	1·3	0·7

References and footnotes: (a) Paege and Gibbs (1961), (b) Altermatt *et al.* (1955a), R.S.A. is the relative specific activity = specific activity of product carbon divided by the specific activity of C-1 or C-2 of glucose × 100.

lactic acids; glucose-2-^{14}C fermentation resulted in carbinol-labeled 2,3-butanediol, carboxyl-labeled acetic acid and carbinol-labeled lactic acid. In addition, these data confirm the head to head condensation of two moles of C_3 to produce 2,3-butanediol.

B. Enzymes of the Embden–Meyerhof Pathway in Bacteria

Many of the enzymes of the Embden–Meyerhof pathway have been detected in enzyme extracts of bacterial cells, and these reports have been collected in the review of Gunsalus *et al.* (1955) and other reviews mentioned in this section.

All hexoses enter the Embden–Meyerhof pathway at the level of glucose or fructose monophosphates (Fig. 6.2). Glucose-1-phosphate is a product of the phosphorolysis of glycogen, starch, sucrose, maltose and cellobiose. Phosphorolysis has an advantage over the hydrolytic cleavage of glycosides because the energy which was used to form the glycoside bond is conserved in the hexose-1-phosphate linkage (see Table 6.6).

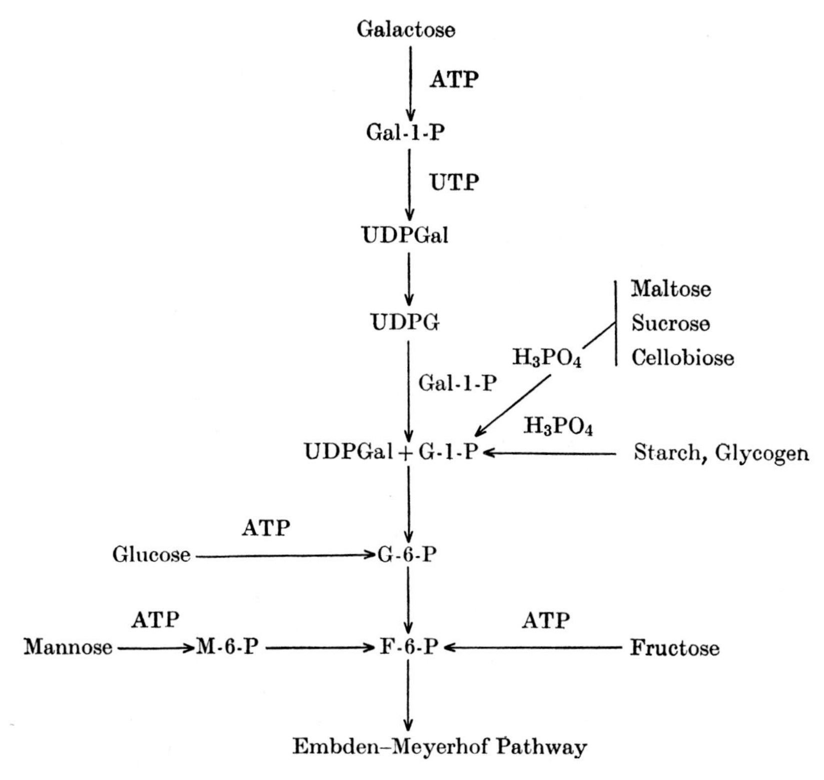

FIG. 6.2. Formation of hexose-6-phosphate from common natural hexoses. Abbreviations: G-1-P = glucose-1-phosphate; G-6-P = glucose-6-phosphate; Gal-1-P = galactose-1-phosphate; F-6-P = fructose-6-phosphate; M-6-P = mannose-6-phosphate; UDPG = uridinediphosphoglucose; UDPGal = uridine-diphosphogalactose.

Phosphoglucomutase (α-D-glucose-1,6-diphosphate:α-D-glucose-1-phosphate phosphotransferase) catalyzes the formation of glucose-6-phosphate from glucose-1-phosphate. Our knowledge of the action of phosphoglucomutase is a result of studies with the muscle and yeast enzymes (Najjar, 1962), but phosphoglucomutase has been reported in bacteria and its existence can be inferred in those organisms known to metabolize glucose-1-phosphate. Glucose-1,6-diphosphate is an inter-

TABLE 6.6. Free energy of hydrolysis of phosphorylated intermediates related to the Embden–Meyerhof pathway (from Atkinson and Morton, 1960).

Phosphoryl derivative	$\Delta G'_0$ kcal/mole
Glucose-1-phosphate	−5·0
Glucose-6-phosphate	−3·3
Fructose-6-phosphate	−3·8
Fructose-1-phosphate	−3·1
2-Phosphoglycerate	−4·2
3-Phosphoglycerate	−3·1
3-Phosphoglyceroyl phosphate (acyl phosphate only)	−11·8
Phosphoenolypruvate	−12·8
Acetyl phosphate	−10·1
Adenosine-5′-triphosphate (terminal phosphate only)	−6·9

$\Delta G'_0$ is the free energy change at unit activities of reactants, usually at or near pH 7·0 and with 0·01M magnesium ion.

mediate in the conversion which involves phosphorylation and dephosphorylation of the enzyme (Najjar, 1962).

$$\text{Glucose-1-phosphate} + \text{phosphoenzyme}$$
$$\rightleftharpoons$$
$$\text{Glucose-1,6-diphosphate} + \text{dephosphoenzyme}$$
$$\rightleftharpoons$$
$$\text{Glucose-6-phosphate} + \text{phosphoenzyme.}$$

Catalytic amounts of glucose-1,6-diphosphate are required for the reaction and are formed by the action of phosphoglucokinase (ATP:D-glucose-1-phosphate 6-phosphotransferase, Paladini *et al.*, 1949) on glucose-1-phosphate and ATP.

Another route to hexose-6-phosphate is by the action of hexokinase (ATP:D-hexose 6-phosphotransferase, Crane, 1962). Muscle and yeast hexokinase phosphorylate glucose, fructose and mannose and, although

$$\text{D-hexose} + \text{ATP} \rightarrow \text{D-hexose-6-phosphate} + \text{ADP}$$

studies with purified bacterial enzymes are lacking, extracts of *S. faecalis* (Moore and O'Kane, 1963) and *E. coli* (Cardini, 1951) also phosphorylate these three sugars. Phosphorylation of glucose to glucose 6-phosphate with adenosine triphosphate by extracts of bacterial cells has been demonstrated several times. Formation of hexose-6-phosphate is an exergonic reaction with a $\Delta G_0'$ of about $-4{\cdot}7$ kcals/mole at pH 7·0, and therefore the equilibrium is considerably in the favor of formation of hexose-6-phosphate (Robbins and Boyer, 1957). In addition to hexokinase, kinases which are specific for individual hexoses are known to occur in species of protozoa (Crane, 1962), and this may be true for bacteria as well.

Galactokinase (ATP:D-galactose 1-phosphotransferase) was first discovered in galactose-adapted yeast (Trucco *et al.*, 1948), and contrasts with hexokinase in that the product is galactose-1-phosphate.

$$\text{D-Galactose} + \text{ATP} \rightarrow \text{D-galactose-1-phosphate} + \text{ADP}$$

Galactokinase has been purified from *E. coli* (Sherman and Adler, 1963).

Galactose-1-phosphate is metabolized by conversion to glucose-1-phosphate with uridine diphosphate glucose as a cofactor in this transformation (Caputto *et al.*, 1950). The first step is the exchange of galactose-1-phosphate with glucose-1-phosphate of uridine disphosphate glucose (UDPG), which is brought about by uridylyl transferase (UDP glucose:α-D-galactose-1-phosphate uridylyl transferase, Kalckar *et al.*,

$$\text{UDPglucose} + \alpha\text{-D-galactose-1-phosphate} \rightleftharpoons \alpha\text{-D-glucose-1-phosphate} + \text{UDPgalactose}$$

1953). The second step is the epimerization of carbon 4 of galactose catalyzed by uridine diphosphate glucose-4-epimerase (Leloir, 1951; Maxwell, 1961). The overall reaction results in the net conversion of one

$$\text{UDPgalactose} \rightleftharpoons \text{UDPglucose}$$

mole of galactose-1-phosphate to glucose-1-phosphate. Uridine diphosphate glucose is required in catalytic amounts and is formed from uridine triphosphate and glucose-1-phosphate by the action of UDPG pyrophosphorylase (UTP:α-D-glucose-1-phosphate uridylyl transferase, Trucco, 1951). The catalytic mechanism of the 4-epimerase is not

$$\text{UTP} + \alpha\text{-D-glucose-1-phosphate} \rightleftharpoons \text{UDPglucose} + \text{pyrophosphate}$$

certain, but purified enzymes contain nicotinamide adenine dinucleotide and it seems likely that some type of oxidation-reduction takes place at carbon 4.

Glucose-6-phosphate and mannose-6-phosphate are converted to fructose-6-phosphate by phosphoglucose isomerase (D-glucose-6-phosphate ketol-isomerase) and phosphomannose isomerase (D-mannose-6-phosphate ketol-isomerase) respectively (Topper, 1961). This conversion is an aldose-ketose transformation, probably with an ene-diol intermediate, and involves very little change in free energy.

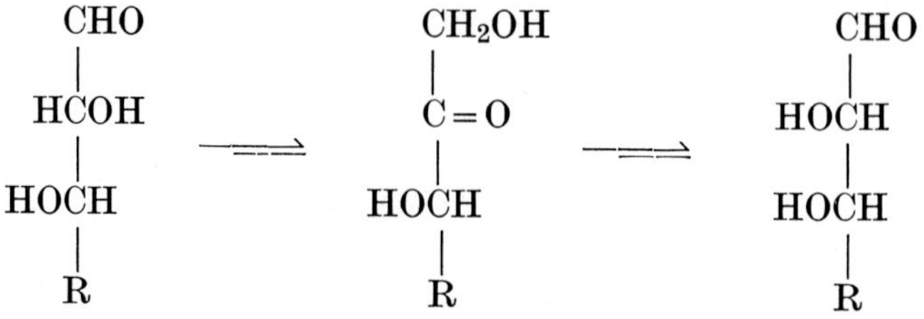

$$
\begin{array}{ccccc}
\text{CHO} & & \text{CH}_2\text{OH} & & \text{CHO} \\
| & & | & & | \\
\text{HCOH} & & \text{C}=\text{O} & & \text{HOCH} \\
| & \rightleftharpoons & | & \rightleftharpoons & | \\
\text{HOCH} & & \text{HOCH} & & \text{HOCH} \\
| & & | & & | \\
\text{R} & & \text{R} & & \text{R}
\end{array}
$$

Glucose-6-phosphate Fructose-6-phosphate Mannose-6-phosphate

Fructose-6-phosphate is the first intermediate common to the metabolism of all the natural sugars metabolized by the Embden–Meyerhof pathway.

From fructose-6-phosphate to pyruvate, the reactions of the Embden–Meyerhof pathway are universal for all organisms which make use of this pathway. Phosphofructokinase (ATP: D-fructose-6-phosphate 1-phosphotransferase) is responsible for the phosphorylation of fructose-6-phosphate with adenosine 5′ triphosphate resulting in fructose-1,6-diphosphate:

D-Fructose-6-phosphate + ATP → D-fructose-1,6-diphosphate + ADP.

This is the only enzyme which is actually unique to the Embden–Meyerhof pathway. In addition to fructose-6-phosphate, highly purified rabbit muscle phosphofructokinase also phosphorylates tagatose-6-phosphate and sedoheptulose-7-phosphate (Lardy, 1962), the latter sugar being the 7 carbon homologue of fructose-6-phosphate. Partially purified *S. faecalis* phosphofructokinase was also active with sedoheptulose-7-phosphate and the product was characterized as sedoheptulose-1,7-diphosphate (Sokatch, 1962).

Aldolase (Ketose-1-phosphate aldehyde lyase) catalyzes the reversible aldol condensation of fructose-1,6-diphosphate to dihydroxyacetone phosphate and glyceraldehyde-3-phosphate. Aldolase is specific for

D-fructose-1,6-diphosphate ⇌ dihydroxyacetone phosphate +
D-glyceraldehyde-3-phosphate

dihydroxyacetone phosphate but a number of aldehydes will substitute for glyceraldehyde-3-phosphate, including unphosphorylated sugars and even aliphatic aldehydes such as acetaldehyde (Rutter, 1961). Sedoheptulose-1,7-diphosphate is nearly as good a substrate for aldolase as is fructose-1,6-diphosphate (Horecker *et al.*, 1955). Aldolase has been reported in a number of species of bacteria, and Bard and Gunsalus (1950) found that aldolase of *Clostridium perfringens* was activated by ferrous iron.

The triose phosphates resulting from the cleavage of fructose-1,6-diphosphate then undergo a rearrangement catalyzed by triose phosphate isomerase (D-glyceraldehyde-3-phosphate ketol-isomerase) which

$$\text{Dihydroxyacetone phosphate} \rightleftharpoons \text{D-glyceraldehyde-3-phosphate}$$

is similar to the isomerization of glucose and fructose-6-phosphates. As a result of this reaction, both halves of the hexose molecule become equivalent and they are both metabolized by the same pathway to pyruvate. This is not the case with some of the non-Embden–Meyerhof pathways of bacteria where the two halves of hexose have different metabolic fates.

A branch of the Embden–Meyerhof pathway leads away from the mainstream through dihydroxyacetone phosphate, which is reduced to glycerolphosphate by glycerolphosphate dehydrogenase (L-glycerol-3-phosphate:NAD oxidoreductase, Baranowski, 1963). Dephosphorylation of glycerol phosphate by a phosphatase produces glycerol, which

$$\text{Dihydroxyacetone phosphate} + \text{NADH} \rightleftharpoons \text{L-glycerol-3-phosphate} + \text{NAD}$$

is frequently a minor product in bacterial fermentations. Most tissues also contain a glycerol phosphate dehydrogenase with FAD as the prosthetic group. Apparently this second enzyme functions in the aerobic oxidation of glycerol phosphate. This is true of *S. faecalis* which has an NAD-linked glycerol phosphate dehydrogenase when grown anaerobically and the FAD enzyme when grown aerobically (Jacobs and VanDemark, 1960).

The mainstream of the Embden–Meyerhof pathway proceeds through glyceraldehyde-3-phosphate which is oxidized by glyceraldehyde-3-phosphate dehydrogenase (D-glyceraldehyde-3-phosphate:NAD oxidoreductase [phosphorylating]). Both the yeast and muscle enzymes have been crystallized and found to contain bound NAD (Velick and Furfine, 1963). The reaction consists of at least two steps:

$$\text{Enzyme} + \text{D-glyceraldehyde-3-phosphate} + \text{NAD} \rightleftharpoons \text{3-phosphoglyceroyl enzyme} + \text{NADH}$$

$$\text{3-Phosphoglyceroyl enzyme} + H_3PO_4 \rightleftharpoons \text{3-phosphoglyceroyl phosphate} + \text{enzyme.}$$

This is a key reaction in the Embden–Meyerhof pathway, since the substrate is now prepared to give up part of its energy to the organism in the form of useful chemical energy. The glyceroyl phosphate bond is a mixed anhydride, and as a class of compounds anhydrides have a high free energy of hydrolysis (Table 6.6). For example, the free energy of hydrolysis of the glyceroyl phosphate linkage is $-11{\cdot}8$ kcals/mole at pH 7·0, while that of the ester phosphate at carbon 3 of glyceric acid is only $-3{\cdot}1$ kcals/mole.

Part of the energy of the acyl phosphate linkage is captured and stored as a phosphate anhydride in ATP as a result of the reaction catalyzed by phosphoglyceric acid kinase (ATP: 3-phospho-D-glycerate

$$\text{1,3-Diphospho-D-glycerate} + \text{ADP} \rightleftharpoons \text{3-phospho-D-glycerate} + \text{ATP}$$

1-phosphotransferase). The equilibrium favors the formation of ATP because the acyl phosphate linkage has a considerably higher free energy of hydrolysis than does the terminal phosphoanhydride of ATP (Table 6.6).

3-Phosphoglyceric acid is converted to 2-phosphoglyceric acid by phosphoglyceric acid mutase, a reaction which involves very little energy change. Two types of phosphoglyceric acid mutase are known; one

$$\text{3-Phospho-D-glycerate} \rightleftharpoons \text{2-phospho-D-glycerate}$$

requires 2,3-diphosphoglycerate as a cofactor (2,3-diphospho-D-glycerate: 2-phospho-D-glycerate phosphotransferase) and the other does not (Grisolia and Joyce, 1959). The former enzyme is the more widely distributed in nature and occurs in *E. coli*. It is tempting to think that this mutase has a phosphoenzyme intermediate similar to that of phosphoglucomutase, but there is no convincing evidence that this is true.

2-Phosphoglycerate is converted to another high energy phosphate compound, phosphoenolpyruvate, by phosphoenolpyruvate hydratase (2-phospho-D-glycerate hydrolyase) or enolase as it was originally named. The reaction is a dehydration of 2-phosphoglycerate resulting in phosphoenol pyruvate, which has the highest free energy of hydrolysis of all the intermediates of the Embden–Meyerhof pathway.

$$\text{2-Phospho-D-glycerate} \rightleftharpoons \text{phosphoenolpyruvate} + H_2O$$

The last reaction of the Embden–Meyerhof pathway common to all organisms using this pathway is the transfer of phosphate from phosphoenolpyruvate to ADP catalyzed by pyruvate kinase (ATP: pyruvate phosphotransferase, Boyer, 1962). The equilibrium of this reaction lies far in the direction of ATP formation. The enol form of pyruvate quickly rearranges to the more stable keto form.

In the fermentation of glucose to the stage of pyruvate, a net of two moles of ADP is converted to ATP per mole of glucose used. From the best estimations available, this means that a total of approximately 14 kcals of useful chemical energy is extracted from glucose. This is only about 2% of the total of 686 kcals available from the complete oxidation of glucose, which makes fermentation a relatively inefficient process. If the organism starts with an oligosaccharide, such as starch, where the initial cleavage is by phosphorylase, then the net ATP yield is three moles per mole hexose.

C. Formation of Fermentation Products

1. *Lactic acid formation*

The products of hexose fermentation after the phosphoenolpyruvate-hydratase reaction are two moles each of pyruvate and NADH. In muscle and lactic acid forming bacteria, pyruvate is reduced to lactic acid with lactic dehydrogenase (L-lactate:NAD oxidoreductase) which completes the pathway. Muscle lactic acid is L(+), but lactic acid produced by bacteria may be either L(+), D(−), or DL (Wood, 1961).

2. *Non-oxidative pyruvate cleavage and condensation reactions*

Pyruvate undergoes cleavage reactions, most of which involve thiamine pyrophosphate, and the initial reaction between pyruvate and thiamine is identical in many of these cases. On the basis of chemical considerations and work with enzymes and enzyme models (Breslow, 1962; Krampitz *et al.*, 1962; Holzer *et al.*, 1962), the first step can be formulated as a reaction of pyruvate with thiamine pyrophosphate bound to the enzyme to form lactyl thiamine pyrophosphate (I, Fig. 6.3). The second step of the thiamine-catalyzed reaction is the decarboxylation of pyruvate resulting in hydroxyethyl thiamine pyrophosphate (II, Fig. 6.3) and carbon dioxide. The fate of the two carbon fragments of pyruvate attached to thiamine depends on the enzyme system involved; for example, in the yeast carboxylase (2-oxo-acid carboxy-lyase) reaction, hydroxyethyl thiamine pyrophosphate, dissociates to produce acetaldehyde and thiamine pyrophosphate. Carboxylase also forms small amounts of acetoin (2-oxo-3-hydroxy butane), which can be explained as the condensation of another mole of acetaldehyde with hydroxyethyl thiamine.

Bacterial formation of acetoin occurs by a different route involving two separate enzymes. The first step is the condensation of two moles of pyruvate forming α-acetolactic acid (Fig. 6.4) and carbon dioxide by a thiamine-containing enzyme, α-acetolactate synthetase (Juni, 1952). Synthetic hydroxyethyl thiamine pyrophosphate has been shown to react with this enzyme prepared from *A. aerogenes* (Krampitz *et al.*,

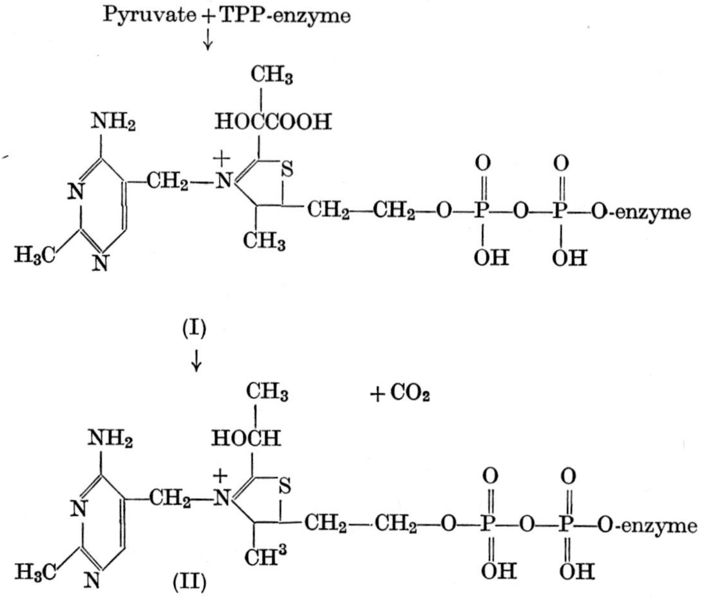

Pyruvate + TPP-enzyme

↓

(I)

↓

$+ CO_2$

(II)

FIG. 6.3. Thiamine pyrophosphate catalyzed decarboxylation of pyruvate.

1962). In this case, hydroxyethyl thiamine pyrophosphate condenses with pyruvate. The second step is catalyzed by acetolactate decarboxylase (2-acetolactate: carboxy-lyase), which requires manganese, but not thiamine, and results in the decarboxylation of α-acetolactate to acetoin and carbon dioxide. In bacterial fermentations, acetoin is usually reduced to 2,3-butanediol with NADH by butylene glycol dehydro-

$$2 \text{ Pyruvate} \rightarrow \alpha\text{-acetolactate} + \text{carbon dioxide}$$

$$\alpha\text{-Acetolactate} \rightarrow \text{acetoin} + \text{carbon dioxide}$$

genase (2,3-butylene-glycol:NAD oxidoreductase), which means that the other mole of NADH formed in hexose fermentation must be used elsewhere.

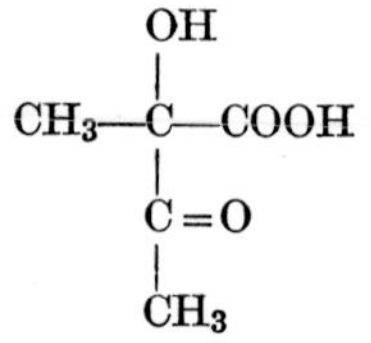

$$CH_3-C(OH)(COOH)-C(=O)-CH_3$$

FIG. 6.4. Structure of α-acetolactic acid.

3. *Pyruvate oxidation*

In a number of bacterial fermentations, pyruvate is oxidized to acetate and carbon dioxide. The electrons removed in the oxidation by pyridine nucleotide-linked enzymes can be used to reduce other fermentation products or can be evolved as hydrogen gas by the complex hydrogenase system (Gest, 1954). The oxidation of pyruvate is an exergonic reaction and thus can serve to increase the yield of ATP from hexose during fermentation. The usual pathway of pyruvate oxidation by both aerobic and anaerobic bacteria, and the sole pathway in animal tissues, requires both thiamine pyrophosphate and lipoic acid. (6,8-dimercaptooctanoic acid). Lipoic acid was discovered in *S. faecalis*, where it is essential for pyruvate oxidation but not for growth (O'Kane and Gunsalus, 1948). Here again the initial reaction is the decarboxyla-

(1) $CH_3COCOOH$ + TPP-enzyme → $CH_3CH(OH)$—TPP-enzyme + CO_2

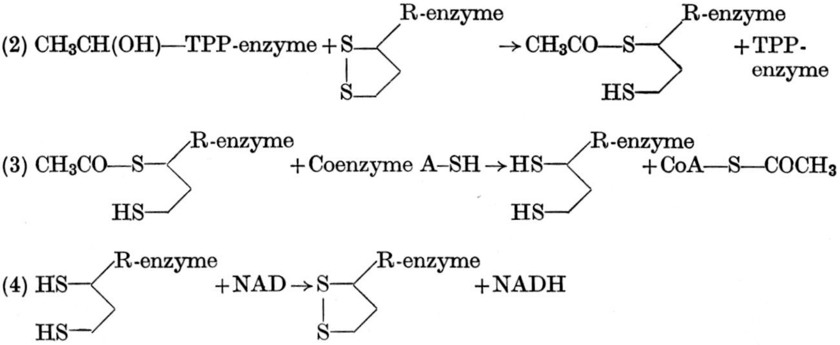

(2) $CH_3CH(OH)$—TPP-enzyme + S(—S)<R-enzyme → CH_3CO—S(HS—)<R-enzyme + TPP-enzyme

(3) CH_3CO—S(HS—)<R-enzyme + Coenzyme A–SH → HS(HS—)<R-enzyme + CoA—S—$COCH_3$

(4) HS(HS—)<R-enzyme + NAD → S(—S)<R-enzyme + NADH

FIG. 6.5. Lipoic acid pathway for oxidation of pyruvate (after Gunsalus, 1954). TPP = Thiamine pyrophosphate, R = $(CH_2)_4COOH$.

tion of pyruvate by a thiamine pyrophosphate-containing enzyme to form hydroxyethyl thiamine pyrophosphate and carbon dioxide (Fig. 6.5). Lipoic acid, which is firmly bound to its enzyme, accepts the hydroxyethyl moiety from thiamine pyrophosphate, resulting in reductive cleavage of the disulfide bond and acetylation of the sulfhydryl on carbon 6 of lipoic acid. Next, the acetyl group is transferred to the sulfhydryl of coenzyme A by lipoic transacetylase (acetyl-CoA: dihydrolipoate s-acetyltransferase), leaving reduced lipoic acid and forming acetyl coenzyme A. Lipoic acid is oxidized to the disulfide by NAD with lipoic dehydrogenase (NADH: lipoamide oxidoreductase), now identified as the old Straub diaphorase (Massey, 1963), and oxidized lipoic acid is now ready to participate in the oxidation of another mole of pyruvate.

4. *Acetate and ethanol formation*

The bacterial system for obtaining the energy of the thioester bond of acetyl coenzyme A in the form of ATP involves the use of the enzyme phosphotransacetylase (acetyl-CoA: orthophosphate acetyltransferase), which was discovered in *C. kluyveri* (Stadtman *et al.*, 1951). This enzyme catalyzes the transfer of the acetyl group from coenzyme A to phosphate, resulting in the formation of acetyl phosphate and free coenzyme A. Acetokinase (ATP: acetate phosphotransferase) catalyzes the transfer

$$\text{Acetyl-CoA} + H_3PO_4 \rightleftharpoons \text{acetylphosphate} + \text{CoA}$$

$$\text{Acetylphosphate} + \text{ADP} \rightleftharpoons \text{acetate} + \text{ATP}$$

of phosphate from acetylphosphate to ADP (Rose, 1962).

CH_3CH_2OH

$+NADH \rightleftharpoons +NAD$

CH_3CHO

$+NADH \rightleftharpoons +NAD + CoA—SH$

$CH_3CO—S—CoA$

$\rightleftharpoons$ $CH_3CO—S—CoA \leftarrow$

$CH_3COCH_2CO—S—CoA + CoA—SH$

$+NAD \rightleftharpoons +NADH$

$CH_3CHOHCH_2CO—S—CoA$

$\rightleftharpoons -H_2O$

$CH_3CH{=}CHCO—S—CoA$

$-2(H) \rightleftharpoons +2(H)$

$CH_3CH_2CH_2CO—S—CoA$

$\rightleftharpoons +CH_3COOH$

$CH_3CH_2CH_2COOH + CH_3CO—S—CoA$

FIG. 6.6. Fermentation of ethanol and acetate by *Clostridium kluyveri* (after Barker, 1956).

Free acetate can be reduced to ethyl alcohol by bacteria, but first must be activated to acetyl coenzyme A. This may be accomplished by coenzyme A transferase, which transfers coenzyme A between acyl compounds (see Fig. 6.6), or by a combination of acetokinase and phosphotransacetylase. *Clostridium kluyveri* contains an aldehyde dehydrogenase which reduces acetyl coenzyme A with NADH to acetaldehyde, and an alcohol dehydrogenase which reduces acetaldehyde to ethanol with NADH (Burton and Stadtman, 1953). Alcohol formation from acetate or acetyl coenzyme A seems to be the rule among bacteria and is a good means of expending NADH formed in fermentation, but wastes the energy available in acetyl coenzyme A. *Zymomonas mobilis* is

$$\text{Acetyl-CoA} + \text{NADH} \rightleftharpoons \text{acetaldehyde} + \text{CoA} + \text{NAD}$$

$$\text{Acetaldehyde} + \text{NADH} \rightleftharpoons \text{ethyl alcohol} + \text{NAD}$$

an exception to this generalization since it possesses a yeast-type carboxylase and forms ethyl alcohol by the reduction of acetaldehyde (Gibbs and DeMoss, 1951).

5. *The phosphoroclastic reaction*

The phosphoroclastic reaction is a decomposition of pyruvate to acetyl phosphate and either formate, as in the case of the coliform bacteria, or hydrogen and carbon dioxide, as in the case of the clostridia. The requirements for both systems include thiamine pyrophosphate,

$$\text{Pyruvate} + H_3PO_4 \begin{cases} \rightarrow \text{Acetylphosphate} + \text{formate} \quad \textit{Coliform bacteria} \\ \rightarrow \text{Acetylphosphate} + H_2 + \text{Carbon dioxide} \quad \textit{Clostridia} \end{cases}$$

coenzyme A and phosphate, and the clostridial system requires ferredoxin as well (Mortenson *et al.*, 1963). The fine points of ferredoxin catalysis are not known, but ferredoxin is reduced during pyruvate oxidation and probably functions as an electron transport factor between "pyruvate dehydrogenase" and hydrogenase. Formate is probably not an intermediate in the clostridial clastic reaction, since it does not exchange with the carboxyl of pyruvate under the same conditions that carbon dioxide can exchange (Wilson *et al.*, 1948; Wolfe and O'Kane, 1955; Mortlock *et al.*, 1959). On the other hand, formate does exchange with the carboxyl of pyruvate in the *E. coli* clastic reaction (Chantrenne and Lipmann, 1950; Strecker, 1951) and in the *S. faecalis* clastic reaction (Wood and O'Kane, 1964).

The term phosphoroclastic is actually a misnomer because the initial product of pyruvate decomposition is acetyl coenzyme A, and acetyl phosphate arises by a secondary reaction with phosphotransacetylase in the presence of substrate amounts of phosphate. The unique aspect of this reaction is the utilization of ferredoxin rather than lipoic acid as an electron transport factor. Ferredoxin has not been found in the coliform bacteria and it seems likely that it does not enter into the phosphorclastic reaction in these organisms.

6. *Formate metabolism*

Formate is oxidized by cell-free extracts of *Methanobacterium omelianskii* and *Clostridium acidi-urici* by a ferredoxin-dependent formic dehydrogenase (formate:NAD oxidoreductase), which results in the reduction of NAD (Brill *et al.*, 1964). In the intact organism, the electrons are presumably passed to the hydrogenase system, ultimately to be evolved as hydrogen gas.

7. *Formation of* C_4 *products*

Butyrate synthesis was studied by Barker and Stadtman and their associates (Barker, 1956) with *C. kluyveri*, which ferments ethanol and acetate to butyric acid plus smaller amounts of higher fatty acids. Oxidation of alcohol to acetyl coenzyme A takes place by a reverse of the pathway described for the reduction of acetyl coenzyme A to alcohol (Fig. 6.6). Acetoacetyl thiolase (Acetyl-CoA:acetyl-CoA C-acetyl transferase) then catalyzes the condensation of two moles of acetyl coenzyme A to form acetoacetyl coenzyme A and one equivalent of free coenzyme A. Acetoacetyl coenzyme A is reduced by 3-hydroxybutyryl coenzyme A dehydrogenase with NADH to 3-hydroxybutyryl coenzyme A. The corresponding animal enzyme (L-3-hydroxyacyl-CoA:NAD oxidoreductase) produces the L(+) isomer of 3-hydroxybutyric acid. Enoyl coenzyme A hydratase (L-3-hydroxyacyl-CoA hydro-lyase) is responsible for the next step, the dehydration of 3-hydroxybutyryl coenzyme A to crotonyl coenzyme A (*trans* 2-butenoyl coenzyme A). Crotonyl coenzyme A is reduced to butyryl coenzyme A by butyryl coenzyme A dehydrogenase of *C. kluyeri*. Coenzyme A transferase catalyzes the transfer of coenzyme A from butyryl coenzyme A to the carboxyl of acetate, forming butyric acid and acetyl coenzyme A, thus activating acetate for the thiolase reaction. Coenzyme A transferase is a non-specific enzyme and can be used to activate a number of fatty acids with acyl coenzyme A bonds generated during metabolism. It is possible that this enzyme provides a general method for activation of fatty acids in those bacteria which possess it.

An interesting problem in the *C. kluyveri* fermentation concerns the source of energy made available to the organism by the fermentation of ethanol and acetate. The scheme outlined in Fig. 6.6 does not provide for a net synthesis of ATP, and since it is obvious that the organism uses this fermentation for energy, *C. kluyveri* must possess unknown methods of ATP formation.

III. Non-Embden–Meyerhof Fermentations

A. Balances and Isotope Studies

1. *The heterolactic fermentation*

The fermentation of glucose by all species of *Leuconostoc* and several species of *Lactobacillus* is quite different from the homolactic fermentation in that equimolar amounts of carbon dioxide, ethyl alcohol and lactic acid are formed per mole of glucose (DeMoss *et al.*, 1951). Because of the mixture of products obtained, this type of fermentation is known as the heterolactic fermentation. Enzyme assays of *Leuconostoc mesenteroides* revealed that this organism lacked aldolase, which made the operation of the Embden–Meyerhof pathway unlikely. The existence of an alternative pathway in *L. mesenteroides* was established by Gunsalus and Gibbs (1952) when they found that fermentation of glucose-1-^{14}C by *L. mesenteroides* resulted in $^{14}CO_2$, unlabeled ethanol and lactic acid (Fig. 6.7). Fermentation of glucose-3,4-^{14}C resulted in ethanol-1-^{14}C, carboxyl-labeled lactic acid and unlabeled carbon dioxide. Gluconate and 2-ketogluconate are also fermented by *L. mesenteroides*, producing in each case one mole of carbon dioxide and lactic acid, but in addition, 0·5 mole of acetic acid and 0·5 mole of ethyl alcohol from gluconate and a mole of acetic acid from 2-ketogluconate (Blackwood and Blakley, 1956). The labeling data indicate that the cleavage is the same as that in the glucose fermentation by this organism (Blakley and Blackwood, 1957).

Fermentation of aldopentoses by both homofermentative and heterofermentative bacteria results in the production of one mole each of acetic and lactic acids (Fred *et al.*, 1921). The mechanism was clearly similar to the heterolactic fermentation of hexoses since fermentation of labeled pentoses by *L. pentosus*, a homofermenter (Lampen *et al.*, 1951; Bernstein, 1953), and *L. pentoaceticus*, a heterofermenter (Rappoport *et al.*, 1951), indicated a cleavage between carbons 2 and 3 of the pentose with the methyl group of acetic acid arising from carbon 1, and the carboxyl group of lactic acid arising from carbon 3 of pentose (Fig. 6.7).

2. Zymomonas mobilis (Pseudomonas lindneri) *fermentation*

Kluyver and Hoppenbrouwers (1931) determined that the products of glucose fermentation by *Z. mobilis* were nearly two moles of each of carbon dioxide and ethanol and trace amounts of lactic acid. Gibbs and DeMoss (1954) confirmed this observation and further found that after glucose-1-^{14}C fermentation, only carbon dioxide was labeled (Fig. 6.7). For comparison, fermentation of glucose-1-^{14}C by yeast resulted in methyl-labeled ethanol and unlabeled carbon dioxide (Koshland and Westheimer, 1950). Glucose-3,4-^{14}C fermentation by *Z. mobilis* resulted in methyl-labeled ethyl alcohol and labeled carbon dioxide. In both the *Zymomonas* and *Leuconostoc* fermentations, carbon 1 of glucose becomes carbon dioxide, but the origin of ethyl alcohol from carbons 2 and 3 is reversed in the two fermentations. Thus, in the *Zymomonas* fermenta-

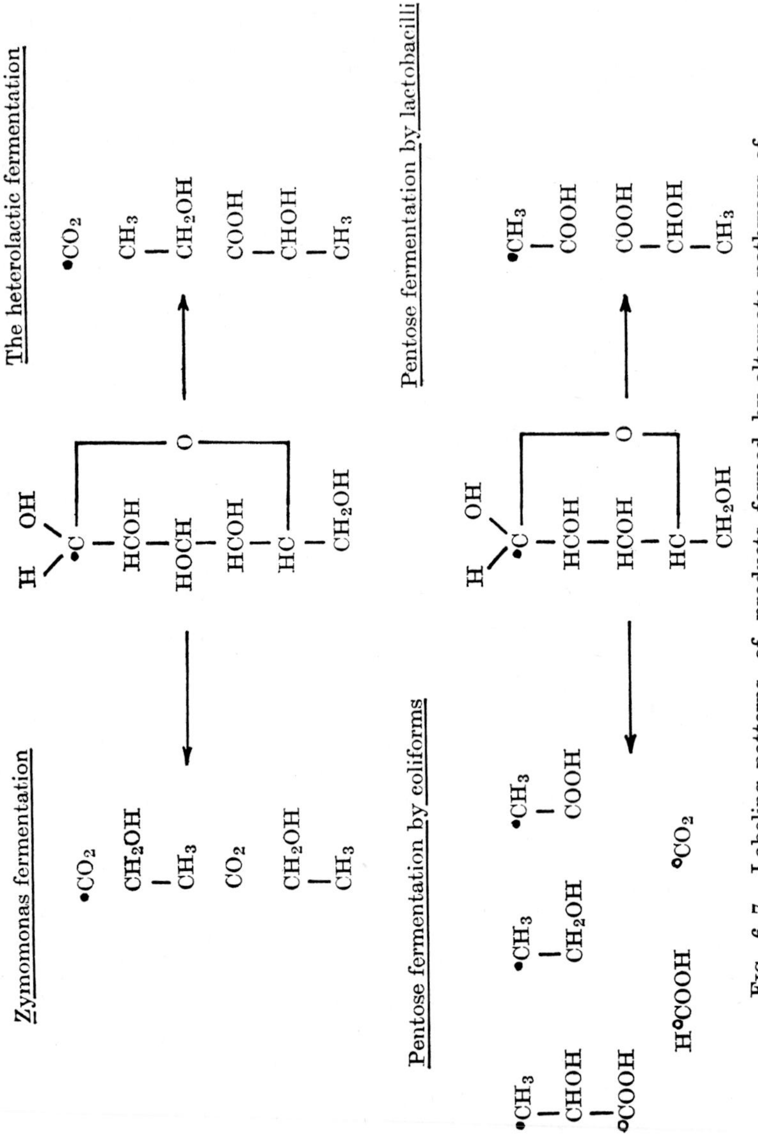

FIG. 6.7. Labeling patterns of products formed by alternate pathways of glucose-1-^{14}C and ribose-1-^{14}C fermentation by bacteria. (•) = One arbitrary unit of radioactivity, (◦) = one-half arbitrary unit of radioactivity.

tion pyruvate is a likely precursor of carbon dioxide and ethanol, but not in the *Leuconostoc* fermentation.

3. *Pentose fermentation by coliform bacteria and clostridia*

In contrast to the clear-cut results with lactobacilli and *Zymomonas*, the fermentation of pentoses by *A. aerogenes* resulted in an array of products similar to that obtained in the glucose fermentation by this organism (Altermatt *et al.*, 1955b). Fermentation of xylose-1-^{14}C and ribose-1-^{14}C resulted in methyl-labeled lactic acid, ethyl alcohol and acetic acid, but the carboxyl carbon of lactic acid, carbon dioxide and formate were also labeled at about one-half the specific activity of the methyl groups (Fig. 6.7). The most reasonable interpretation of these data was that pentose was entirely converted to pyruvate, probably by way of hexose. Similar results were obtained with pentose fermentation by *E. coli* (Gibbs and Paege, 1961) and several species of *Clostridium* (Cynkin and Gibbs, 1958).

B. Enzymes of Non-Embden–Meyerhof Pathways in Bacteria

1. *Enzymes of the heterolactic pathway*

The first evidence of a non-Embden–Meyerhof pathway was obtained by Warburg and Christian (1931) when they discovered that glucose-6-phosphate was oxidized by red blood cell preparations with a cofactor later identified as NADP. The pathway through glucose-6-phosphate was considered to be a branch leading off the main pathway of sugar metabolism and came to be known as the hexosemonophosphate shunt. Warburg *et al.* (1935) later identified the product of glucose-6-phosphate oxidation as 6-phosphogluconate and named the enzyme *zwischenferment*, now known as glucose-6-phosphate dehydrogenase (D-glucose-6-phosphate:NADP oxidoreductase). Cori and Lipmann (1952) showed that the actual product of the reaction was 6-phosphogluconolactone which was hydrolyzed to 6-phosphogluconate by a lactonase (Brodie and Lipmann, 1955, Fig. 6.8), although spontaneous hydrolysis does occur at physiological pH's. Glucose-6-phosphate dehydrogenase is wide-spread in bacteria, including homofermentative and heterofermentative organisms (Gunsalus *et al.*, 1955), and indeed, in most living cells. As a rule, the enzyme functions with NADP, although glucose-6-phosphate dehydrogenase from *L. mesenteroides* will function with either NAD or NADP (DeMoss *et al.*, 1953).

6-Phosphogluconate is oxidized to ribulose-5-phosphate and carbon dioxide by 6-phosphogluconate dehydrogenase (6-phosphogluconate: NADP oxidoreductase [decarboxylating] Horecker *et al.*, 1951). This

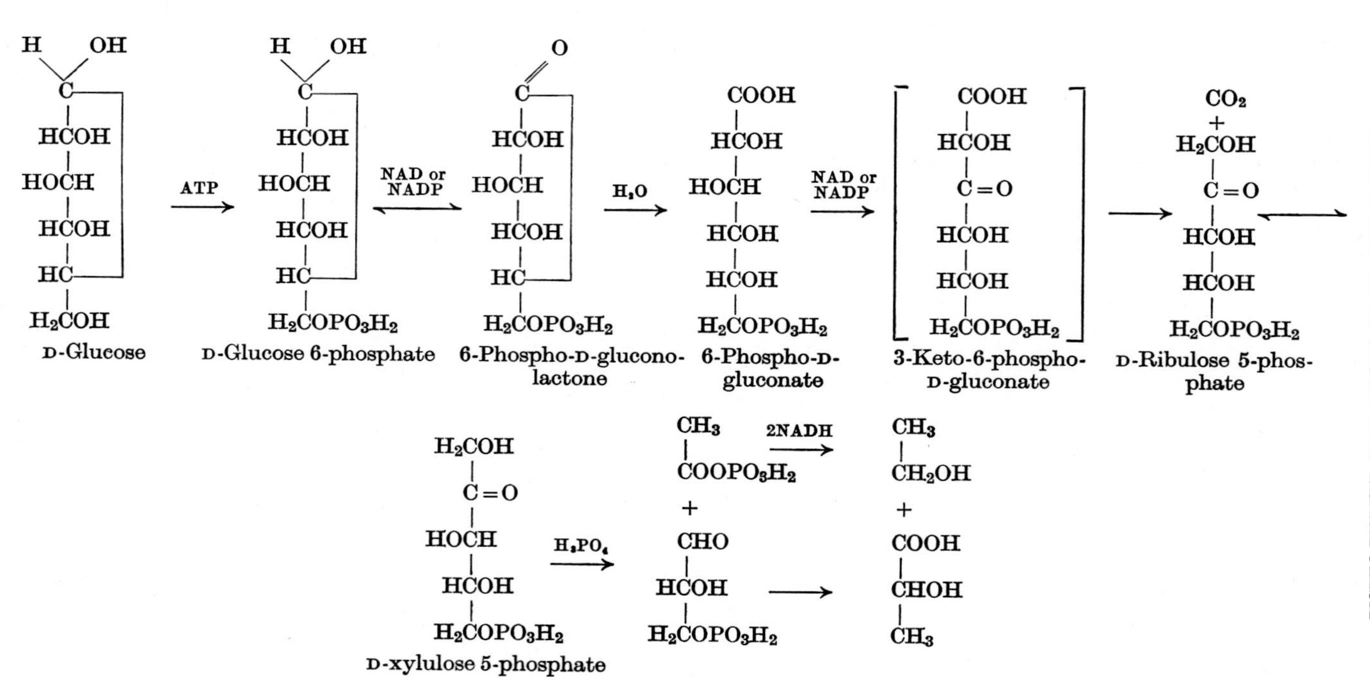

FIG. 6.8. Pathway for fermentation of D-glucose by heterolactic acid bacteria.

enzyme usually functions with NADP, although the enzyme from *L. mesenteroides* requires NAD (DeMoss, 1955). This is a single enzyme reaction and the probable 3-keto intermediate is not stable enough to be isolated. 6-Phosphogluconate dehydrogenase is also wide-spread in bacteria and other living tissues. D-ribulose-5-phosphate undergoes epimerization at carbon 3 by D-ribulose-5-phosphate-3-epimerase (Stumpf and Horecker, 1956) to become D-xylulose-5-phosphate. This reaction differs from those involved in the epimerization of galactose in that uridine diphosphate sugar derivatives are not involved.

The unique reaction of the heterolactic fermentation pathway is the cleavage of D-xylulose-5-phosphate by the enzyme phosphoketolase (D-xylulose-5-phosphate D-glyceraldehyde-3-phosphate-lyase [phosphate-acetylating], Heath *et al.*, 1958a) with phosphate to produce acetyl phosphate and glyceraldehyde-3-phosphate. Phosphoketolase contains thiamine pyrophosphate and, although the mode of action is not certain, it is probably similar to other thiamine-catalyzed reactions (Breslow, 1962). Phosphoketolase is found only in bacteria and is limited to lactic acid bacteria with the exception of a similar enzyme in *Acetobacter xylinum* which acts on fructose-6-phosphate, the 6 carbon homologue of xylulose-5-phosphate (Schramm *et al.*, 1958). Acetyl phosphate is reduced to ethyl alcohol by the reactions outlined in section II.C.4 of this chapter, using electrons removed in the oxidation of glucose-6-phosphate. Glyceraldehyde-3-phosphate is metabolized to pyruvate and lactic acid by the same reactions outlined for the Embden–Meyerhof pathway.

The common naturally-occurring pentoses are D-xylose, D-ribose and L-arabinose, all of which are fermented by *L. pentosus* via the phosphoketolase pathway after conversion to D-xylulose-5-phosphate (Fig. 6.9). D-Xylose is isomerized by xylose isomerase (D-xylose ketol-isomerase, Mitsuhashi and Lampen, 1953) to the ketopentose D-xylulose, which is then phosphorylated by xylulokinase (ATP:D-xylulose-5-phosphotransferase, Stumpf and Horecker, 1956). L-Arabinose is isomerized to L-ribulose by a specific L-arabinose isomerase (L-arabinose ketol-isomerase, Heath *et al.*, 1958b) and L-ribulose is phosphorylated by ribulokinase which uses either D- or L-ribulose (ATP:L (or D) ribulose 5-phosphotransferase, Burma and Horecker, 1958a). L-Ribulose-5-phosphate is converted to D-xylulose-5-phosphate by L-ribulose-5-phosphate-4-epimerase (Burma and Horecker, 1958b). As in the case of the 3-epimerase, uridine diphosphate derivatives are not intermediates. Ribose is fermented by phosphorylation with ribokinase (ATP:D-ribose 5-phosphotransferase, Heath *et al.*, 1958a) forming ribose-5-phosphate, which is then converted to ribulose-5-phosphate by phosphoribose isomerase (D-ribose-5-phosphate ketol-isomerase, Horecker *et al.*, 1951) and then to xylulose-5-phosphate by the 3-epimerase.

In the heterolactic fermentation of glucose, ATP is generated in the oxidation of glyceraldehyde-3-phosphate and in the phosphoenol-

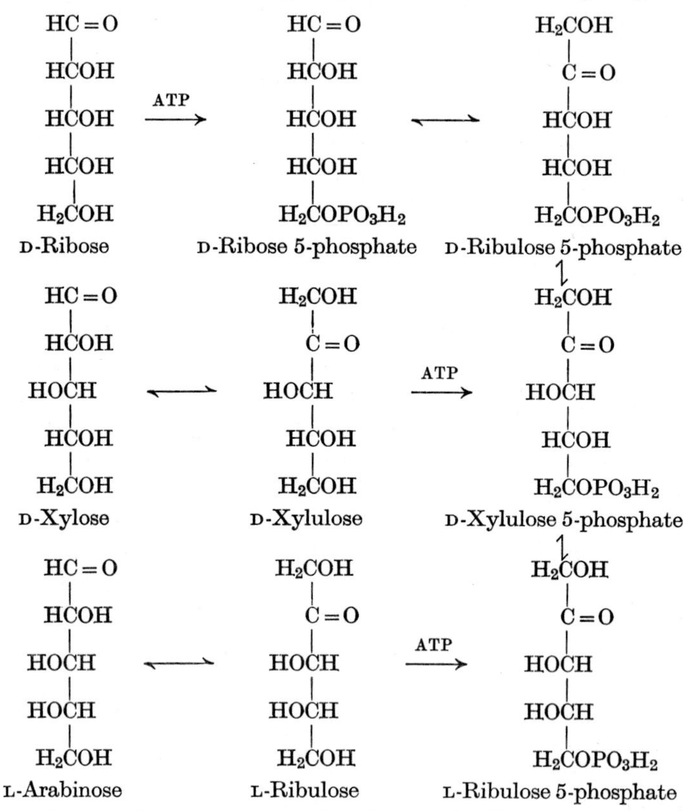

FIG. 6.9. Pathways of pentose metabolism in *Lactobacillus pentosus* leading to D-xylulose 5-phosphate.

pyruvate hydratase reaction. Since one ATP is required for the phosphorylation of glucose, there is a net of one mole of ATP produced in this fermentation. A similar set of calculations reveal that a net of two moles of ATP per mole of pentose fermented by the phosphoketolase pathway is expected, since the energy of the acetyl phosphate linkage should be obtained as ATP.

2. *Enzymes of the* Zymomonas mobilis *pathway*

The *Z. mobilis* pathway of glucose fermentation is similar to a type of glucose oxidation discovered in *Pseudomonas saccharophila* by

Entner and Doudoroff (1952, Fig. 6.10). Glucose is metabolized to 6-phosphogluconate by a kinase and glucose-6-phosphate dehydrogenase. A novel reaction in this pathway is the dehydration of 6-phosphogluconate by 6-phosphogluconate dehydrase (6-phospho-D-gluconate hydro-lyase, MacGee and Doudoroff, 1954), which is then cleaved to pyruvate and glyceraldehyde-3-phosphate by 2-keto-3-deoxy-6-phosphogluconate aldolase (6-phospho-2-keto-3-deoxy-D-gluconate D-glyceraldehyde-3-phosphate-lyase, Kovachevich and Wood, 1955). Glyceraldehyde-3-phosphate is metabolized by the usual triose phosphate pathway to pyruvate, which is decarboxylated by a yeast-type carboxylase to acetaldehyde and then reduced by an alcohol dehydrogenase (Gibbs and DeMoss, 1951). This fermentation pathway would be expec-

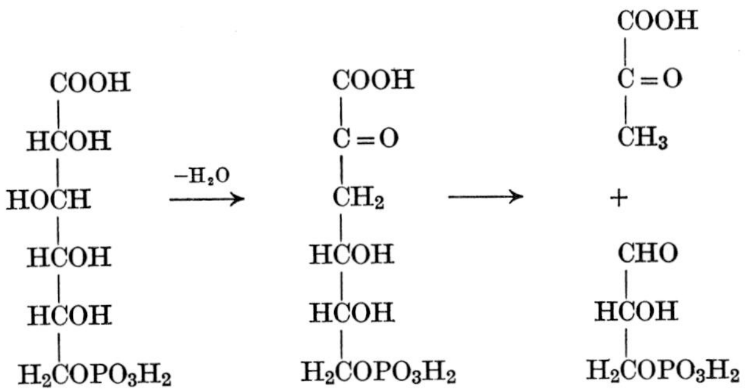

$$\begin{array}{ccccc} & & & & COOH \\ & & & & | \\ COOH & & COOH & & C{=}O \\ | & & | & & | \\ HCOH & & C{=}O & & CH_3 \\ | & \xrightarrow{-H_2O} & | & & \\ HOCH & & CH_2 & \longrightarrow & + \\ | & & | & & \\ HCOH & & HCOH & & CHO \\ | & & | & & | \\ HCOH & & HCOH & & HCOH \\ | & & | & & | \\ H_2COPO_3H_2 & & H_2COPO_3H_2 & & H_2COPO_3H_2 \end{array}$$

6-Phosphogluconate — 2-Keto-3-deoxy-6-phosphogluconate

FIG. 6.10. The Entner–Doudoroff pathway to pyruvate and glyceraldehyde 3-phosphate from 6-phosphogluconate.

ted to yield only one mole of ATP per mole of glucose fermented. This expectation appears to be borne out by the growth yield studies of Bauchop and Elsden (1960), who found that *Z. mobilis* attained only about a half the final cell crop on glucose compared to *S. faecalis*.

3. *Enzymes of pentose fermentation by coliform bacteria and clostridia*

Fermentation of pentoses by *A. aerogenes* involves hexose synthesis from pentose as outlined in Fig. 6.11, at least in the case of L-arabinose fermentation (Simpson and Wood, 1958; Simpson *et al.*, 1958; Wolin *et al.*, 1958). The pathway from xylulose-5-phosphate to hexose phosphate was originally discovered by Horecker and Racker and their associates working with yeast and plant enzymes and verified for *A. aerogenes*. Xylulose-5-phosphate and ribose-5-phosphate are produced by the pathways already outlined for *L. pentosus* and are the substrates

				•CH₂OH										
				CO				•CH₂OH				•CH₂OH		
•CH₂OH		•CHO		HO•CH				CO				CO		
CO		HCOH		HCOH				HO•CH		CHO		HOCH		
HOCH	+	HCOH	⇌ (1)	HCOH	+	CHO	⇌ (2)	HCOH	+	HCOH	⇌ Xu-5-P (3)	HCOH	+	CHO
HCOH		HCOH		HCOH		HCOH		HCOH		HCOH		HCOH		HCOH
$H_2COPO_3H_2$		$H_2COPO_3H_2$		$H_2COPO_3H_2$		$H_2COPO_3H_2$		$H_2COPO_3H_2$		$H_2COPO_3H_2$		$H_2COPO_3H_2$		$H_2COPO_3H_2$
D-Xylylose 5-phosphate		D-Ribose 5-phosphate		D-Sedoheptulose 7-phosphate		D-Glyceraldehyde 3-phosphate		D-Fructose-6-phosphate		D-Erythrose-4-phosphate		D-Fructose-6-phosphate		D-Glyceraldehyde 3-phosphate

FIG. 6.11. Hexose phosphate synthesis from labeled pentose phosphate with transketolase, reactions (1) and (3), and transaldolase, reaction (2). In the second transketolase reaction, xylulose 5-phosphate is labeled in carbon 1. Carbon atoms with an • are radioactive.

for transketolase (D-sedoheptulose-7-phosphate: D-glyceraldehyde-3-phosphate glycolaldehyde transferase, Horecker, *et al.*, 1953; Srere *et al.*, 1955) resulting in sedoheptulose-7-phosphate and glyceraldehyde-3-phosphate. Transketolase is a thiamine pyrophosphate containing enzyme which catalyzes the reversible transfer of carbons 1 and 2 of xylulose-5-phosphate to the acceptor aldopentose, ribose-5-phosphate. The products of the transketolase reaction are the substrates for the next enzyme, transaldolase (D-sedoheptulose-7-phosphate: D-glyceraldehyde-3-phosphate dihydroxyacetone transferase, Horecker and Smyrniotis, 1955), which catalyzes the transfer of carbons 1-3 of sedoheptulose-7-phosphate to glyceraldehyde-3-phosphate resulting in fructose-6-phosphate and erythrose-4-phosphate. Erythrose-4-phosphate is utilized by a second transketolase reaction with a second mole of xylulose-5-phosphate producing fructose-6-phosphate and glyceraldehyde-3-phosphate. The overall reaction results in 2·5 moles of hexose phosphate from 3 moles of pentose phosphate. Fructose-6-phosphate and glyceraldehyde-3-phosphate are fermented by the usual Embden–Meyerhof enzymes to pyruvate. It is interesting to note that the adjacent hydroxyls produced in the transketolase and transaldolase reactions have the *trans* configuration which is also true in the case of the aldolase reaction.

Starting with 1-^{14}C pentose, one mole of fructose-6-phosphate becomes labeled in carbons 1 and 3, resulting in pyruvate labeled in the methyl and carboxyl carbons (Fig. 6.11). The other mole of fructose-6-phosphate is labeled only in carbon 1, producing pyruvate labeled in the methyl carbon. Hence the methyl carbon of pyruvate will have twice the specific activity of the carboxyl carbon, which agrees with the isotope distribution of the fermentation products in both the *Aerobacter* and *Clostridium* fermentations.

The transketolase–transaldolase sequence of reactions is frequently referred to as the pentose phosphate pathway, and has considerable importance in such processes as photosynthetic and chemolithotrophic carbon dioxide fixations, carbohydrate oxidation, pentose phosphate formation for DNA and RNA formation and aromatic amino acid biosynthesis.

IV. Mixed Pathways

A. Propionic Fermentation by Propionibacterium: Products and Enzymic Reactions

The fermentation of glucose and lactic acid by *Propionibacterium* species results in the formation of propionic acid as the major product with lesser amounts of acetic acid, carbon dioxide and succinate (Wood *et al.*, 1937). A similar fermentation of lactic acid takes place with *Veillonella alcalescens* (*Micrococcus lactilyticus*, *Veillonella gazogenes*, Foubert and Douglas, 1948) with the exception that hydrogen is produced. Early *Propionibacterium* fermentation balances always exhibited an excess of oxidized products and high carbon recoveries (Table 6.7). Eventually it was discovered that calcium carbonate, which was added to the medium as a buffer, disappeared during the fermentation. When the amount of carbonate utilized was taken into account, both the oxidized products and carbon recovery were satisfactorily accounted

TABLE 6.7. Evidence for carbon dioxide fixation into products of glycerol (O/R value = 1·0) fermentation by *Propionibacterium*.

	moles/mole glycerol
Products	
Propionic acid	0·593
Acetic acid	0·020
Succinic acid	0·345
Oxidized products	+0·345
Reduced products	−0·593
Net:	−0·148
Expected	−1·00
Moles carbon in products	3·219
Carbon recovery	107%

In this particular fermentation, 0·432 moles of carbon dioxide disappeared from the medium per mole of glycerol fermented. Data taken from Wood and Werkman (1936).

for. However, it was not until the heavy isotope of carbon became available that Wood *et al.* (1941) were able to establish conclusively that $^{13}CO_2$ was fixed into the carboxyls of succinate during glycerol fermentation by *Propionibacterium*. This was the first clear-cut demonstration of heterotrophic carbon dioxide fixation.

Studies of the fermentation of radioactive glucose by *Propionibacterium* indicated that the fermentation pathway was complex (Table 6.8). Isotope from glucose-1-^{14}C appeared mainly in the reduced carbons of the products; carbons 2 and 3 of propionate, the methylene carbons of succinate and the methyl carbon of acetate suggesting an Embden–Meyerhof pathway. However, carbon dioxide also contained an appreciable amount of radioactivity, as did the carboxyls of the fermentation acids, suggesting a hexosemonophosphate pathway as well. Glucose-2-^{14}C fermentation resulted in propionate and succinate labeled as in the case of glucose-1-^{14}C fermentation, but acetate was labeled more heavily in the carboxyl carbon. In the case of glucose-3,4-^{14}C fermentation, isotope was found in carbon dioxide and in the carboxyls of succinate and propionate, which again argues for an Embden–Meyerhof pathway. These data are best explained by the operation of a mixed pathway in *Propionibacterium*; part of the glucose is fermented

TABLE 6.8. Isotope distribution of products of radioactive glucose fermentation by *Propionibacterium* (Wood *et al.*, 1955). R.S.A. = relative specific activity (see Table 6.5).

	G-1-^{14}C	G-2-^{14}C	G-3,4-^{14}C
Products	R.S.A.	R.S.A.	R.S.A.
Propionic acid			
CH_3	14·2	20·2	6·59
| CH_2	15·7	19·4	8·05
| COOH	8·09	3·07	55·4
Succinic acid			
| CH_2	15·5	13·5	—
| COOH	4·5	5·17	41·4
Acetic acid			
CH_3	15·1	13·5	6·50
| COOH	5·23	24·1	5·05
Carbon dioxide	19·7	3·14	48·1

by the Embden–Meyerhof pathway and part is fermented by the pentose phosphate pathway, but with the Embden–Meyerhof pathway predominating.

It is apparent from the data in Table 6.8 that succinate and propionate are related to each other by the similarity of their labeling patterns. The point is clearly demonstrated by the fermentation of lactate-2- and -3-^{14}C by *Propionibacterium*, which results in randomization of the isotope between carbons 2 and 3 of propionate (Leaver *et al.*, 1955). This type of evidence suggested that a symmetrical C_4 intermediate, such as succinate, was a precursor of propionate.

Wood *et al.* (1956) ruled out carbon dioxide fixation into propionate as the primary means of succinate formation when they found that

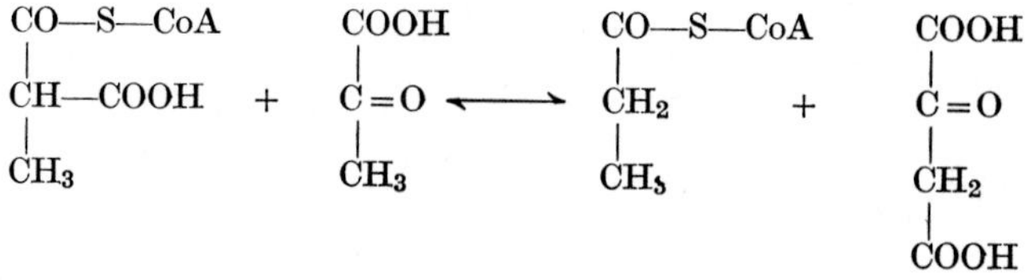

$$\begin{array}{l} CO{-}S{-}CoA \\ | \\ CH{-}COOH \\ | \\ CH_3 \end{array} + \begin{array}{l} COOH \\ | \\ C{=}O \\ | \\ CH_3 \end{array} \rightleftarrows \begin{array}{l} CO{-}S{-}CoA \\ | \\ CH_2 \\ | \\ CH_3 \end{array} + \begin{array}{l} COOH \\ | \\ C{=}O \\ | \\ CH_2 \\ | \\ COOH \end{array}$$

FIG. 6.12. Transcarboxylation between methylmalonyl-CoA and pyruvate.

resting cells of *Propionibacterium* randomized isotope between carbons 2 and 3 of propionate with negligible carbon dioxide fixation. The key enzyme in propionate formation is a biotin-containing enzyme named methylmalonyl transcarboxylase (methylmalonyl-CoA:pyruvate carboxyltransferase, Swick and Wood, 1960) which catalyzes the transfer of the carboxyl group of methylmalonyl coenzyme A to pyruvate with the formation of oxaloacetate and propionyl coenzyme A (Fig. 6.12). Free carbon dioxide is not an intermediate in this reaction.

1·5 Glucose → 3 pyruvate + 6(H)
Pyruvate → acetate + carbon dioxide + 2(H)

Sum: 1·5 Glucose → 2 Pyruvate + acetate + carbon dioxide + 8(H)

2 Pyruvate + 2 methylmalonyl-CoA ⇌ 2 Propionyl-CoA + 2 oxaloacetate
2 Oxaloacetate + 8(H) ⇌ 2 Succinate
2 Succinate + 2 propionyl-CoA ⇌ 2 Succinyl-CoA + 2 propionate
2 Succinyl-CoA ⇌ 2 Methylmalonyl-CoA

Sum: 2 Pyruvate + 8(H) → 2 Propionate

Overall: 1·5 Glucose → 2 Propionate + acetate + carbon dioxide

FIG. 6.13. Reactions in the formation of propionate by *Propionibacterium*.

The pathway of propionate formation by *Propionibacterium* is outlined in Fig. 6.13 (Wood and Stjernholm, 1962). The reducing power required for the reduction of pyruvate to propionate is supplied by the metabolism of 1·5 moles of glucose to 3 moles of pyruvate, one of the latter being oxidized to acetate and carbon dioxide. By means of the transcarboxylase, pyruvate is converted to oxaloacetic acid and then reduced to succinate by reactions similar to those of the tricarboxylic acid cycle. Reduction of oxaloacetate is the step which brings about the randomization of carbons 2 and 3 of propionate, since succinate is a symmetrical compound. Coenzyme A is transferred to succinate from propionyl coenzyme A by coenzyme A transferase, and then succinate is isomerized to methylmalonyl coenzyme A by methylmalonyl mutase (methylmalonyl-CoA CoA-carbonyl mutase, Stjernholm and Wood, 1961). Methyl malonyl mutase contains a form of vitamin B_{12} as the prosthetic group.

Carbon dioxide fixation apparently serves the function of providing oxaloacetic acid which is required in catalytic amounts in the propionic fermentation. The elusive fixation reaction was finally discovered by Siu and Wood (1962) and found to be a reversible fixation of carbon dioxide into phosphoenolpyruvate with the transfer of phosphate from phosphoenolpyruvate to inorganic orthophosphate; they named the enzyme phosphoenolpyruvate carboxyl transphosphorylase (pyrophosphate:oxaloacetate carboxy-lyase [phosphorylating]).

$$CO_2 + \text{Phosphoenolpyruvate} + P_i \rightleftharpoons \text{oxaloacetate} + PP_i$$

If only the Embden–Meyerhof pathway were used to produce pyruvate in the propionic fermentation, then 1·5 moles of glucose would yield 3 moles of ATP plus 1 mole of ATP from pyruvate oxidation to acetate via acetokinase. The overall ATP yield would be 4 moles per 1·5 moles glucose or 2·67 moles ATP per mole glucose, assuming that no ATP is produced during the formation of propionate from pyruvate. Glucose which is oxidized to pentose phosphate and carbon dioxide followed by conversion of pentose phosphate to hexose monophosphate yields slightly less than two moles of ATP per mole hexose:

$$3\ \text{Glucose} \rightarrow 3CO_2 + 2{\cdot}5\ \text{fructose-6-phosphate} + 12(H)$$

$$2{\cdot}5\ \text{Fructose-6-phosphate} \rightarrow 5\ \text{pyruvate} + 5ATP + 10(H)$$

$$\text{Sum:}\quad 3\ \text{Glucose} \rightarrow 3CO_2 + 5\ \text{pyruvate} + 5ATP + 22(H)$$

This is a more oxidative pathway than the Embden–Meyerhof pathway and would probably result in sparing pyruvate oxidation to acetate. The 22(H) evolved in the oxidation of 3 moles of glucose is more than enough to reduce the resulting 5 moles of pyruvate to propionate. The overall ATP yield in this case would then be 1·67 moles ATP per mole glucose. In their studies, Bauchop and Elsden obtained a growth yield of 37·5 g of *Propionibacterium pentosaceum* per mole of glucose, which means nearly 4 moles of ATP per mole glucose, and they predicted that a mole of ATP is formed for each mole of propionate produced.

B. Gluconate Fermentation by *Streptococcus faecalis*

Streptococcus faecalis ferments gluconic acid with the production of 0·5 mole carbon dioxide, 1·5 moles lactic acid and the rest a mixture of small amounts of acetic acid, formic acid and ethyl alcohol. In the presence of arsenite, 1·75 moles of lactic acid were produced with negligible amounts of the minor products (Sokatch and Gunsalus, 1957). Fermentation of gluconate-1-^{14}C resulted in radioactive carbon dioxide and carboxyl-labeled lactic acid. Fermentation of gluconate-2-^{14}C resulted in unlabeled carbon dioxide and lactic acid labeled heavily in carbon 2, but also appreciably labeled in carbons 1 and 3. The best explanation for the isotope distribution is that one mole of gluconate is oxidized via 6-phosphogluconate dehydrogenase to carbon dioxide and pentose phosphate; pentose phosphate is then converted to hexose phosphate and fermented to the usual products of hexose fermentation

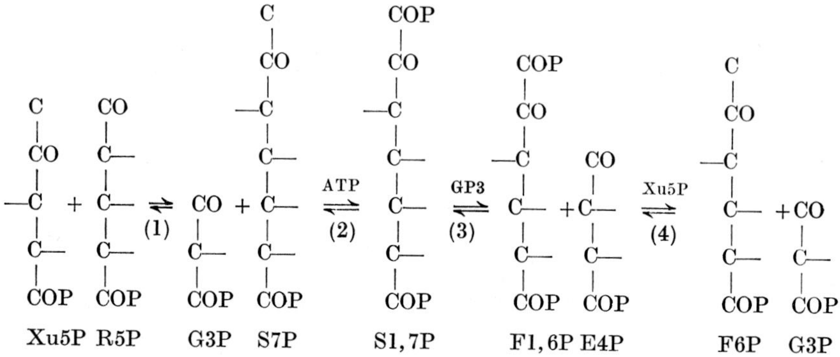

Xu5P + R5P ⇌ (1) G3P + S7P ⇌ ATP (2) S1,7P ⇌ GP3 (3) F1,6P + E4P ⇌ Xu5P (4) F6P + G3P

Sum: 3 Pentose phosphate = 2·5 hexose phosphate

Fig. 6.14. Hexose phosphate synthesis by *S. faecalis*. Reaction (1) is catalyzed by transketolase, (2) by phosphofructokinase, (3) by aldolase and (4) by transketolase (Sokatch, 1962). Compare with Fig. 6.11.

by this organism. An equal amount of 6-phosphogluconate is converted to pyruvate and glyceraldehyde-3-phosphate by way of the Entner–Doudoroff pathway, which would account for the label in the carboxyl carbon of lactic acid after gluconate-1-^{14}C fermentation. Many of the expected enzymes were detected in *S. faecalis*, including 2-keto-3-deoxy-6-phospho-D-gluconate aldolase (Sokatch *et al.*, 1956), which was present in increased amounts in cells grown on gluconate as opposed to cells grown on glucose. Although transketolase was present transaldolase could not be detected in this organism, but it is possible to bypass transaldolase by the use of a combination of phosphofructokinase and aldolase, which are known to function with sedoheptulose phosphates (Fig. 6.14). Fermentation of 2-ketogluconate by *S. faecalis* produced 1·0 mole carbon dioxide and a mixture of ethyl alcohol, acetic acid and formic acid, or in the presence of arsenite, 1·67 moles of lactic acid. The labeling data indicate that 2-ketogluconate fermentation involves only the pentose phosphate pathway, as it occurs in *S. faecalis* and not the Entner–Doudoroff pathway (Goddard and Sokatch, 1964). *Streptococcus faecalis* is unique among the lactic acid bacteria in utilizing the pentose phosphate pathway for hexonic acid fermentation as opposed to the phosphoketolase pathway.

V. Miscellaneous Pathways

Two other organisms which produce propionate as a fermentation product are *Clostridium propionicum* (Cardon and Barker, 1947) and *Peptostreptococcus elsdenii* (Elsden *et al.*, 1956), the latter organism originally being referred to as LC, for "large coccus", until the generic designation was made. *Clostridium propionicum* ferments lactate to propionate, acetate and carbon dioxide in the proportions 2:1:1, while *P. elsdenii* forms a mixture of acetic, propionic, butyric and valeric

Lactic acid → Acetic acid + carbon dioxide + 4(H)

2 Lactic acid + 2 propionyl-CoA → 2 Lactyl-CoA + 2 propionic acid

2 Lactyl-CoA → 2 Acrylyl-CoA + $2H_2O$

2 Acrylyl-CoA + 4(H) → 2 Propionyl-CoA

Sum: 3 Lactic acid → 2 propionic acid + acetic acid + carbon dioxide

FIG. 6.15. Acrylate pathway for the production of propionate from lactic acid by *Clostridium propionicum* and *Peptostreptococcus elsdenii.*

acids from lactate. Both organisms differ from *Propionibacterium* in that lactic acid is converted to propionate without randomization of carbons 2 and 3 (Leaver *et al.*, 1955; Ladd, 1959).

The probable route for conversion of lactic acid to propionic acid by both *C. propionicum* and *P. elsdenii* is shown in Fig. 6·15. Coenzyme A transferase of *P. elsdenii* is responsible for the transfer of coenzyme A from propionycol enzyme A to lactic acid to produce lactyl coenzyme A. Vagelos *et al.* (1959) discovered lactyl coenzyme A dehydrase, the enzyme which removes water from lactyl coenzyme A to produce acrylyl coenzyme A, in a propionate-oxidizing pseudomonad but were unable to find the enzyme in *C. propionicum*. Baldwin *et al.* (1965) purified lactyl coenzyme A dehydrase several fold from *P. elsdenii*. Both *Pseudomonas* and *Peptostreptococcus* enzymes are specific for L(+) lactic acid. Acrylyl coenzyme A is then reduced to propionyl coenzyme A by an acyl coenzyme A dehydrogenase present in both *C. propionicum* and *P. elsdenii* (Stadtman and Vagelos, 1957; Baldwin *et al.*, 1965).

References

Altermatt, H. A., Simpson, F. J. and Neish, A. C. (1955a). *Can. J. Microbiol.* **1**, 473.

Altermatt, H. A., Simpson, F. J. and Neish, A. C. (1955b). *Can. J. Biochem. Physiol.* **33**, 615.

Atkinson, M. R. and Morton, R. K. (1960). *In* "Comparative Biochemistry" (M. Florkin and H. S. Mason, eds.), Academic Press, New York, Vol. II, p. 1.

Baldwin, R. L., Wood, W. A. and Emery, R. S. (1965). *Biochim. Biophys. Acta* **97**, 202.

Baranowski, T. (1963). *In* "The Enzymes" (P. D. Boyer, H. Lardy and K. Myrback, eds.), Academic Press, New York, Vol. 7, 85.

Bard, R. C. and Gunsalus, I. C. (1950). *J. Bacteriol* **59**, 387.

Barker, H. A. (1956). "Bacterial Fermentations", John Wiley and Sons, New York.

Barker, H. A. and Kamen, M. D. (1945). *Proc. Natl Acad. Sci. U.S.* **31**, 219.

Bauchop, T. and Elsden, S. R. (1960). *J. Gen. Microbiol.* **23**, 457.

Bernstein, I. A. (1953). *J. Biol. Chem.* **205**, 309.

Bhat, J. V. and Barker, H. A. (1947). *J. Bacteriol.* **54**, 381.

Blackwood, A. C. and Blakley, E. R. (1956). *Can. J. Microbiol.* **2**, 741.

Blackwood, A. C., Neish, A. C. and Ledingham, G. A. (1956). *J. Bacteriol.* **72**, 497.

Blakley, E. R. and Blackwood, A. C. (1957). *Can. J. Microbiol.* **3**, 741.

Boyer, P. D. (1962). *In* "The Enzymes" (P. D. Boyer, H. Lardy and K. Myrback, eds.), Academic Press, New York, Vol. 6, p. 95.

Breslow, R. (1962). *Ann. N.Y. Acad. Sci.* **98**, 445.

Brill, W. A., Wolin, E. A. and Wolfe, R. S. (1964). *Science* **144**, 297.

Brodie, A. F. and Lipmann, F. (1955). *J. Biol. Chem.* **212**, 677.

Burma, D. P. and Horecker, B. L. (1958a). *J. Biol. Chem.* **231**, 1039.

Burma, D. P. and Horecker, B. L. (1958b). *J. Biol. Chem.* **231**, 1053.

Burton, R. M. and Stadtman, E. R. (1953). *J. Biol. Chem.* **202**, 873.

Caputto, R., Leloir, L. F., Cardini, C. E. and Paladini, A. C. (1950). *J. Biol. Chem.* **184**, 333.

Cardini, C. E. (1951). *Enzymologia* **14**, 362.

Cardon, B. P. and Barker, H. A. (1947). *Arch. Biochem. Biophys.* **12**, 165.

Chantrenne, H. and Lipmann, F. (1950). *J. Biol. Chem.* **187**, 757.

Cori, O. and Lipmann, F. (1952). *J. Biol. Chem.* **194**, 417.

Crane, R. K. (1962). *In* "The Enzymes" (P. D. Boyer, H. Lardy and K. Myrback, eds.), Academic Press, New York, Vol. 6, p. 47.

Cynkin, M. A. and Gibbs, M. (1958). *J. Bacteriol.* **75**, 335.

DeMoss, R. D. (1955). *In* "Methods in Enzymology" (S. P. Colowick and N. O. Kaplan, eds.), Academic Press, New York, Vol. I, p. 328.

DeMoss, R. D., Bard, R. C. and Gunsalus, I. C. (1951). *J. Bacteriol.* **62**, 499.

DeMoss, R. D., Gunsalus, I. C. and Bard, R. C. (1953). *J. Bacteriol.* **66**, 10.

Elsden, S. R., Volcani, B. E., Gilchrist, F. M. C. and Lewis, D. (1956). *J. Bacteriol.* **72**, 681.

Entner, N. and Doudoroff, M. (1952). *J. Biol. Chem.* **196**, 853.

Foubert, E. L. and Douglas, H. C. (1948). *J. Bacteriol.* **56**, 35.

Fred, E. B., Peterson, W. H. and Anderson, J. P. (1921). *J. Biol. Chem.* **48**, 385.

Friedemann, T. E. (1939). *J. Biol. Chem.* **130**, 757.

Fukuyama, T. T. and O'Kane, D. J. (1962). *J. Bacteriol.* **84**, 793.

Gest, H. (1954). *Bacteriol. Rev.* **18**, 43.

Gibbs, M. and DeMoss, R. D. (1951). *Arch. Biochem. Biophys.* **34**, 478.
Gibbs, M. and DeMoss, R. D. (1954). *J. Biol. Chem.* **207**, 689.
Gibbs, M. and Paege, L. M. (1961). *J. Biol. Chem.* **236**, 6.
Gibbs, M., Dumrose, R., Bennett, F. A. and Bubeck, M. R. (1950). *J. Biol. Chem* **184**, 545.
Gibbs, M., Sokatch, J. R. and Gunsalus, I. C. (1955). *J. Bacteriol.* **70**, 572.
Goddard, J. L. and Sokatch, J. R. (1964). *J. Bacteriol.* **87**, 844.
Grisolia, S., and Joyce, B. K. (1959). *J. Biol. Chem.* **234**, 1335.
Gunsalus, I. C. (1954). *In* "The Mechanism of Enzyme Action" (W. D. McElroy and B. D. Glass, eds.), The Johns Hopkins Press, Baltimore, p. 545.
Gunsalus, I. C. and Gibbs, M. (1952). *J. Biol. Chem.* **194**, 871.
Gunsalus, I. C. and Niven, C. F. (1942). *J. Biol. Chem.* **145**, 131.
Gunsalus, I. C., Horecker, B. L. and Wood, W. A. (1955). *Bacteriol. Rev.* **19**, 79.
Heath, E. C., Hurwitz, J., Horecker, B. L. and Ginsburg, A. (1958a). *J. Biol. Chem.* **231**, 1009.
Heath, E. C., Horecker, B. L., Smyrniotis, P. Z. and Takagi, Y. (1958b). *J. Biol. Chem.* **231**, 1031.
Holzer, H., da Fonseca-Wollheim, F., Kohlhaw, C. and Woenckhaus, C. W. (1962). *Ann. N.Y. Acad. Sci.* **98**, 453.
Horecker, B. L. and Smyrniotis, P. Z. (1955). *J. Biol. Chem.* **212**, 811.
Horecker, B. L., Smyrniotis, P. Z. and Seegmiller, J. E. (1951). *J. Biol. Chem.* **193**, 383.
Horecker, B. L., Smyrniotis, P. Z. and Klenow, H. (1953). *J. Biol. Chem.* **205**, 661.
Horecker, B. L., Smyrniotis, P. Z., Hiatt, H. and Marks, P. (1955). *J. Biol. Chem.* **212**, 827.
Jacobs, N. J. and VanDemark, P. J. (1960). *J. Bacteriol.* **79**, 532.
Juni, E. (1952). *J. Biol. Chem.* **195**, 715.
Kalckar, H. M., Braganca, B. and Munch-Peterson, A. (1953). *Nature* **172**, 1038.
Kluyver, A. J. and Hoppenbrouwers, W. J. (1931). *Arch. Mikrobiol.* **2**, 245.
Koshland, D. E. and Westheimer, F. H. (1950). *J. Am. Chem. Soc.* **72**, 3383.
Kovachevich, R. and Wood, W. A. (1955). *J. Biol. Chem.* **213**, 757.
Krampitz, L. O., Suzuki, I. and Greull, G. (1962). *Ann. N.Y. Acad. Sci.* **98**, 466.
Ladd, J. N. (1959). *Biochem. J.* **71**, 16.
Lampen, J. O., Gest, H. and Sowden, J. C. (1951). *J. Bacteriol.* **61**, 97.
Lardy, H. A. (1962). *In* "The Enzymes" (P. D. Boyer, H. Lardy and K. Myrback, eds.), Academic Press, New York, Vol. 6, p. 67.
Leaver, F. W., Wood, H. G. and Stjernholm, R. (1955). *J. Bacteriol.* **70**, 251.
Leloir, L. F. (1951). *Arch. Biochem. Biophys.* **33**, 186.
MacGee, J. and Doudoroff, M. (1954). *J. Biol. Chem.* **210**, 617.
Massey, V. (1963). *In* "The Enzymes" (P. D. Boyer, H. Lardy and K. Myrback, eds.), Academic Press, New York, Vol. 7, p. 275.
Maxwell, E. S. (1961). *In* "The Enzymes" (P. D. Boyer, H. Lardy and K. Myrback, eds.), Academic Press, New York, Vol. 5, p. 433.
Mitsuhashi, S. and Lampen, J. O. (1953). *J. Biol. Chem.* **204**, 1011.
Moore, L. D. and O'Kane, D. J. (1963). *J. Bacteriol.* **86**, 766.
Mortenson, L. E., Valentine, R. C. and Carnahan, J. E. (1963). *J. Biol. Chem.* **238**, 794.
Mortlock, R. P., Valentine, R. C. and Wolfe, R. S. (1959). *J. Biol. Chem.* **234**, 1653.

Najjar, V. A. (1962). *In* "The Enzymes" (P. D. Boyer, H. Lardy and K. Myrback, eds.), Academic Press, New York, Vol. 6, p. 161.
O'Kane, D. J. and Gunsalus, I. C. (1948). *J. Bacteriol.* **56**, 499.
Osburn, O. L., Brown, R. W. and Werkman, C. H. (1937). *J. Biol. Chem.* **121**, 685.
Paege, L. M., and Gibbs, M. (1961). *J. Bacteriol.* **81**, 107.
Paege, L. M., Gibbs, M. and Bard, R. C. (1956). *J. Bacteriol.* **72**, 65.
Paladini, A. C., Caputto, R., Leloir, L. F., Trucco, R. E. and Cardini, C. E. (1949). *Arch. Biochem. Biophys.* **23**, 55.
Prescott, S. C. and Dunn, C. G. (1959). "Industrial Microbiology", McGraw-Hill, New York.
Rappoport, D. A., Barker, H. A. and Hassid, W. Z. (1951). *Arch. Biochem. Biophys.* **31**, 326.
Reynolds, H. and Werkman, C. H. (1937). *J. Bacteriol.* **33**, 603.
Robbins, E. A. and Boyer, P. D. (1957). *J. Biol. Chem.* **224**, 121.
Rose, I. A. (1962). *In* "The Enzymes" (P. D. Boyer, H. Lardy and K. Myrback, eds.), Academic Press, New York, Vol. 6, p. 115.
Rutter, W. J. (1961). *In* "The Enzymes" (P. D. Boyer, H. Lardy and K. Myrback, eds.), Academic Press, New York, Vol. 5, p. 341.
Schramm, M., Klybas, V. and Racker, E. (1958). *J. Biol. Chem.* **233**, 1283.
Sherman, J. (1937). *Bacteriol. Rev.* **1**, 3.
Sherman, J. R. and Adler, J. (1963). *J. Biol. Chem.* **238**, 873.
Simpson, F. J. and Wood, W. A. (1958). *J. Biol. Chem.* **230**, 473.
Simpson, F. J., Wolin, M. J. and Wood, W. A. (1958). *J. Biol. Chem.* **230**, 457.
Siu, P. M. L. and Wood, H. G. (1962). *J. Biol. Chem.* **237**, 3044.
Sokatch, J. R. (1962). *Arch. Biochem. Biophys.* **99**, 401.
Sokatch, J. R. and Gunsalus, I. C. (1957). *J. Bacteriol.* **73**, 452.
Sokatch, J. R., Prieto, A. P. and Gunsalus, I. C. (1956). *Bact. Proc.* p. 112.
Srere, P. A., Cooper, J. R., Klybas, V. and Racker, E. (1955). *Arch. Biochem. Biophys.* **59**, 535.
Stadtman, E. R. and Vagelos, P. R. (1957). *Proc. Intern. Symp. Enzyme Chem. Tokyo Kyoto*, p. 86.
Stadtman, E. R., Novelli, G. D. and Lipmann, F. (1951). *J. Biol. Chem.* **191**, 354.
Steele, R. H., White, A. G. C. and Pierce, W. A. (1954). *J. Bacteriol.* **67**, 86.
Stephenson, M. and Stickland, L. H. (1932). *Biochem. J.* **26**, 712.
Stjernholm, R. and Wood, H. G. (1961). *Proc. Natl Acad. Sci. U.S.* **47**, 303.
Strecker, H. J. (1951). *J. Biol. Chem.* **189**, 815.
Stumpf, P. K. and Horecker, B. L. (1956). *J. Biol. Chem.* **218**, 753.
Swick, R. W. and Wood, H. G. (1960). *Proc. Natl Acad. Sci. U.S.* **46**, 28.
Topper, Y. J. (1961). *In* "The Enzymes" (P. D. Boyer, H. Lardy and K. Myrback, eds.), Academic Press, New York, Vol. 5, p. 429.
Trucco, R. E. (1951). *Arch. Biochem. Biophys.* **34**, 482.
Trucco, R. E., Caputto, R., Leloir, L. F. and Mittleman, N. (1948). *Arch. Biochem. Biophys.* **18**, 137.
Vagelos, P. R., Earl, J. M. and Stadtman, E. R. (1959). *J. Biol. Chem.* **234**, 765.
Velick, S. F. and Furfine, C. (1963). *In* "The Enzymes" (P. D. Boyer, H. Lardy and K. Myrback, eds.), Academic Press, New York, Vol. 7, p. 243.
Warburg, O. and Christian, W. (1931). *Biochem. Z.* **242**, 206.
Warburg, O., Christian, W. and Griese, A. (1935). *Biochem. Z.* **282**, 157.
Wilson, J., Krampitz, L. O. and Werkman, C. H. (1948). *Biochem. J.* **42**, 598.
Wolfe, R. S. and O'Kane, D. J. (1955). *J. Biol. Chem.* **215**, 637.

Wolin, M. J., Simpson, F. J. and Wood, W. A. (1958). *J. Biol. Chem.* **232**, 559.
Wood, H. G. (1952). *J. Biol. Chem.* **199**, 579.
Wood, H. G. and Stjernholm, R. (1962). *In* "The Bacteria" (I. C. Gunsalus and R. Y. Stanier, eds.), Academic Press, New York, Vol. III, p. 41.
Wood, H. G. and Werkman, C. H. (1936). *Biochem. J.* **30**, 48.
Wood, H. G., Stone, R. W. and Werkman, C. H. (1937). *Biochem. J.* **31**, 349.
Wood, H. G., Werkman, C. H., Hemingway, A. and Wier, A. O. (1941). *J. Biol. Chem.* **139**, 377.
Wood, H. G., Stjernholm, R. and Leaver, F. W. (1955). *J. Bacteriol.* **70**, 510.
Wood, H. G., Stjernholm, R. and Leaver, F. W. (1956). *J. Bacteriol.* **72**, 142.
Wood, N. P. and O'Kane, D. J. (1964). *J. Bacteriol.* **87**, 97.
Wood, W. A. (1961). *In* "The Bacteria" (I. C. Gunsalus and R. Y. Stanier, eds.), Academic Press, New York, Vol. II, p. 59.

7. AEROBIC METABOLISM OF CARBOHYDRATES

I. Hexose Oxidation

A. Oxidation by Way of the Embden–Meyerhof Pathway

Probably the most universal means of hexose oxidation is metabolism by way of the Embden–Meyerhof pathway to pyruvate followed by oxidation of pyruvate to acetyl-CoA and oxidation of acetyl-CoA by means of the tricarboxylic acid cycle. Energy is obtained both by substrate-level phosphorylation, the sole energy mechanism in fermentation, and by oxidative phosphorylation resulting from electron transport through the cytochrome system.

It is not a simple matter to decide which of the several pathways of hexose oxidation outlined in this chapter is the major pathway unless the enzymes of only one pathway exist in a given organism. Frequently, however, enzymes of the Embden–Meyerhof pathway occur alongside those of another pathway, usually those of the oxidative pentose phosphate pathway. All pathways which result in the complete oxidation of hexose to carbon dioxide will theoretically yield respiratory carbon dioxide of the same specific activity from specifically labeled glucose. In practice, however, labeled substrates are of some value in determining the major pathway of hexose oxidation because of assimilation of part of the substrate, in particular, those carbons of glucose farthest removed from oxidative attack. Since the first step which results in carbon dioxide release from glucose to the Embden–Meyerhof pathway is the oxidative decarboxylation of pyruvate, the carbons most readily assimilated are carbons 2 and 3 of pyruvate, originally carbons 1, 2 and 5, 6 of glucose. Assimilation is especially important in the case of aerobic organisms grown in a glucose-salts medium where glucose is the sole source of carbon and energy. Approximately half the glucose molecule is oxidized and half assimilated under these circumstances. The existence of other pathways of glucose catabolism complicates the labeling picture, which makes a precise judgement of the relative importance of simultaneous pathways difficult. In the final analysis, the decision of which is the major pathway depends on many pieces of circumstantial evidence, but mainly on the labeling data and enzymic analysis of the organism.

The results of a number of studies where the total yield of $^{14}CO_2$ from the positions of specifically labeled glucose was measured in a Warburg respirometer were summarized by Cheldelin *et al.* (1962) (Table 7.1).

TABLE 7.1. Relative Importance of Metabolic Pathways of Glucose Oxidation by Microorganisms as Determined by Oxidation of Labeled Glucose.

Organism	Pathway of hexose oxidation		
	Embden–Meyerhof	Hexosemonophosphate oxidation	Entner–Doudoroff
Saccharomyces cerevisiae	Major	Minor	
Candida utilis	Major	Minor	
Streptomyces griseus	Major	Minor	
Penicillium chrysogenum	Major	Minor	
Penicillium digitatum	Major	Minor	
Fusarium lini	Major	Minor	
Escherichia coli	Major	Minor	
Sarcina lutea	Major	Minor	
Bacillus subtilis	Major	Minor	
Pseudomonas reptilovora		Minor	Major
Pseudomonas aeruginosa		Minor	Major
Acetobacter suboxydans		Sole	
Zymomonas mobilis			Sole
Pseudomonas saccharophila			Sole

Major = major pathway, minor = minor pathway, sole = sole pathway. From Cheldelin *et al.*, 1962.

Facultative bacteria such as *E. coli* and *B. subtilis*, as well as several species of fungi, were found to oxidize glucose with the Embden–Meyerhof pathway as the major catabolic pathway; that is, more CO_2 was evolved from positions 3 and 4 of glucose than from the other carbons.

B. THE OXIDATIVE PENTOSE CYCLE

When the enzymes of the pentose phosphate pathway were first characterized, the function of these enzymes in oxidative organisms was not immediately certain for the reasons outlined in the previous section. Couri and Racker (1959) proposed a cyclic pathway of hexose oxidation using the enzymes of the pentose phosphate pathway (Table 7.2). For convenience of discussion, three moles of glucose are taken and oxidized to three moles of pentose phosphate, reactions (1) and (6), Fig. 7.1. Two moles of pentose phosphate are converted to fructose-6-phosphate and tetrose phosphate via sedoheptulose-7-phosphate, reactions (2) and (3). Fructose-6-phosphate is converted to glucose-6-phosphate by phosphoglucose isomerase, reaction (4), completing the first turn of the cycle for this molecule. Tetrose phosphate, formed in reaction (3), is converted to fructose-6-phosphate by transketolase using one of the three moles

TABLE 7.2. The Oxidative Pentose Phosphate Pathway

Reaction	Enzyme
3 G + 3 ATP → 3 G-6-P + 3 ADP	Hexokinase
3 G-6-P → 3 6-PG + 6(H)	Glucose-6-phosphate dehydrogenase
3 6-PG → 3 Ru-5-P + 3 CO_2 + 6(H)	6-Phosphogluconate dehydrogenase
Ru-5-P → R-5-P	Phosphoribose isomerase
2 Ru-5-P → 2 Xu-5-P	Ribulose-5-phosphate-3-epimerase
R-5-P + Xu-5-P → S-7-P + G-3-P	Transketolase
S-7-P + G-3-P → F-6-P + E-4-P	Transaldolase
E-4-P + Xu-5-P → F-6-P + G-3-P	Transketolase
Net: 3G + 3ATP → 3ADP + 12(H) + 2 F-6-P + G-3-P + 3 CO_2	

Abbreviations: G = glucose, G-6-P = glucose-6-phosphate, 6-PG = 6-phosphogluconate, Ru-5-P = ribulose-5-phosphate, R-5-P = ribose-5-phosphate, Xu-5-P = xylulose-5-phosphate, S-7-P = sedoheptulose-7-phosphate, G-3-P = glyceraldehyde-3-phosphate, F-6-P = fructose-6-phosphate, E-4-P = erythrose-4-phosphate.

of pentose phosphate formed in the initial oxidation of glucose, reaction (5). Fructose-6-phosphate formed in reaction (5) is isomerized to glucose-6-phosphate, reaction (7), and the triose phosphate formed in reaction (5) is converted to pyruvate, reaction (8), and oxidized via the tricarboxylic acid cycle. In those organisms which do not possess a tricarboxylic acid cycle, fructose-1,6-diphosphate can be formed from triose phosphate by aldolase, dephosphorylated by fructose-1,6-diphosphatase (D-fructose-1,6-diphosphate-1-phosphohydrolase) and oxidized by the usual oxidative pentose cycle reactions.

After one turn of the oxidative pentose cycle, all of carbon 1 of glucose has been evolved as carbon dioxide. Carbon 2 of glucose has moved into carbons 1 and 3 of hexosemonophosphate and carbon 3 of glucose has moved into carbons 2 and 3 of hexosemonophosphate. As the cycle continues, carbons 2 and 3 of glucose equilibrate with carbons 1 to 3 of hexosemonophosphate; carbons 4 to 6 of glucose do not equilibrate unless resynthesis of hexose occurs by way of hexosediphosphate. Katz

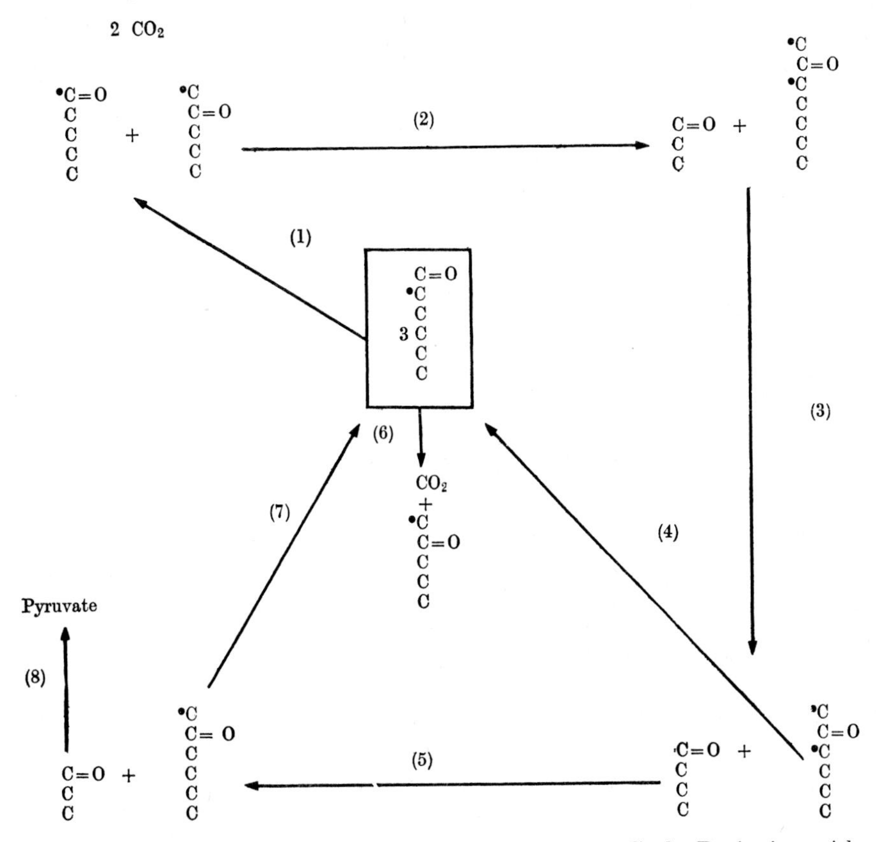

FIG. 7.1. Flow of Carbon in the Oxidative Pentose Cycle Beginning with Glucose-2-^{14}C. The numbers refer to reactions discussed in the text.

and Wood (1960) showed mathematically that carbons 1 to 3 of hexosemonophosphate will assume a characteristic labeling pattern from glucose-2-^{14}C or glucose-3-^{14}C, providing that radioactive glucose is taken up as fast as it is oxidized. However, it should be emphasized that a number of factors affect the labeling pattern of hexosemonophosphate, in particular, the extent of recycling, the occurrence of other oxidative pathways, the drainage of intermediates for biosynthesis, and the contribution of hexosemonophosphate from other metabolic pathways.

Organisms which use the oxidative pentose cycle evolve more carbon dioxide from carbons 1 to 3 of glucose than from carbons 4 to 6, and by this criterion the oxidative pentose cycle is widely distributed in bacteria, at least as a minor pathway (Table 7.1). On the basis of enzyme content, the oxidative pentose cycle is the major pathway of hexose oxidation in *Acetobacter* and the closely related *Gluconobacter* (De Ley, 1960). *Acetobacter suboxydans* uses solely the pentose cycle for hexose oxidation (Table 7.1) and is an interesting organism because it does not have a functional tricarboxlic acid cycle (Rao, 1957). Hence, triose phosphate must be either assimilated or converted to hexosediphosphate and hexosemonophosphate, and then oxidized.

C. The Entner–Doudoroff Pathway

The Entner–Doudoroff pathway is the third major pathway of hexose oxidation (Fig. 7.2), and the enzymes of this pathway are described in

Glucose

↓ +ATP

Glucose-6-phosphate

↓ +NADP

6-Phosphogluconate

↓ $-H_2O$

2-Keto-3-deoxy-6-phosphogluconate

↓

Pyruvate + glyceraldehyde-3-phosphate

Fig. 7.2. Entner–Doudoroff Pathway of Glucose Oxidation.

connection with the discussion of the glucose fermentation by *Z. mobilis* in Chapter 6. The pathway was discovered in *P. saccharophila* by Entner and Doudoroff (1952), who found that glucose was oxidized by this organism to two moles of pyruvate with the carboxyl carbon of one mole of pyruvate derived from carbon 1 of glucose. Subsequent studies by MacGee and Doudoroff (1954) led to the isolation and characterization of the sodium salt of 2-keto-3-deoxy-6-phosphogluconate. A similar set of studies by Kovachevich and Wood (1955a, b) established the occurrence of the same pathway in *P. fluorescens*.

The Entner–Doudoroff pathway is the major pathway of hexose oxidation by organisms of the genus *Pseudomonas* and related genera, excluding *Acetobacter* (De Ley, 1960). Glyceraldehyde-3-phosphate, which is formed in the cleavage of 2-keto-3-deoxy-6-phosphogluconate, is oxidized to pyruvate by the triose phosphate pathway and pyruvate is then oxidized by way of the tricarboxylic acid cycle.

D. Direct Hexose Oxidation

Many species of *Pseudomonas* accumulate high yields of 2-ketogluconate when grown with high concentrations of glucose and a limiting nitrogen source (Stubbs *et al.*, 1940). Stokes and Campbell (1951) showed that dried cells of *P. aeruginosa* converted glucose and gluconate to 2-ketogluconate without phosphorylation. Wood and Schwerdt (1953) extended these observations by demonstrating the presence of particulate, cytochrome-containing enzymes in extracts of *P. fluorescens* which catalyzed the oxidation of glucose and gluconate without the

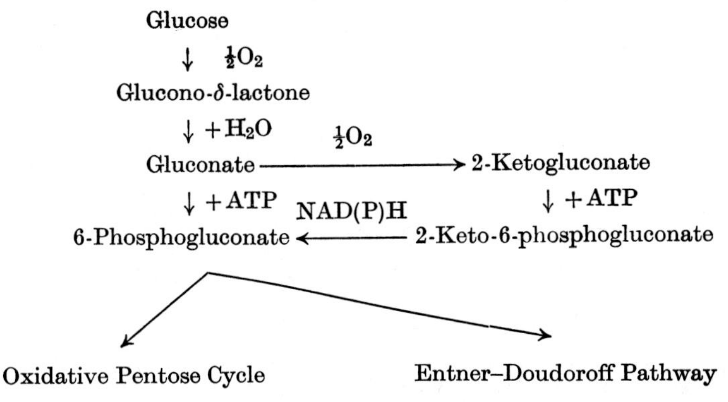

Fig. 7.3. Direct glucose oxidation by *Pseudomonas* and *Aerobacter*.

requirement for a soluble cofactor (Fig. 7.3). It is probable that gluconate-δ-lactone is an intermediate in the oxidation of glucose to gluconate by these organisms, as is the case with the corresponding enzymes from other sources (Cori and Lipmann, 1952).

Although the production of 2-ketogluconate can account for a considerable portion of the glucose added, the conditions used are artificial and the process is really a metabolic side issue off the main pathway of glucose metabolism by these organisms. De Ley (1953) discovered the first step in the pathway for the utilization of 2-ketogluconate using extracts of *Aerobacter cloacae* where the pathway for 2-ketogluconate catabolism is inducible (Fig. 7.3). 2-Ketogluconate was phosphorylated by ATP: 2-keto-D-gluconate 6-phosphotransferase to form 2-keto-6-phosphogluconate, which was reduced with either NADH or NADPH by 6-phospho-D-gluconate:NAD(P) oxidoreductase to form 6-phosphogluconate (De Ley and Verhofstede, 1955). *Aerobacter cloacae* metabolizes 6-phosphogluconate by the oxidative pentose cycle. The same pathway for 2-ketogluconate metabolism to 6-phosphogluconate was established for *P. fluorescens* by Narrod and Wood (1956) and Frampton and Wood (1961 a, b). This organism can metabolize 6-phosphogluconate by either the Entner–Doudoroff pathway or the oxidative pentose cycle (Fig. 7.3).

De Ley and Doudoroff (1957) studied galactose oxidation by *P. saccharophila* (Fig. 7.4). The enzyme responsible for galactose oxidation, D-galactose:NAD oxidoreductase, produced a lactone, presumably

Galactose
↓ +NAD
Galactono-δ-lactone
↓ $+H_2O$
Galactonate
↓ $-H_2O$
2-Keto-3-deoxygalactonate
↓ +ATP
2-Keto-3-deoxy-6-phosphogalactonate
↓
Pyruvate + glyceraldehyde-3-phosphate

FIG. 7.4. Metabolism of galactose by *Pseudomonas saccharophila*.

D-galactono-δ-lactone, from galactose. Chemically synthesized D-galactono-δ-lactone was hydrolyzed by a lactonase from *P. saccharophila* grown on galactose. D-Galactonic acid was dehydrated by D-galactonate hydro-lyase to form 2-keto-3-deoxygalactonate. In the presence of ATP, 2-keto-3-deoxygalactonate was further metabolized to pyruvate and triose phosphate. Doudoroff and Shuster (1962) subsequently separated two aldolases from *P. saccharophila*, one of which was specific for 2-keto-3-deoxy-6-phosphogalactonic acid and the other was the previously described enzyme active on 2-keto-3-deoxy-6-phosphogluconic acid.

2-Ketogluconate is formed by many species of *Acetobacter* during growth on glucose; 5-ketogluconate is formed by *Acetobacter* species of the suboxydans and mesoxydans groups. 2,5-Diketogluconate is formed only by *Acetobacter melanogenum* (Katznelson *et al.*, 1953) and its further metabolism is uncertain. 2-And 5-ketogluconates are on metabolic side paths in the *Acetobacter* since both are metabolized by way of 6-phos-

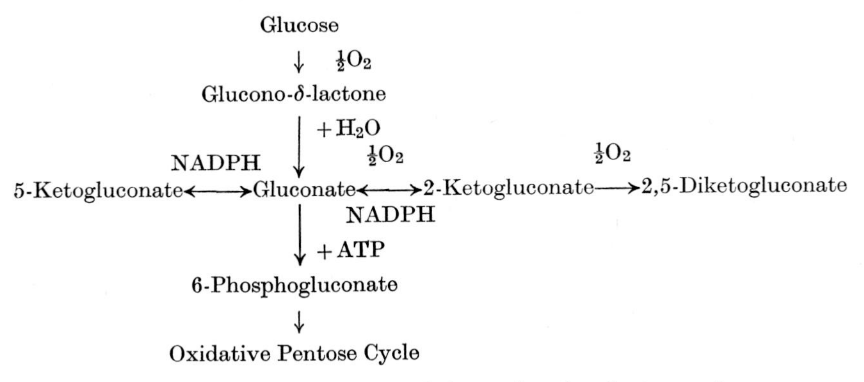

FIG. 7.5. Direct oxidation of glucose by *Acetobacter* species.

phogluconate (Fig. 7.5). *Acetobacter suboxydans* contains particulate enzymes for the oxidation of glucose to 2-ketogluconate (De Ley and Stouthamer, 1959). In addition, there are two soluble dehydrogenases, active with NADP, which reduce 2-ketogluconate and 5-ketogluconate to gluconate. The latter enzyme is apparently also responsible for the formation of 5-ketogluconate in the medium from glucose and gluconate. Gluconate is metabolized via the oxidative pentose cycle after phosphorylation to 6-phosphogluconate.

II. Hexuronic Acid Oxidation

The pathways for hexuronic acid oxidation in bacteria are inducible systems leading to the formation of 2-keto-3-deoxy-6-phosphogluconate. This contrasts with the animal pathways for glucuronate metabolism which lead to the formation of ascorbic acid and, by decarboxylation at C-6, to L-xylulose. The first step in glucuronate and galacturonate metabolism by *E. coli* is isomerization which results in fructuronic acid and tagaturonic acid, respectively (Fig. 7.6; Ashwell *et al.*, 1960). Both reactions are apparently catalyzed by the same enzyme, D-glucuronate ketol-isomerase, which is induced by growth on either glucuronic or galacturonic acid. Fructuronic and tagaturonic acids are reduced at carbon 2 with NADH by separate enzymes, D-mannonate dehydrogenase and D-altronate dehydrogenase (D-mannonate:NAD oxidoreductase, D-altronate:NAD oxidoreductase), respectively (Hickman and Ashwell, 1960). Since the products of these dehydrogenases, mannonic and altronic acids, are known hexonic acids, they are discussed by their usual names rather than as derivatives of fructuronic acid and tagaturonic acid. In this terminology, however, carbon 6 of fructuronic acid

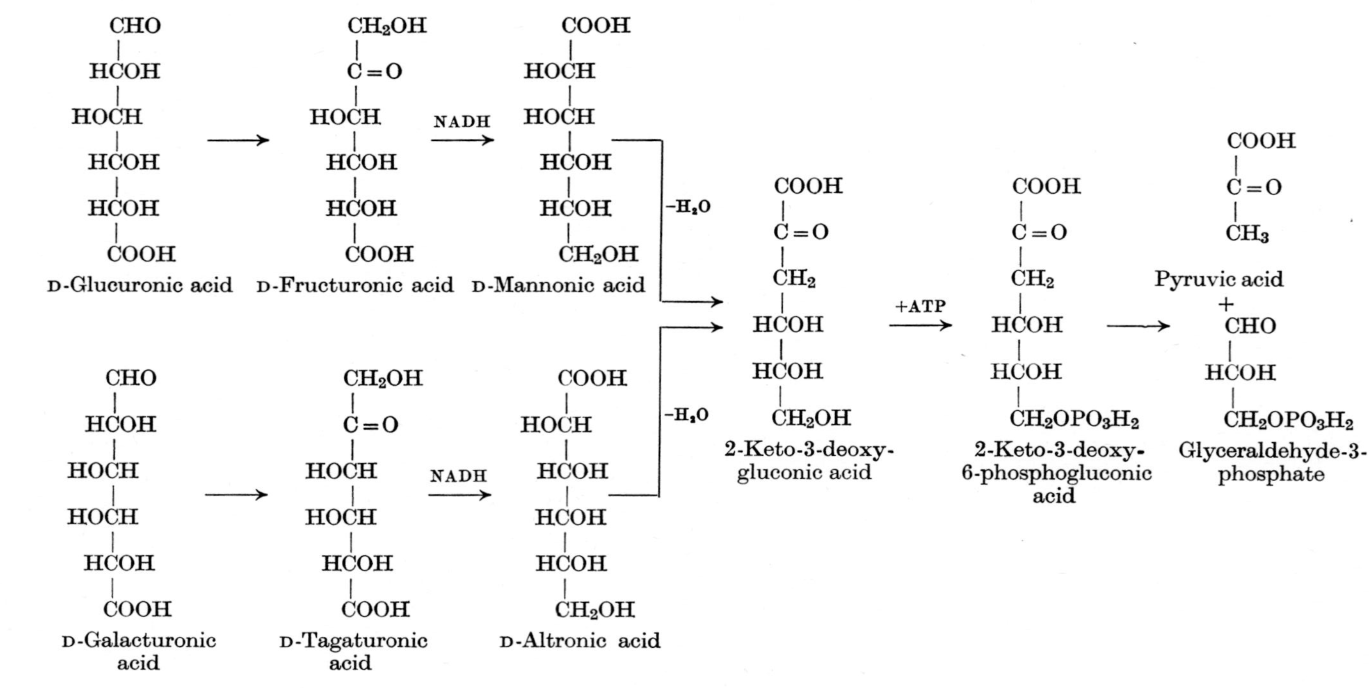

FIG. 7.6. Metabolism of hexuronic acids by *Escherichia coli*.

and tagaturonic acid corresponds to carbon 1 of the hexonic acids. The next step in the metabolism of both hexonic acids is dehydration by specific mannonic and altronic dehydrases (D-mannonate hydro-lyase, D-altronate hydro-lyase), resulting in the formation of 2-keto-3-deoxy-D-gluconate in both cases (Smiley and Ashwell, 1960). *Escherichia coli* forms an inducible kinase for 2-keto-3-deoxy-D-gluconate (ATP: 2-keto-3-deoxy-D-gluconate 6-phosphotransferase) which is cleaved to pyruvate and glyceraldehyde-3-phosphate by an aldolase similar to that of the pseudomonads (Cynkin and Ashwell, 1960).

The metabolism of polygalacturonic acid was studied with a soil pseudomonad originally isolated from an enrichment culture with alginic acid as the source of carbon and energy (Preiss and Ashwell, 1963a). This organism hydrolyzed part of the polygalacturonic acid to galacturonic acid which was then metabolized by the tagaturonic acid pathway. Preiss and Ashwell also isolated a sugar from enzyme digests of polygalacturonate which they characterized as 4-deoxy-L-*threo*-5-hexoseulose uronic acid (Fig. 7.7). It seems most likely that this acid arose as a result of successive enzyme-catalyzed elimination reactions on polygalacturonate resulting in a series of unsaturated oligosaccharides, and eventually yielding the unstable unsaturated monosaccharide designated as compound I, Fig. 7.7. Compound I rearranged non-enzymically to 4-deoxy-L-*threo*-hexoseulose uronic acid, which is the same compound isolated by Linker *et al.* (1960) from the hydrolysis of the unsaturated disaccharide produced by a *Flavobacterium* species from hyaluronic acid (see Fig. 4.8). Further metabolism of 4-deoxy-L-*threo*-5-hexoseulose uronic acid occurred by an isomerase, resulting in 3-deoxy-D-*glycero*-2,5-hexodiulosonic acid, and a dehydrogenase which functions with either NADPH or NADH, resulting in 2-keto-3-deoxy-D-gluconic acid. This latter compound was then metabolized by the action of a kinase and aldolase, leading to the formation of pyruvate and glyceraldehyde-3-phosphate (Preiss and Ashwell, 1963b).

Alginic acid is a polymer of uncertain composition, but which contains D-mannuronic acid and L-guluronic acid. In a study previous to their work on polygalacturonate metabolism, Preiss and Ashwell (1962) found that the same pseudomonad produced 4-deoxy-L-*erythro*-5-hexoseulose uronic acid from alginic acid (Fig. 7.7). The route of formation of this deoxyhexoseulose uronic acid is also presumed to involve a series of unsaturated oligosaccharides with compound II, Fig. 7.7, being the immediate precursor to 4-deoxy-L-*erythro*-5-hexoseulose uronic acid. A dehydrogenase reduces this latter sugar with NADPH to 2-keto-3-deoxygluconic acid which is then metabolized via the Entner–Doudoroff pathway.

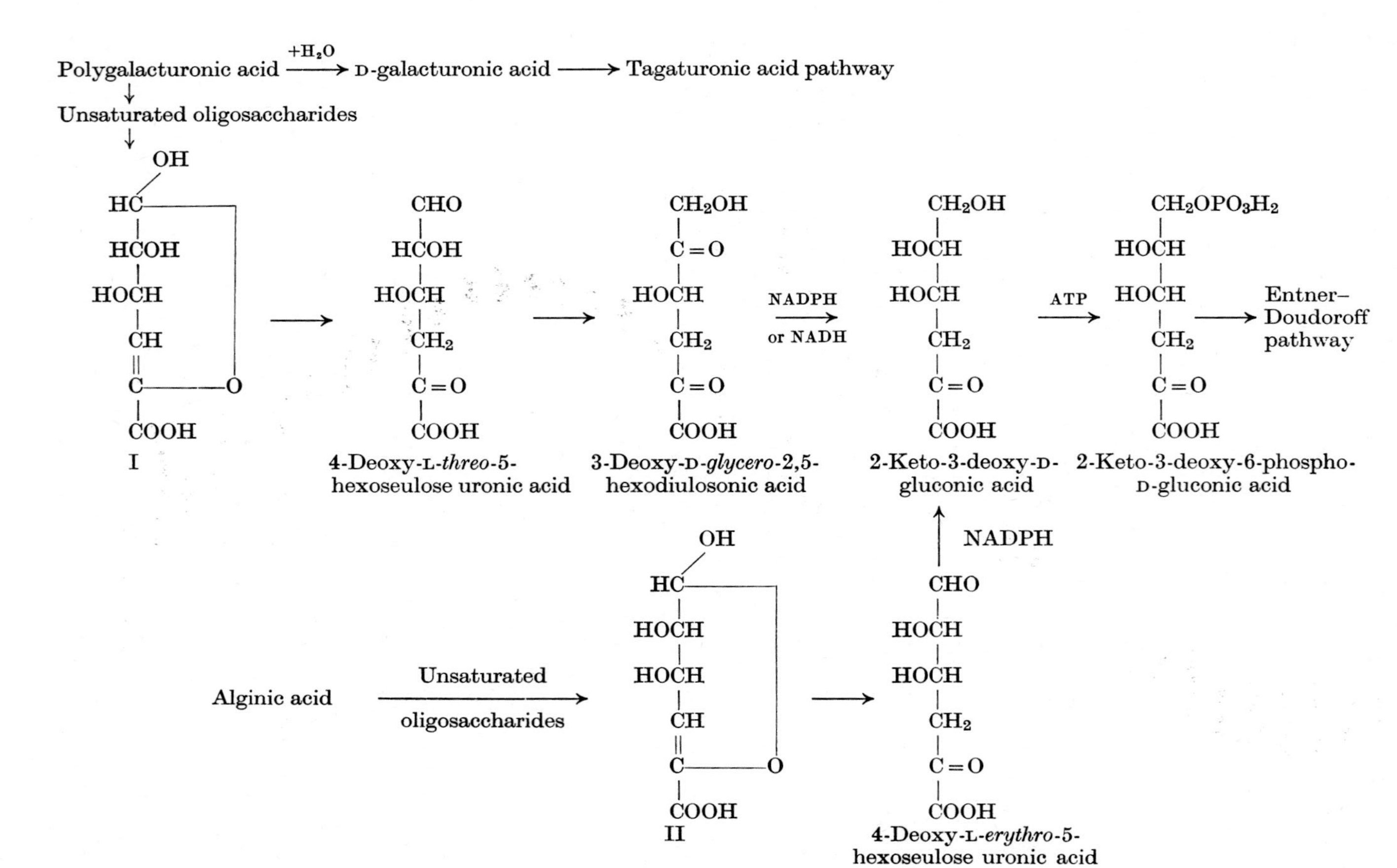

FIG. 7.7. Metabolism of polygalacturonic acid and alginic acid by bacteria.

III. Pentose Oxidation

Pentose oxidation by *A. aerogenes* follows the same scheme outlined in Chapter 6 for conversion of the pentose to D-xylulose-5-phosphate which is then oxidized by the oxidative pentose cycle (Mortlock and Wood, 1964; see also Fig. 7.11).

However, pentose oxidation by *P. saccharophila* follows a unique pathway. The first step in the oxidation of L-arabinose is oxidation to L-arabino-γ-lactone by L-arabinose:NAD oxidoreductase (Fig. 7.8; Weimberg and Doudoroff, 1955). L-Arabino-γ-lactone hydrolase is responsible for the hydrolysis of L-arabino-γ-lactone to L-arabonic acid. D-Arabinonate hydro-lyase is the enzyme which converts L-arabonic acid to L-2-keto-3-deoxy-4,5-dihydroxyvaleric acid (Weimberg, 1959). The ketodeoxy acid is transformed to 2-oxoglutarate by an enzyme system from *P. saccharophila* which requires NAD. The conversion also represents the withdrawal of a mole of water but the number of enzymes involved in this transformation is unknown.

The metabolism of D-arabinose by *P. saccharophila* is similar in its initial stages to that of L-arabinose (Fig. 7.8). D-Arabinose is oxidized by an NAD-linked enzyme to D-arabonic acid which is dehydrated to D-2-keto-3-deoxy-4,5-dihydroxyvaleric acid by D-arabinonate hydro-lyase. This latter compound has a different metabolic fate from the corresponding acid in L-arabinose metabolism. D-2-Keto-3-deoxy-4,5-dihydroxyvaleric acid is oxidized by another NAD-linked enzyme to one mole each of pyruvate and glycolate (Palleroni and Doudoroff, 1956, 1957). The identity of the intermediates, if any, between the ketodeoxy acid and pyruvate and glycolate is unknown.

IV. Polyol Oxidation

Acetobacter species accumulate keto sugars from many polyols under conditions which permit limited growth of the organism, that is, conditions similar to those which result in the accumulation of 2-ketogluconate by *Pseudomonas* species. Bertrand, working with *A. xylinum*, and later Hudson and his associates, working with *A. suboxydans*, were the first to stipulate the structural requirements of the polyol which would allow the organism to oxidize the substrate. The Bertrand–Hudson rule states that *Acetobacter* species oxidize polyols which have *cis* hydroxyl groups adjacent to a primary (or secondary) alcohol, and that the carbon immediately adjacent to the alcohol is the position which is oxidized (see Fig. 7.9; Hann *et al.*, 1938). They further pointed out that *A. suboxydans* is somewhat more specific in that the *cis*

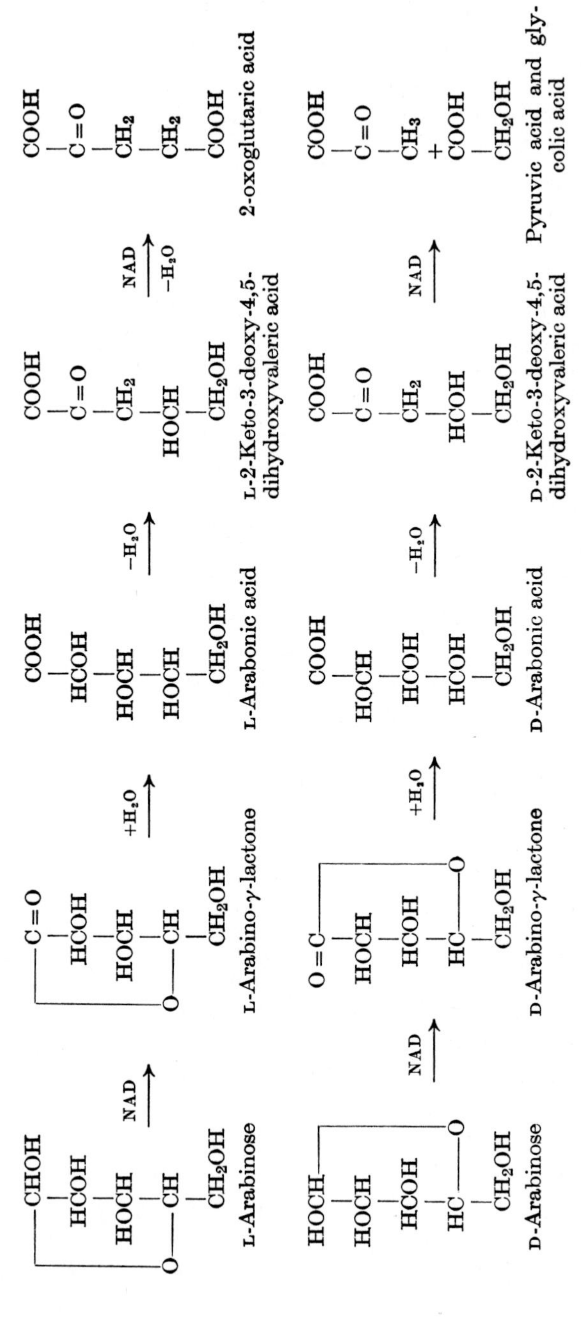

FIG. 7.8. Metabolism of D- and L-arabinose by *Pseudomonas saccharophila.*

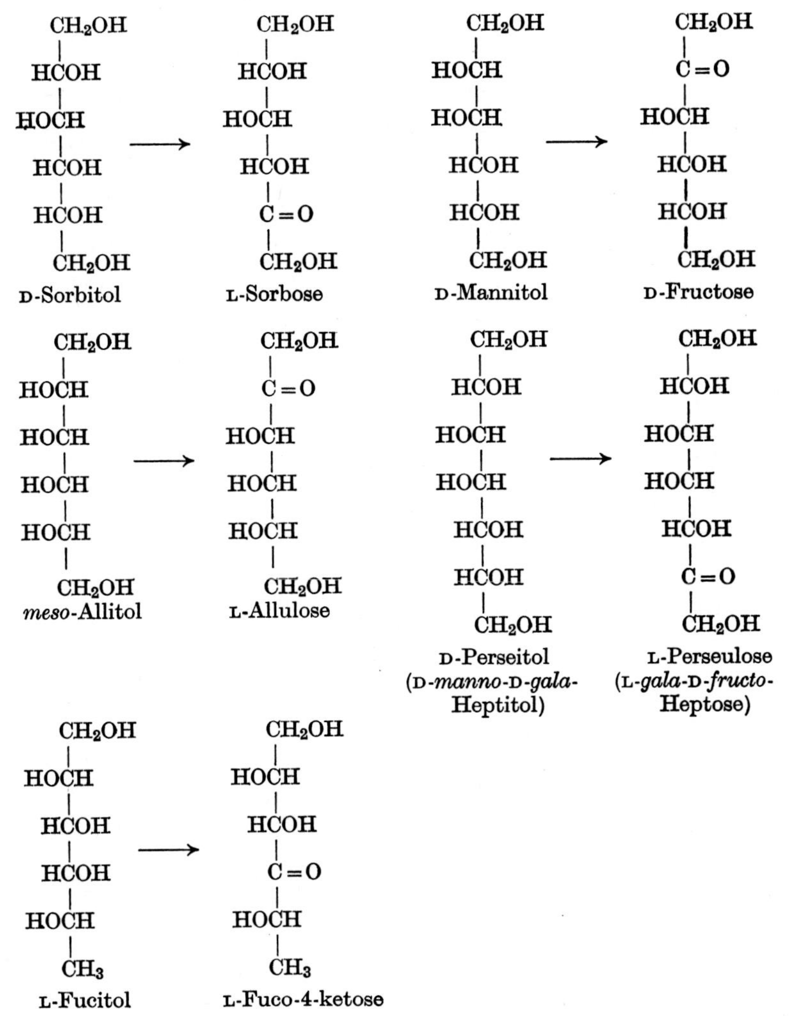

FIG. 7.9. Representative polyol oxidations by *Acetobacter* species (Cheldelin *et al.*, 1962).

hydroxyls must be of the D configuration. Apparently these interesting oxidations do not result in an energy benefit for the organism, but they are a useful means of providing many unique keto sugars for biochemical studies.

Arcus and Edson (1956) discovered the enzymic basis for these oxidations when they found that *A. suboxydans* contained a particulate enzyme which did not require a soluble coenzyme and which was responsible for the oxidation of several polyols to keto sugars as predicted by the Bertrand–Hudson rule. Arcus and Edson also studied an NAD-linked polyol dehydrogenase active on xylitol, sorbitol, mannitol, sedoheptitol and other polyols. The product of both mannitol and sorbitol oxidation was fructose, and therefore, in the case of sorbitol, the oxidation did not correspond to the Bertrand–Hudson rule.

Cummins *et al.* (1957) found that their strain of *Acetobacter suboxydans* formed both NAD- and NADP-linked sorbitol dehydrogenases, the former leading to fructose, as observed by Arcus and Edson, and the latter leading to sorbose. The NADP and particulate enzymes provide this strain of *A. suboxydans* with two means for oxidation of sorbitol to sorbose.

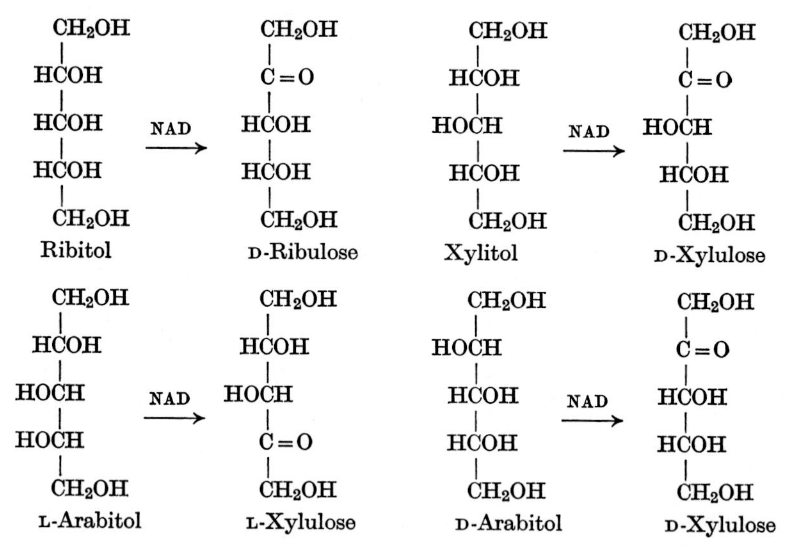

FIG. 7.10. Oxidation of pentitols to ketopentoses by NAD-linked enzymes from *Aerobacter aerogenes*. References: ribitol dehydrogenase (ribitol: NAD oxidoreductase), Fromm (1958), Wood *et al.* (1961); xylitol dehydrogenase (xylitol: NAD oxidoreductase, D-xylulose forming), Fossitt *et al.* (1964); L-arabitol dehydrogenase (L-arabinitol:NAD oxidoreductase, L-xylulose forming), Fossit *et al.* (1964); D-arabitol dehydrogenase (D-arabinitol:NAD oxidoreductase), Lin (1961), Wood *et al.* (1961).

Aerobacter aerogenes is able to use all four pentitols, ribitol, xylitol, L-arabitol and D-arabitol, as an energy source and metabolizes these compounds by means of NAD-linked dehydrogenases to ketopentoses (Fig. 7.10). The configurations of the D and L forms of xylitol and ribitol are identical and therefore there is only one isomer of each of these two pentitols. Pentitol oxidation by *Aerobacter aerogenes* leads to D-xylulose, L-xylulose and D-ribulose which are then metabolized by way of D-

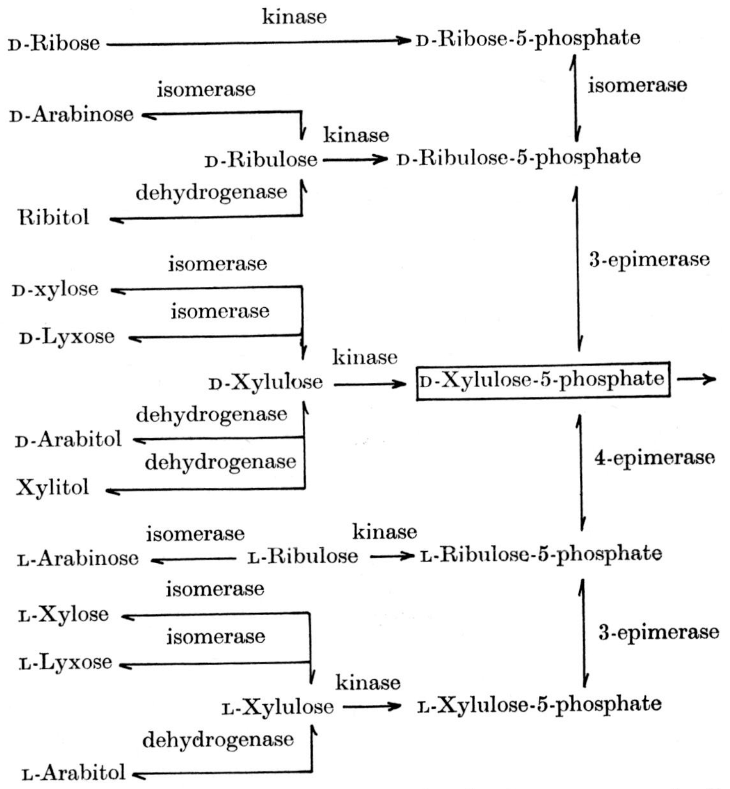

FIG. 7.11. Pentose and pentitol metabolism by *Aerobacter aerogenes* leading to D-xylulose-5-phosphate. (From Mortlock and Wood, 1964).

xylulose-5-phosphate and the pathways of pentose metabolism outlined in Chapter 6 (see Fig. 7.11).

V. Miscellaneous Oxidations

The oxidation of maltose and lactose to the corresponding bionic acids by growing cultures of several species of *Pseudomonas* was discovered by Stodola and Lockwood (1947) (Fig. 7.12). Bentley and Slechta (1960) isolated a particulate enzyme from *Pseudomonas quercitopyrogallica* which catalyzed the oxidation of melibiose, lactose, maltose and cellobiose to their respective bionic acids. None of the disaccharides tested stimulated the yield of organism above that obtained with the basal peptone medium; hence, it seems likely that the routes

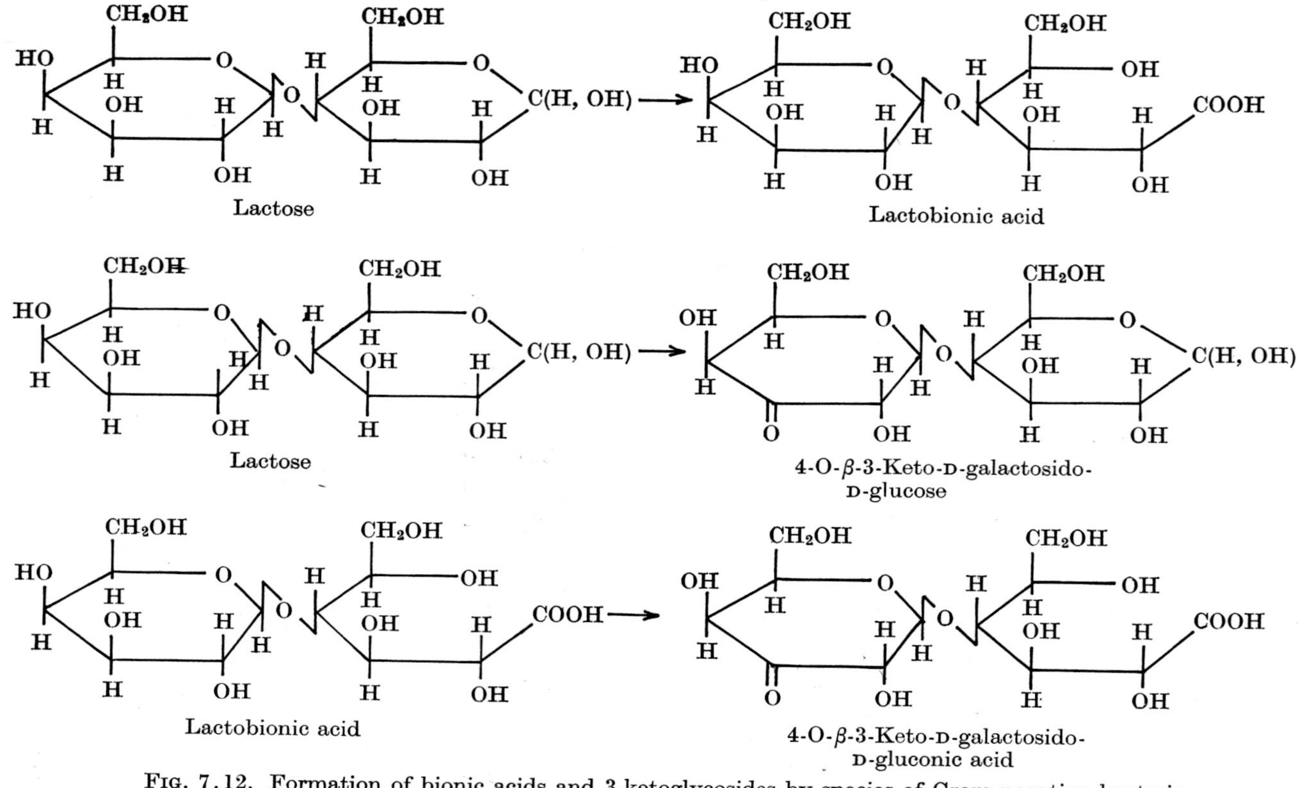

FIG. 7.12. Formation of bionic acids and 3-ketoglycosides by species of Gram negative bacteria (reproduced with permission from De Ley, 1960).

to the formation of the bionic acids are metabolic dead-ends without energy production.

Bernaerts and De Ley (1960a,b) isolated a strain of *Alcaligenes faecalis* from river water which oxidized lactose, maltose, lactobionic acid and maltobionic acid to the corresponding 3-ketoglycosides (Fig. 7.12).

References

Arcus, A. C. and Edson, N. L. (1956). *Biochem. J.* **64**, 385.

Ashwell, G., Wahba, A. and Hickman, J. (1960). *J. Biol. Chem.* **235**, 1559.

Bentley, R. and Slechta, L. (1960). *J. Bacteriol.* **79**, 346.

Bernaerts, M. J. and De Ley, J. (1960a). *J. Gen. Microbiol.* **22**, 129.

Bernaerts, M. J. and De Ley, J. (1960b). *J. Gen. Microbiol.* **22**, 137.

Cheldelin, V. H., Wang, C. H. and King, T. E. (1962). *In* "Comparative Biochemistry" (M. Florkin and H. S. Mason, eds.), Academic Press, New York, Vol. III, p. 427.

Cori, O. and Lipmann, F. (1952). *J. Biol. Chem.* **194**, 417.

Couri, D. and Racker, E. (1959). *Arch. Biochem. Biophys.* **83**, 195.

Cummins, J. T., Cheldelin, V. H. and King, T. E. (1957). *J. Biol. Chem.* **226**, 301.

Cynkin, M. A. and Ashwell, G. (1960). *J. Biol. Chem.* **235**, 1576.

De Ley, J. (1953). *Enzymologia* **16**, 99.

De Ley, J. (1960). *J. Appl. Bacteriol.* **23**, 400.

De Ley, J. and Doudoroff, M. (1957). *J. Biol. Chem.* **227**, 745.

De Ley, J. and Stouthamer, A. J. (1959). *Biochim. Biophys. Acta* **34**, 171.

De Ley, J. and Verhofstede, S. (1955). *Naturwissenschaften* **42**, 584.

Doudoroff, M. and Shuster, C. W. (1962). *Bacteriol. Proc.* p. 112.

Entner, N. and Doudoroff, M. (1952). *J. Biol. Chem.* **196**, 853.

Fossitt, D., Mortlock, R. P., Anderson, R. L. and Wood, W. A. (1964). *J. Biol. Chem.* **239**, 2110.

Frampton, E. W. and Wood, W. A. (1961a). *J. Biol. Chem.* **236**, 2571.

Frampton, E. W. and Wood, W. A. (1961b). *J. Biol. Chem.* **236**, 2578.

Fromm, H. J. (1958). *J. Biol. Chem.* **238**, 1049.

Hann, R. M., Tilden, E. B. and Hudson, C. S. (1938). *J. Am. Chem. Soc.* **60**, 1201.

Hickman, J. and Ashwell, G. (1960). *J. Biol. Chem.* **235**, 1566.

Katz, J. and Wood, H. G. (1960). *J. Biol. Chem.* **235**, 2165.

Katznelson, H., Tanenbaum, S. W. and Tatum, E. L. (1953). *J. Biol. Chem.* **204**, 43.

Kovachevich, R. and Wood, W. A. (1955a). *J. Biol. Chem.* **213**, 745.

Kovachevich, R. and Wood, W. A. (1955b). *J. Biol. Chem.* **213**, 757.

Lin, E. C. C. (1961). *J. Biol. Chem.* **236**, 31.

Linker, A., Hoffman, P., Meyer, K., Sampson, P. and Korn, E. D. (1960). *J. Biol. Chem.* **235**, 3061.

MacGee, J. and Doudoroff, M. (1954). *J. Biol. Chem.* **210**, 617.

Mortlock, R. P. and Wood, W. A. (1964). *J. Bacteriol.* **88**, 838.

Narrod, S. A. and Wood, W. A. (1956). *J. Biol. Chem.* **220**, 45.

Palleroni, N. J. and Doudoroff, M. (1956). *J. Biol. Chem.* **223**, 499.

Palleroni, N. J. and Doudoroff, M. (1957). *J. Bacteriol.* **74**, 180.

Preiss, J. and Ashwell, G. (1962). *J. Biol. Chem.* **237**, 317.
Preiss, J. and Ashwell, G. (1963a). *J. Biol. Chem.* **238**, 1571.
Preiss, J. and Ashwell, G. (1963b). *J. Biol. Chem.* **238**, 1577.
Rao, M. R. R. (1957). *Ann. Rev. Microbiol.* **11**, 317.
Smiley, J. D. and Ashwell, G. (1960). *J. Biol. Chem.* **235**, 1571.
Stodola, F. H. and Lockwood, L. B. (1947). *J. Biol. Chem.* **171**, 213.
Stokes, F. N. and Campbell, J. J. R. (1951). *Arch. Biochem. Biophys.* **30**, 121.
Stubbs, J. J., Lockwood, L. B., Roe, E. T., Tabenkin, B. and Ward, G. E. (1940). *Industr. Eng. Chem.* (*Industr.*) **32**, 1626.
Weimberg, R. (1959). *J. Biol. Chem.* **235**, 727.
Weimberg, R. and Doudoroff, M. (1955). *J. Biol. Chem.* **217**, 607.
Wood, W. A. and Schwerdt, R. F. (1953). *J. Biol. Chem.* **201**, 501.
Wood, W. A., McDonough, M. J. and Jacobs, L. B. (1961). *J. Biol. Chem.* **236**, 2190.

8. OXIDATION OF ORGANIC ACIDS

I. The Tricarboxylic Acid Cycle

A. Tricarboxylic Acid Cycle Enzymes in Bacteria

Acetyl-CoA is a product of carbohydrate metabolism by the oxidation of pyruvic acid, a product of fatty acid oxidation and amino acid oxidation, and is formed from acetate of the medium by the action of acetokinase and phosphotransacetylase. The most important mechanism for the oxidation of acetyl-CoA is the tricarboxylic acid cycle formulated by Krebs and his colleagues (Fig. 8.1).

Citrate synthetase (citrate oxaloacetate-lyase [CoA-acetylating]) is the first enzyme of the tricarboxylic acid cycle and is a key enzyme in the complete oxidation of acetate. This enzyme was first crystallized from pigeon liver (Stern and Ochoa, 1951) and catalyzes the following reaction (Stern, 1961):

$$\text{Acetyl-CoA} + \text{oxaloacetate} + H_2O \rightarrow \text{citrate} + \text{CoA} + H^+$$

Citrate synthetase has been measured in unfractionated extracts of several species of bacteria (Table 8.1) and purified from *Mycobacterium tuberculosis* (Goldman, 1957).

Aconitase (citrate [isocitrate] hydro-lyase) is the second enzyme of the tricarboxylic acid cycle and catalyzes the reversible hydration of *cis*-aconitate (Dickman, 1961). The levels of aconitase in extracts of several bacterial species are shown in Table 8.1.

$$\begin{array}{c} \text{Citrate} \\ \rightleftharpoons \\ \textit{cis}\text{-Aconitate} + H_2O \\ \rightleftharpoons \\ \text{Isocitrate} \end{array}$$

In the early studies of the tricarboxylic acid cycle, the belief was generally held that citric acid formed from labeled precursors should have identical isotope content in the two -CH_2COOH groups because citrate was a symmetrical molecule. However, Ogston (1948) pointed

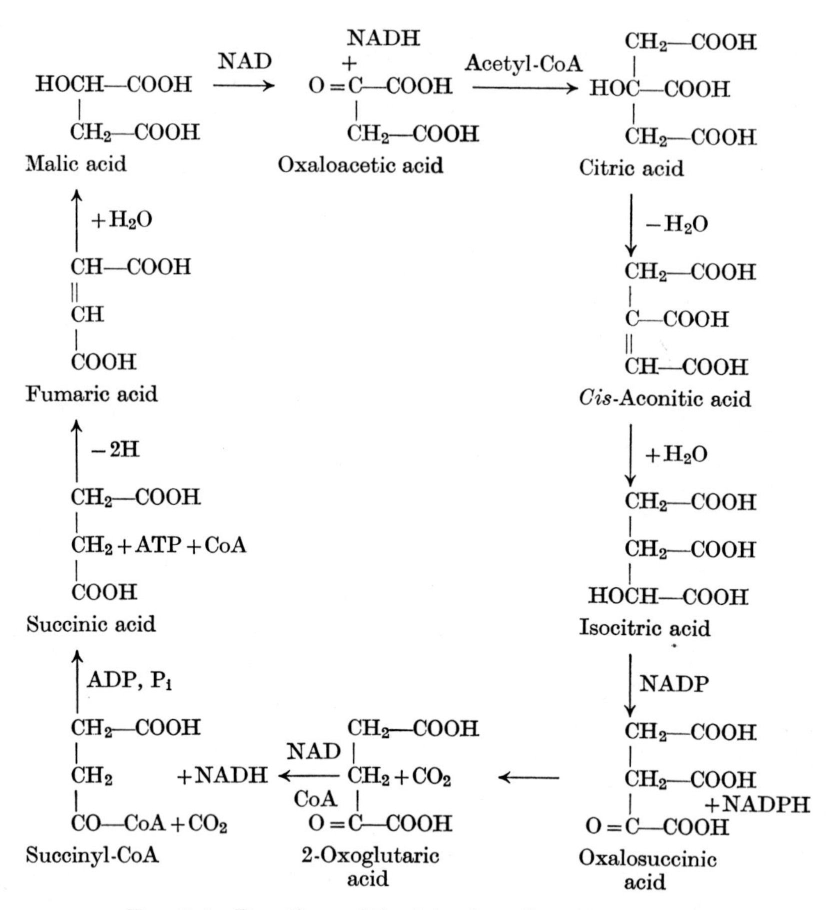

FIG. 8.1. Reactions of the tricarboxylic acid cycle.

out that this was not necessarily so; if citrate were attached to the enzyme at three sites, then aconitase should be able to distinguish between the two -CH_2COOH groups of citrate. This concept is illustrated in Fig. 8.2, where it will be noted that citrate is attached to aconitase by 3 of the 4 substituents of the central carbon atom at points a, b and c on the enzyme surface. In Fig. 8.2(a) the order of attachment of citrate to the enzyme from a to c is -OH, -CH_2COOH and -COOH. In Fig. 8.2(b), citrate is attached to the enzyme by the opposite -CH_2COOH, in which case the order of attachment is -COOH, -CH_2COOH and -OH. Citrate attached at one -CH_2COOH (Fig. 8.2(a)) is the mirror image of itself attached to the enzyme at the opposite -CH_2COOH, even though the molecule does not possess an asymmetric

TABLE 8.1. Activity of tricarboxylic acid enzymes in extracts prepared from several species of bacteria.

Organism	Citrate synthetase	Aconitase	Isocitrate dehydrogenase	2-Oxoglutarate-dehydrogenase	Succinate dehydrogenase	Fumarase	Malate dehydrogenase
	μMoles product formed/hr/mg protein						
Acetobacter acidum-mucosum (a)	29·0	16·4	16·7	0·32*	1·58	8·7	1·48*
Acetobacter rancens (a)	30·7	11·8	7·9	0·38*	2·61	10·7	0·51*
Acetobacter gluconicum (a)	1·1	0·4	0·0	0·10*	1·32	0·5	0·01*
Mycoplasma hominis (b)	0·0	—	0·42	1·2	3·8	2·0	1·42
Mycobacterium tuberculosis (c)	1·02	9·22	5·22	0·21	0·42	38·0	2·58
Hydrogenomonas H16G+ (d)	2·16	0·77	9·0	0·86	2·71	3·52	288

The data from the original literature have all been converted to the same units for comparison. (a) Williams and Rainbow (1964), (b) VanDemark and Smith (1964), (c) Murthy *et al.* (1962), (d) Truper (1965). * These preparations were assayed without a pyridine nucleotide cofactor.

carbon atom in the conventional sense of the term. Carbon atoms which possess the type of asymmetry exhibited by the central carbon atom of citric acid have been termed *meso* carbon atoms by Schwartz and Carter (1954), and their article should be consulted for a further discussion of the subject.

Ogston's theory was first substantiated by Potter and Heidelberger (1949), who synthesized citrate with a rat liver homogenate using $^{14}CO_2$ and then oxidized the citrate to 2-oxoglutarate with another rat liver homogenate. 2-Oxoglutarate was labeled solely in carbon 1, the

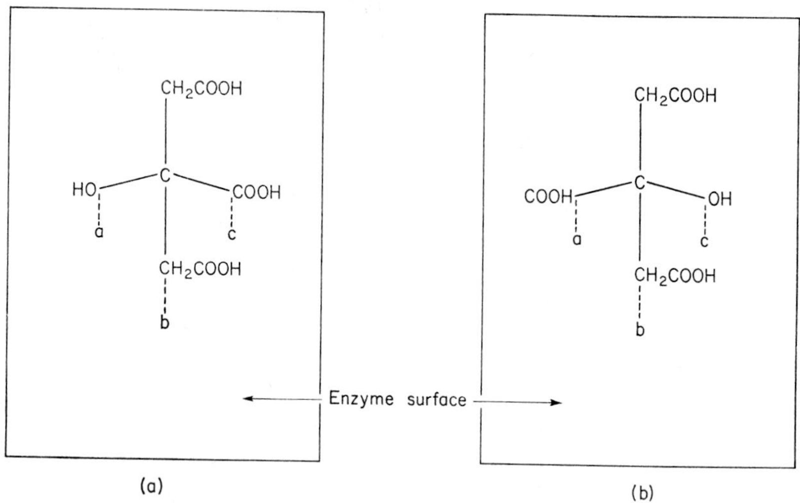

FIG. 8.2. Schematic representation of Ogston's theory for the attachment of citrate to aconitase.

carboxyl group adjacent to the carbonyl group (Fig. 8.3). Under the conditions used by Potter and Heidelberger, $^{14}CO_2$ was fixed into the terminal carboxyl of oxaloacetate, and hence it follows that the portion of citrate dehydrated and rehydrated by aconitase is the portion derived from oxaloacetate. This end of the citrate molecule is also referred to as the aconitase-active portion, as opposed to the aconitase-inactive group derived from acetyl-CoA.

Two kinds of isocitrate dehydrogenases are known; one NADP-linked and the other NAD-linked (Plaut, 1963). In both cases, the reaction catalyzed is:

$$\text{Isocitrate} + \text{NAD(P)} \rightarrow \text{2-oxoglutarate} + CO_2 + \text{NAD(P)H}$$

The natural substrate for both types of dehydrogenase is *threo*-D-isocitrate. The enzyme which requires NADP decarboxylates oxalo-

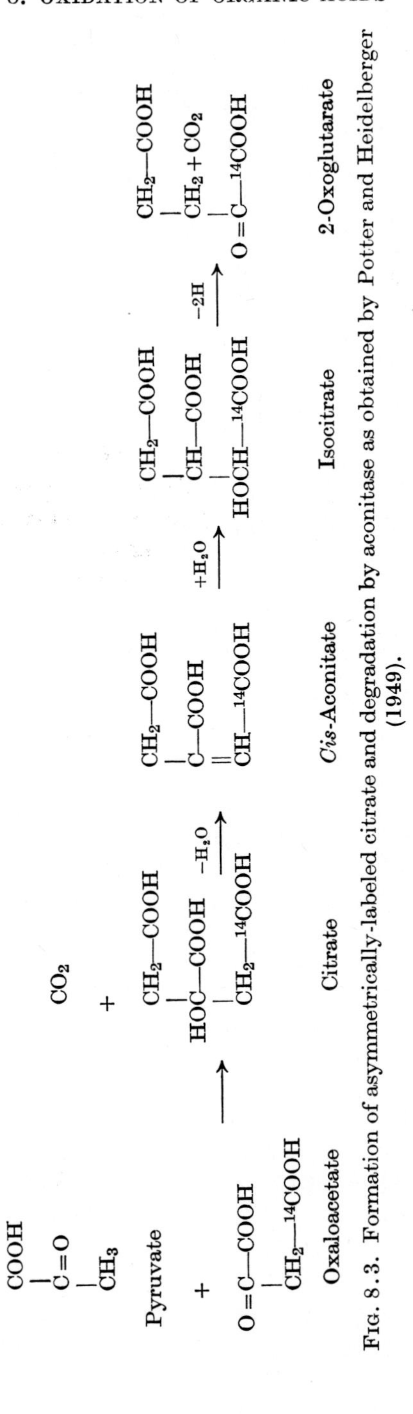

FIG. 8.3. Formation of asymmetrically-labeled citrate and degradation by aconitase as obtained by Potter and Heidelberger (1949).

succinate and reduces 2-oxoglutarate and carbon dioxide with NADPH, resulting in isocitrate. The NAD enzyme does not catalyze the decarboxylation of oxalosuccinate nor the reduction of 2-oxoglutarate with reduced pyridine nucleotide. Little work has been done with bacterial enzymes, but those which have been studied are NADP-linked (*threo*-D_s-isocitrate: NADP oxidoreductase [decarboxylating]; Barban and Ajl, 1952; Goldman, 1956a; see also Table 8.1).

Oxidation of 2-oxoglutarate to succinyl-CoA occurs by reactions directly analogous to those described for pyruvate oxidation in Chapter 6. The enzymes which catalyze the first three reactions of 2-oxoglutarate oxidation are located together on a respiratory particle of *E. coli* (Gunsalus, 1954; Koike *et al.*, 1960):

$$\text{2-Oxoglutarate} + \text{TPP} \rightarrow \text{succinate semialdehyde-TPP} + CO_2$$
$$\text{Succinate semialdehyde-TPP} + \text{lipoate}_{(ox)} \rightarrow \text{succinyl-lipoate} + \text{TPP}$$
$$\text{Succinyl-lipoate} + \text{CoA} \rightarrow \text{succinyl-CoA} + \text{lipoate}_{(red)}$$
$$\text{Lipoate}_{(red)} + \text{NAD} \rightarrow \text{lipoate}_{(ox)} + \text{NADH}$$

$$\text{Sum: 2-Oxoglutarate} + \text{NAD} + \text{CoA} \rightarrow \text{succinyl-CoA} + \text{NADH} + CO_2$$

The last reaction, catalyzed by lipoate dehydrogenase, is common to both pyruvate and 2-oxoglutarate pathways.

Succinyl-CoA is probably deacylated under physiological conditions by the action of succinate thiokinase (succinate:CoA ligase [ADP]):

$$\text{Succinyl-CoA} + P_i + \text{ADP} \rightarrow \text{succinate} + \text{CoA} + \text{ATP}$$

The enzyme from *E. coli* functions with ADP (Hager, 1962), but the mammalian enzyme requires GDP or IDP. Mammalian tissues contain succinyl-CoA deacylase (Gergely *et al.*, 1952) which catalyzes the hydrolytic removal of CoA and thus provides an alternate pathway for succinyl-CoA deacylation, but in this case the energy of the thioester linkage is lost to the cell.

The next reaction in the tricarboxylic acid cycle is catalyzed by succinate dehydrogenase (succinate: [acceptor] oxidoreductase):

$$\text{Succinate} + \text{FAD} \rightarrow \text{fumarate} + \text{FADH}$$

Succinate dehydrogenase from various sources usually reduces artificial electron acceptors such as phenazine methosulfate, 2,6-dichlorophenolindophenol and ferricyanide, but the natural electron acceptor is FAD (Singer and Kearney, 1963). The bacterial succinate dehydrogenase which has been most thoroughly characterized is from an anaerobe, *Micrococcus lactilyticus*. Like the animal and yeast enzymes, succinate dehydrogenase of *M. lactilyticus* contains both FAD and iron (Warringa and Giuditta, 1958), but differs in that fumarate is reduced more rapidly than succinate is oxidized (Warringa *et al.*, 1958) and probably functions under physiological conditions as a fumarate reductase. *Escherichia coli*, a facultative anaerobe, contains two enzymes which catalyze the oxidation of succinate (Hirsch *et al.*, 1963); with one enzyme, succinate oxidation is the most rapid reaction, and with the other enzyme, fumarate reduction is the most rapid reaction. Both enzymes of *E. coli* are associated with oxidative particles, presumably pieces of the cell membrane, and this seems to be true of most preparations of succinate dehydrogenase from bacteria with the exception of the enzyme from *M. lactilyticus*.

Fumarase or fumarate hydratase (L-malate hydro-lyase) catalyzes the hydration of fumaric acid to L-malic acid:

$$\text{Fumarate} + H_2O \rightarrow \text{L-malate}$$

Englard (1960) discovered that the addition of water to fumarate was *trans* and further established that the proton added by fumarase was the same one removed by aconitase (Fig. 8.4). Fumarase has been detected in extracts prepared from several species of bacteria (Table 8.1) and has been purified from *Propionibacterium pentosaceum* (Ayres and Lara, 1962), a fermentative organism which produces propionate by a pathway which involves the reduction of oxaloacetate to succinate (Fig. 6.13).

Malate dehydrogenase (L-malate:NAD oxidoreductase) catalyzes the following reaction:

$$\text{L-malate} + \text{NAD} \rightarrow \text{oxaloacetate} + \text{NADH}$$

Malate dehydrogenase has been partially purified from *Mycobacterium tuberculosis* (Goldman, 1956b) and crystallized from *Bacillus subtilis* (Yoshida, 1965). Enzyme from both sources required NAD as a cofactor; however, several malate dehydrogenases of bacteria which did not require a soluble cofactor have been reported (Table 8.1).

B. Evidence for the Occurrence of the Tricarboxylic Acid Cycle in Bacteria

Evidence for the pathway of pyruvate oxidation via acetate and the tricarboxylic acid cycle was first obtained by Krebs and his associates working with mammalian enzyme preparations, and Krebs' review (Krebs, 1943) should be consulted for details of the early work on this problem. Experiments with bacterial systems came some years later and the initial experiments dealt with the oxidation of acetate and inter-

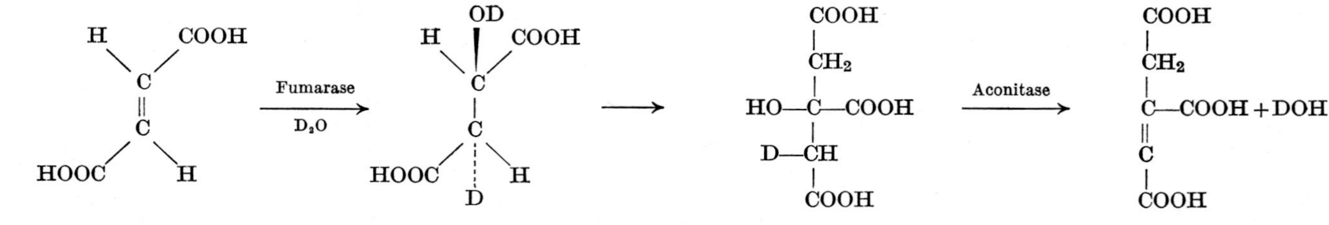

FIG. 8.4. Trans addition of water to fumarate catalyzed by fumarase and detected with the use of D_2O (Englard, 1960). The bold line represents orientation upward from the plane of the paper, the dashed line represents orientation downward from the plane of the paper. Deuterium attached to the —OD substituent of citrate is lost by exchange with water.

mediates of the tricarboxylic acid cycle by whole cells. In many cases it was found that whole cells were unable to oxidize key intermediates of the cycle, and this was regarded as evidence against the occurrence of the cycle in those species of bacteria. Eventually, Barrett and Kallio (1953) found that enzyme preparations of *Pseudomonas fluorescens* were able to oxidize TCA cycle intermediates, while the cells from which such extracts were made were unable to do so, and these workers thereby showed that the difficulty lay in the penetration of the highly polar tricarboxylic acid cycle intermediates into the cell. This observation was made subsequently with other species of bacteria. Krampitz's review (1961) contains an excellent discussion of the early problems involved in the study of the tricarboxylic acid cycle in bacteria.

Deciding if the tricarboxylic acid cycle is involved in the oxidation of acetate by bacteria is similar to deciding the major pathway of carbohydrate oxidation as discussed in Chapter 7. Criteria for making such a decision are also similar, namely, detection of the enzymes of the tricarboxylic acid cycle at a specific activity high enough to account for the growth of the organism, and oxidation of radioactive acetate yielding tricarboxylic acid cycle intermediates labeled as predicted on the basis of the known reactions. Detection of all the enzymes of the tricarboxylic acid cycle is tedious but straightforward, and such data are illustrated in Table 8.1. It can be reasonably concluded that the cycle exists in those organisms for which such data are available.

Results from isotope experiments were ambiguous, however. Early isotope experiments were done by incubating whole bacterial cells or enzyme preparations with acetate-1 or -2-^{14}C. The latter compound was the most useful substrate, since isotope from acetate-1-^{14}C is rapidly lost to carbon dioxide. After a suitable period of time, carrier compounds were added to the incubation mixture and then reisolated in order to facilitate detection of the labeled compounds. Data obtained in many of these experiments appeared to argue against the occurrence of the tricarboxylic acid cycle, since potential intermediates were labeled at too low a specific activity to be derived from acetate-^{14}C. These results were actually due to lack of equilibration of substrate bound to the enzyme with the carrier substrate in solution (Krampitz, 1961). Saz and Krampitz (1954) overcame this problem when they incubated large amounts of cells with acetate-2-^{14}C and isolated the cellular acids without the addition of carrier. Tricarboxylic acid cycle acids isolated from *Micrococcus lysodeikticus* by this technique were labeled as predicted from the enzymes of the tricarboxylic acid cycle (Table 8.2). Most important was the finding that the specific activity of the individual carbons of the acids was always no more, and usually a bit less, than

TABLE 8.2. Isotope Distribution in Intermediates of Tricarboxylic Acid Cycle Following Acetate-2-^{14}C Oxidation by *Micrococcus lysodeikticus* (Saz and Krampitz, 1954).

	Specific activity Counts/min/mMole Carbon × 10^6
Acetate	
CH_3	3·62
COOH	0
Citrate	
COOH	0·44
CH_2	1·28
HOC	1·12
COOH	0·52
CH_2	1·28
COOH	0·44
2-Oxoglutarate	
COOH	0·45
CH_2	0·97
CH_2	0·97
C=O	0·45
COOH	0·33
Succinate	
COOH	0·45
CH_2	0·92
CH_2	0·92
COOH	0·45

The method of degradation released the following pairs of carbons simultaneously and the values reported are averages for the two positions: primary carboxyl carbons of citrate, methylene carbons of citrate, methylene carbons of 2-oxoglutarate, carbonyl carbon and terminal carboxyl carbon of 2-oxoglutarate, methylene carbons of succinate, and carboxyl carbons of succinate. (The values for the methylene carbons of 2-oxoglutarate and succinate given in the original publication were sums of these pairs of carbons and have been divided by 2 in order to simplify presentation.)

that of the compound preceding it in the cycle. This result would be expected because some dilution of isotope would result from the entry of acids from unlabeled endogenous sources.

The method of degradation used in Saz's and Krampitz's experiments was chemical and both primary carboxyl carbons of citrate were obtained simultaneously, as was true of the adjacent methylene carbons. The value for these positions are averages for the two carbons obtained and it is not possible to tell if the two ends of citrate had different isotope contents. However, the carbonyl carbon of 2-oxoglutarate had a lower isotope concentration than did the methylene carbon adjacent to the terminal carboxyl group. This labeling pattern is similar to results of studies with mammalian systems, and agrees with the concept that the carbonyl group is derived from the actonitase active portion of citrate, while the methylene carbon adjacent to the terminal carboxyl group is derived from acetate (compare with Fig. 8.3). The greater the extent of recycling, however, the more equal becomes isotope concentration of the aconitase-active and aconitase-inactive portions of citrate. Swim and Krampitz (1954) reported similar data for studies with *E. coli*.

II. Metabolism of Other Organic Acids by Bacteria

Many organisms, under fermentative conditions with citrate added to the medium, form an inducible enzyme, which also cleaves citrate to acetate and oxaloacetate without the intervention of coenzyme A.

$$\text{Citrate} \rightarrow \text{acetate} + \text{oxaloacetate}$$

Citritase (citrate oxaloacetate-lyase) is the enzyme which catalyzes this reaction, and it has been studied in *S. faecalis* (Gillespie and Gunsalus, 1953) and purified from *A. aerogenes* (Raman, 1961; Dagley and Dawes, 1955), *E. coli* (Wheat and Ajl, 1955) and *S. diacetilactis* (Harvey and Collins, 1963). Citritase from all sources required a divalent cation, usually supplied as magnesium or manganese.

Kornberg and Gotto (1961) studied the metabolism of glycollate by a species of *Pseudomonas* and obtained evidence for the following pathway of glycollate oxidation:

$$2\ \text{Glycollate} + \tfrac{1}{2}O_2 \rightarrow 2\ \text{glyoxylate} + H_2O$$

$$2\ \text{Glyoxylate} \rightarrow C_3\ \text{acid} + \text{carbon dioxide}$$

$$\text{NADH} + C_3\ \text{acid} \rightarrow \text{glycerate} + \text{NAD}$$

Glycollate oxidase (Glycollate: oxygen oxidoreductase), the enzyme which catalyzed the first reaction, was present in extracts of cells grown on glycollate but not in extracts of those grown on succinate. The condensation of 2 moles of glyoxylate to form a C_3 acid, tentatively identified as tartronic semialdehyde, was catalyzed by glyoxylate carboligase also induced by growth on glycollate. Glyoxylate carboligase was discovered by Krakow and Barkulis (1956) working with extracts of *E. coli* grown on glycollate, and they also demonstrated the participation of magnesium and diphosphothiamine in the reaction. Kornberg and Gotto showed the conversion of the C_3 acid to glycerate and then to pyruvate by properly fortified extracts. This particular strain of *Pseudomonas* also contained malate synthetase when grown on glycollate, and therefore was able to catalyze the condensation of glyoxylate and acetyl-CoA to form malate, a reaction discussed further in the section on biosynthetic metabolism dealing with the role of the tricarboxylic acid cycle in supplying intermediates for biosynthesis of amino acids.

Quayle *et al.* (1961) elucidated the pathway of oxalate metabolism by *Pseudomonas oxalaticus*, an organism which can grow with oxalate as the sole carbon and energy source (Table 8.3). Oxalyl-CoA was formed

TABLE 8.3. Reactions in the metabolism of oxalate by *Pseudomonas oxalaticus* (Quayle *et al.*, 1961)

$$\text{Oxalate} + \text{succinyl-CoA} \rightarrow \text{oxalyl-CoA} + \text{succinate}$$
$$\text{Oxalyl-CoA} \rightarrow \text{formyl-CoA} + CO_2$$
$$\text{Succinate} + \text{formyl-CoA} \rightarrow \text{formate} + \text{succinyl-CoA}$$
$$\text{Formate} + \text{NAD} \rightarrow CO_2 + \text{NADH}$$

Sum: $\text{Oxalate} + \text{NAD} \rightarrow 2\ CO_2 + \text{NADH}$

by a transferase reaction with succinyl-CoA. Oxalyl-CoA was decarboxylated to carbon dioxide and formyl-CoA by a decarboxylase activated by diphosphothiamine and magnesium ion. This reaction was studied earlier by Jakoby *et al.* (1956) with an unidentified soil organism grown on oxalate. By means of another transferase reaction, succinyl-CoA was regenerated from formyl-CoA and succinate. The final step in the oxalate pathway of *P. oxalaticus* was the oxidation of formate to carbon dioxide with formate dehydrogenase which required NAD as the cofactor. *Pseudomonas oxalaticus* may contain an alternate pathway of oxalate metabolism, since an enzyme was present in these extracts which reduced oxalyl-CoA to glyoxylate and free coenzyme A with NADPH. Quayle and Keech (1960) had earlier found the conversion of glyoxylate to glycerate by extracts of this organism, probably via glyoxylate carboligase.

Krampitz and Lynen (1964) studied tartrate metabolism by a Gram positive soil organism isolated from an enrichment culture with (+) tartrate as the carbon and energy source. There are three optical isomers of tartaric acid, dextrorotatory, levorotatory and optically inactive. A satisfactory relation of optical activity to absolute configuration has not yet been made for tartaric acid, and therefore these compounds are discussed here as (+), (−) and *meso*, respectively. Krampitz and Lynen found that extracts of their organism metabolized (+) tartrate by a specific dehydrase not able to attack (−) or *meso*-tartrate. Their enzyme preparations decarboxylated oxaloacetate to pyruvate and carbon dioxide and converted pyruvate to lactate, acetate and carbon dioxide. Shilo (1957) isolated strains of *Pseudomonas* by enrichment culture, using all three forms of tartrate, and found that these isolates produced a specific dehydrase for the form of tartrate added to the growth medium. Oxaloacetate was identified as the product of this enzyme and was further metabolized by decarboxylation to pyruvate and carbon dioxide.

References

Ayres, G. C. deM. and Lara, J. F. S. (1962). *Biochim. Biophys. Acta* **62**, 435.

Barban, S. and Ajl, S. (1952). *J. Bacteriol.* **64**, 443.

Barrett, J. T. and Kallio, R. E. (1953). *J. Bacteriol.* **66**, 517.

Dagley, S. and Dawes, E. A. (1955). *Biochim. Biophys. Acta* **17**, 177.

Dickman, S. R. (1961). *In* "The Enzymes" (P. D. Boyer, H. Lardy and K. Myrback, eds.), Academic Press, New York, Vol. 5, 495.

Englard, S. (1960). *J. Biol. Chem.* **235**, 1510.

Gergely, J., Hele, P. and Ramakrishnan, C. V. (1952). *J. Biol. Chem.* **198**, 323.

Gillespie, D. C. and Gunsalus, I. C. (1953). *Bacteriol. Proc.* p. 80.

Goldman, D. S. (1956a). *J. Bacteriol.* **71**, 732.

Goldman, D. S. (1956b). *J. Bacteriol.* **72**, 401.

Goldman, D. S. (1957). *J. Bacteriol.* **73**, 602.

Gunsalus, I. C. (1954). *In* "The Mechanism of Enzyme Action" (W. D. McElroy and B. Glass, eds.), The Johns Hopkins Press, p. 545.

Hager, L. P. (1962). *In* "The Enzymes" (P. D. Boyer, H. Lardy and K. Myrback, eds.), Academic Press, New York, Vol. 6, 387.

Harvey, R. J. and Collins, E. B. (1963). *J. Biol. Chem.* **235**, 2648.

Hirsch, C. A., Rasminsky, M., Davis, B. D. and Lin, E. C. C. (1963). *J. Biol. Chem.* **238**, 3770.

Jakoby, W. B., Ohmura, E. and Hayaishi, O. (1956). *J. Biol. Chem.* **222**, 435.

Koike, M., Reed, L. J. and Carroll, R. M. (1960). *J. Biol. Chem.* **235**, 1924.

Kornberg, H. L. and Gotto, A. M. (1961). *Biochem. J.* **78**, 69.

Krakow, G. and Barkulis, S. S. (1956). *Biochim. Biophys. Acta* **21**, 593.

Krampitz, L. O. (1961). *In* "The Bacteria" (I. C. Gunsalus and R. Y. Stainer, eds.), Academic Press, New York, Vol. II, p. 209.

Krampitz, L. O. and Lynen, F. (1964). *Biochem. Z.* **341**, 97.

Krebs, H. A. (1943). *Advan. Enzymol.* **3**, 191.

Murthy, P. S., Sirsi, M. and Ramakrishnan, T. (1962). *Biochem. J.* **84**, 263.

Ogston, A. G. (1948). *Nature* **162**, 963.
Plaut, G. W. E. (1963). *In* "The Enzymes" (P. D. Boyer, H. Lardy and K. Myrback, eds.), Academic Press, New York, Vol. 7, p. 105.
Potter, V. R. and Heidelberger, C. (1949). *Nature* **164**, 180.
Quayle, J. R. and Keech, D. B. (1960). *Biochem. J.* **75**, 515.
Quayle, J. R., Keech, D. B. and Taylor, G. A. (1961). *Biochem. J.* **78**, 225.
Raman, C. S. (1961). *Biochim. Biophys. Acta* **52**, 212.
Saz, H. J. and Krampitz, L. O. (1954). *J. Bacteriol.* **67**, 409.
Schwartz, P. and Carter, H. E. (1954). *Proc. Natl Acad. Sci. U.S.* **40**, 499.
Shilo, M. J. (1957). *Gen. Microbiol.* **16**, 472.
Singer, T. P. and Kearney, E. B. (1963). *In* "The Enzymes" (P. D. Boyer, H. Lardy and K. Myrback, eds.), Academic Press, New York, Vol. 7, p. 384.
Stern, J. R. (1961). *In* "The Enzymes" (P. D. Boyer, H. Lardy and K. Myrback, eds.), Academic Press, New York, Vol. 5, p. 367.
Stern, J. R. and Ochoa, S. (1951). *J. Biol. Chem.* **191**, 161.
Swim, H. E. and Krampitz, L. O. (1954). *J. Bacteriol.* **67**, 419.
Truper, H. G. (1965). *Biochim. Biophys. Acta* **111**, 565.
VanDemark, P. J. and Smith, P. F. (1964). *J. Bacteriol.* **88**, 1602.
Warringa, M. G. P. J. and Giuditta, A. (1958). *J. Biol. Chem.* **230**, 111.
Warringa, M. G. P. J., Smith, O. H., Giuditta, A. and Singer, T. P. (1958). *J. Biol. Chem.* **230**, 97.
Wheat, R. W. and Ajl, S. J. (1955). *J. Biol. Chem.* **217**, 897.
Williams, P. J. and Rainbow, C. (1964). *J. Gen. Microbiol.* **35**, 237.
Yoshida, A. (1965). *J. Biol. Chem.* **240**, 1113.

9. ELECTRON TRANSPORT

I. Enzymes of the Cytochrome System

The International Union of Biochemistry recognizes four major groups of cytochromes, *a*, *b*, *c* and *d* (Table 9.1). The basis for separation into groups is the absorption spectrum of the reduced enzyme, which is the result of reduction of the iron porphyrin prosthetic group. Reduced cytochrome enzymes have three characteristic absorption maxima designated α, β and γ or Soret bands, from the longer to the shorter wavelengths respectively. Cytochromes *a* have an α-band in the region

TABLE 9.1. Classification of cytochrome enzymes proposed by the International Union of Biochemistry (*Enzyme Nomenclature*, 1965). Methods for the identification of the haem group are also discussed in this publication.

Group	α-Band of pyridine ferrohaemochrome in alkali mμ	Ether solubility of haemin after treatment of cytochrome with acetone-HCl	Nature of haem
a	580–590	Soluble	Contains formyl side-chain (*e.g.* haem *a*)
b	556–558	Soluble	Protohaem
c	549–551	Insoluble	Haem covalently bound to protein by side-chains
d	600–620	Soluble	Iron dihydroporphyrin

of 590 mμ, cytochromes *b* have an α-band in the region of 558 mμ, and cytochromes *c* have an α-band in the region of 550 mμ. It is common practice to further differentiate between cytochromes of any group by subscripts such as cytochrome b_1, when spectral properties indicate that a newly discovered cytochrome belongs to the group, but it is different from the other members. This terminology is not informative and, except for the cases where such nomenclature is firmly entrenched, new cytochromes of a given group are named by the addition of the absorption maximum of the reduced α-band as a subscript to the group; for example, cytochrome c_{551}. Both types of nomenclature exist in the literature and will be used here for that reason.

There is also a functional correlation with the group designation. Cytochromes *a* are usually terminal oxidases, that is their physiological electron acceptor is oxygen and they are the most electropositive of the cytochrome enzymes with an E_0' in the range of $+0{\cdot}3$ volts. Cytochromes of the *b* type have an E_0' near to 0·0 or are slightly electronegative. The physiological electron donor for *b*-type cytochromes is usually a reduced flavoprotein and the physiological electron acceptor is cytochrome *c*. Cytochromes *c* function as intermediate electron carriers between *b* and *a* with an E_0' of about $+0{\cdot}25$ volts. All true cytochrome enzymes are one-electron acceptors as a result of the oxidation and reduction of iron at the center of the prophyrin nucleus.

Several cytochromes of the *b*-type have been purified from bacterial

TABLE 9.2. Cytochromes which have been purified from bacterial sources.

	Reference
Cytochromes *b*	
Escherichia coli	Deeb and Hager, 1964
Micrococcus lysodeikticus	Jackson and Lawton, 1959
Rhodopseudomonas spheroides	Orlando and Horio, 1961
Cytochromes *c*	
Micrococcus species	Hori, 1961
Pseudomonas aeruginosa	Horio *et al.*, 1960
Pseudomonas saccharophila	Yamanaka *et al.*, 1963
Chromatium species	Bartsch and Kamen, 1960
Rhosopseudomonas palustris	Morita, 1960
Rhodopseudomonas spheroides	Orlando, 1962
Desulfovibrio desulfuricans	Postgate, 1961
Cytochromes *a* or cytochrome oxidase	
Pseudomonas aeruginosa	Horio *et al.*, 1961; Yamanaka and Okunuki, 1963a
Azotobacter vinelandii	Layne and Mason, 1958
Micrococcus denitrificans	Vernon and White, 1957.

sources (Table 9.2). One such enzyme is cytochrome b_1 of *E. coli* which was crystallized by Deeb and Hager (1964). In their study, cytochrome b_1 was found to be the physiological electron acceptor for a flavin pyruvate oxidase of *E. coli*. Cytochrome b_1 of *E. coli* was associated with the cell membrane, which is frequently true of bacterial cytochromes. The molecular weight was 62,000 and the prosthetic group was identified as iron protoporphyin IX. The difference spectrum, the spectrum of the reduced enzyme minus the spectrum of the oxidized enzyme, revealed α, β and γ bands at 557·5, 527·5 and 427·5 mμ, respectively (Fig. 9.1). The E_0' of the crystalline enzyme was determined to be $-0{\cdot}34$ volts,

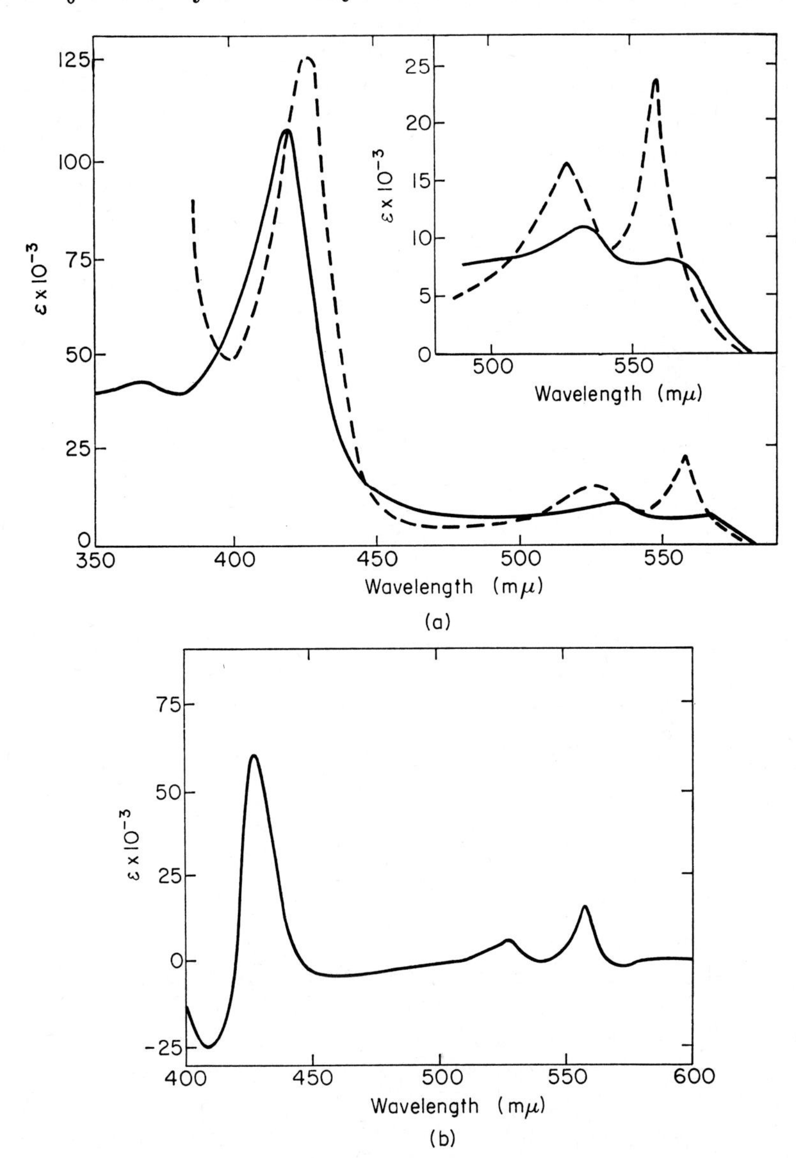

FIG. 9.1. Spectra of crystalline cytochrome b_1 of *Escherichia coli* (Deeb and Hager, 1964). (a) Oxidized (———) and reduced (------) spectra. (b) Difference spectrum = spectrum of oxidized enzyme minus spectrum of reduced enzyme.

which is extremely electronegative for cytochromes *b*. However, Deeb and Hager also isolated a protein associated with cytochrome b_1 in crude enzyme preparations which had the ability to raise the E_0' of cytochrome b_1 to more electropositive values near to $-0{\cdot}1$.

Horio *et al.* (1960) crystallized a *c*-type cytochrome from *P. aeruginosa* which they designated *Pseudomonas* cytochrome c_{551}. The crystalline

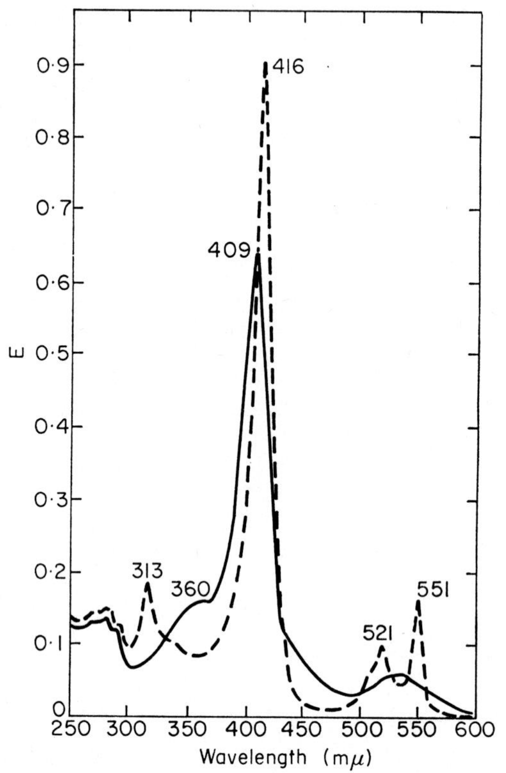

FIG. 9.2. Oxidized (———) and reduced (– – – –) spectra of crystalline *Pseudomonas* cytochrome c_{551} (Horio *et al.*, 1960).

enzyme had properties very similar to those of cytochromes *c* from other sources. The reduced enzyme had an α-band at 551 mμ (Fig. 9.2), had a low molecular weight (between 7600 and 8100 g/mole), and it was not possible to split the haem from the enzyme with acetone-HCl. The E_0' was $+0{\cdot}286$ volts, which is in the range of values obtained with cytochromes *c* from other sources. A partial list of cytochromes *c* which have been purified from bacteria is contained in Table 9.2. Several *c*-type cytochromes have been described in anaerobically grown photosynthetic bacteria. *Desulfoviobrio desulfuricans*, an anaerobic sulfate reducer, contains a single haem pigment, cytochrome c_3 (Postgate, 1961).

Not many studies of cytochromes *a* from bacterial sources have been reported, but a somewhat unusual cytochrome oxidase has been crystallized from enzyme preparations of *P. aeruginosa* (Horio *et al.*, 1961; Yamanaka and Okunuki, 1963a, b). The crystalline enzyme rapidly oxidized reduced cytochrome c_{551} prepared from the same organism and combined with cyanide and carbon monoxide, as do classical cytochrome oxidases. In addition, reduced *Pseudomonas* cytochrome oxidase was able to reduce nitrite and may be responsible for nitrite reduction under physiological conditions. Probably the most unusual characteristic of this enzyme is the possession of two different haem groups, haem a_2, which can be split from the enzyme by acetone-HCl, and haematohaem, a haem characteristic of *c*-type cytochromes and which cannot be split from the enzyme by treatment with acetone-HCl. The spectrum of the reduced *Pseudomonas* cytochrome oxidase is shown in Fig. 9.3.

Smith's review (1961) should be consulted for further details on the topic of bacterial cytochromes.

II. Respiratory Chains of Bacteria

The procedures used to determine the chain of events occurring between dehydrogenation of substrate and reduction of oxygen are measurement of reduced cytochromes following addition of substrate and use of inhibitors whose mode of action is known. One complication is that more than one pathway to oxygen may exist, and such is the case with the respiratory chain of *Mycobacterium phlei* illustrated in Fig. 9.4. Asano and Brodie (1964) fractionated enzyme preparations of this organism by high speed centrifugation into a particulate and a soluble fraction. The particulate fraction contained the bulk of the respiratory enzymes and cofactors, including a naphthoquinone identified as vitamin K_9H. The soluble fraction contained most of the dehydrogenases such as β-hydroxybutyate dehydrogenase as well as the bulk of the NAD. Spectral studies of the particles disclosed the presence of cytochromes *a*, a_3, *b*, c_1 and *c*. The particles produced a characteristic spectrum in the reduced state when a stream of carbon monoxide gas was bubbled through the preparation, indicating the presence of a terminal oxidase which was also inhibited by cyanide and sulfide. Treatment of the reduced particle with 2-*n*-nonylhydroxyquinoline N-oxide (NOQNO), an inhibitor known to act between cytochromes *b* and *c*, resulted in an increased reduction of cytochrome *b*, but oxidation of cytochromes *c* and *a*, indicating that the latter enzymes are beyond cytochrome *b* in the respiratory chain.

The respiratory particle of *M. phlei* contained three pathways to oxygen, one from the oxidation of malate, one from the oxidation of

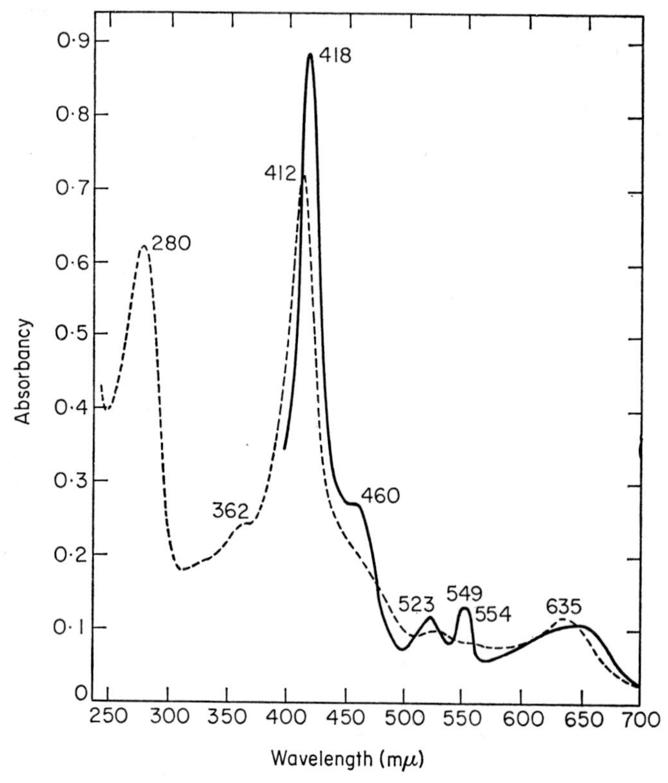

FIG. 9.3. Oxidized (– – – – –) and reduced (———) spectra of crystalline *Pseudomonas* cytochrome oxidase (Yamanaka and Okunuki, 1963a).

NADH and one from the oxidation of succinate (Fig. 9.4, Asano and Brodie, 1964). All three substrates were capable of reducing cytochromes *b* through *a*. Malate oxidation was catalyzed by two separate enzyme systems present in the soluble fraction. One enzyme was identified as a malate-vitamin K reductase (Asano *et al.*, 1965) which was activated by FAD and which catalyzed the reduction of vitamin K_9H of the particle. The other enzyme (indicated as the soluble NAD^+-linked factor of Fig. 9.4) required NAD for malate oxidation and was probably the conventional malate dehydrogenase of the tricarboxylic acid cycle. All three pathways converged at the stage of cytochrome *b* since NOQNO had the same effect with all three substrates. When the particles were irradiated at 360 mμ to destroy endogenous K_9H, addition of vitamin K_1, used instead of K_9H because of the ready availability of K_1, restored the ability of the particles to oxidize NADH and malate, but not succinate. This observation suggested that the former two pathways have a common site at the level of K_9H, but not the latter. The nature of the

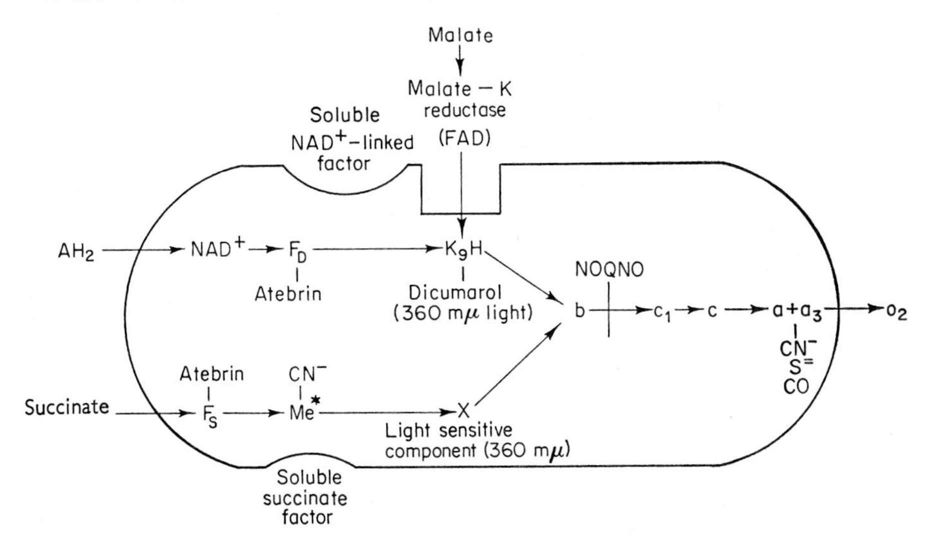

FIG. 9.4. Diagram of the respiratory chains of *Mycobacterium phlei* (Asano and Brodie, 1964). NOQNO = 2-*n*-nonylhydroxyquinoline N-oxide.

factor in the succinate pathway destroyed by light is still unknown. The succinate pathway apparently involves the participation of a metal somewhere before the stage of cytochrome *b*, since cyanide and other metal chelates inhibit succinate oxidation. The particles themselves were able to oxidize succinate, but succinate oxidation was stimulated by the addition of the soluble fraction and NAD, which could be a result of the action of fumarase and the soluble NAD-linked malate dehydrogenase. Atabrin inhibited all three pathways, although it proved much less effective with the NAD pathway.

Pathways for oxidation of NADH and NADPH exist in the soluble fraction prepared from *M. phlei* and are termed bypass pathways because they cause electrons to be shunted around the particulate respiratory chain. These enzymes react either with oxygen or transfer electrons back into the respiratory chain near the level of cytochrome *c*. Two enzymes which catalyzed NADH oxidation were separated from the soluble fraction of *M. phlei*, both of which were flavoproteins, one active with FAD and the other with FMN (Asano and Brodie, 1965a). The NADH bypass enzymes were assayed with menadione as an artificial electron acceptor but the physiological electron acceptor is unknown. A bypass pathway for NADPH was reported by Murthy and Brodie (1964) in *M. phlei*. NADPH was also oxidized by the NADH pathway of the respiratory particle since pyridine nucleotide transhydrogenase (reduced-NADP:NAD oxidoreductase), which catalyzed the reduction of NAD by NADPH, was present in *M. phlei*.

III. Oxidative Phosphorylation

A. Phosphorylative Sites

Withdrawal of a pair of electrons from one mole of NADH with the reduction of oxygen results in approximately 52 kcals of energy, enough to form several moles of ATP per atom of oxygen reduced. However, the experimentally obtained values of ATP produced per atom of oxygen consumed, the P:O ratios, do not come near to the theoretical values, either with bacterial or mammalian preparations. The reasons for this are two-fold: firstly, there are not enough phosphorylative sites between NADH and oxygen to utilize all the energy produced, and secondly, interfering reactions and loss of phosphorylative sites during fractionation reduce the yield of ATP. Several reports of oxidative phosphorylation with bacterial preparations have been made with P:O values lower, as a rule, than those obtained with mammalian preparations, very likely because of the second reason stated in the preceding sentence.

One method of detecting oxidative phosphorylation is to incubate the enzyme preparation with ADP, inorganic phosphate and hexokinase and to determine glucose-6-phosphate at the end of the reaction as a measure of the amount of ATP formed. The sensitivity of this method can be increased by the use of ^{32}P and measurement of radioactive glucose-6-phosphate or ATP. In order to identify sites of oxidative phosphorylation, the electron transport pathway of the organism must be known. Substrates tested must not be oxidized beyond the first step, otherwise the number of electrons transported is uncertain.

The preceding requirements were met by the phosphorylating system prepared from *M. phlei* by Asano and Brodie (1965b), and the phosphorylative sites identified in their studies are illustrated in Fig. 9.5. The span between cytochrome *c* and oxygen, common to all three respiratory chains, was tested by the use of reduced dyes, such as reduced phenazine methosulfate, which connect into the respiratory chain at the level of cytochrome *c*, resulting in electron transport between cytochrome *c* and oxygen only. Phosphorylation at this level

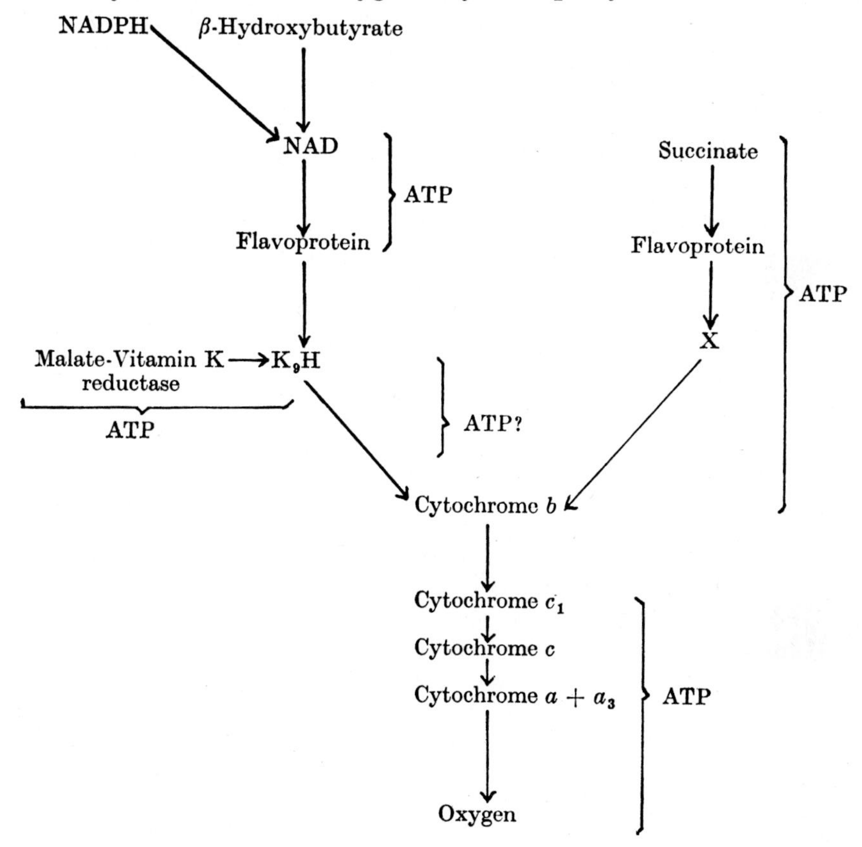

FIG. 9.5. Phosphorylative sites on the respiratory chain of *Mycobacterium phlei*. X is the light sensitive component of Fig. 9.4.

required a factor present in the soluble fraction, possibly similar to the coupling factors operative in oxidative phosphorylation by mitochondria.

Phosphorylation associated with malate oxidation was unusual because this site was soluble and did not require the participation of the particulate electron transport chain. In addition to the malate-vitamin

K reductase, FAD, vitamin K suspended in phospholipid and an artificial electron acceptor, thiazolyl blue tetrazolium, were required for phosphorylation. Malate was not oxidized beyond the stage of oxaloacetate, thereby satisfying the prerequisite for a one-step oxidation.

In order to isolate the segment between succinate or NAD and cytochrome *c*, the particles were incubated anaerobically with thiazolyl blue tetrazolium, and without the soluble factor necessary for phosphorylation in the span between cytochrome *c* and oxygen. Thiazolyl blue tetrazolium acts as an "electron sink" by siphoning off electrons at the level of cytochrome *c* and thereby preventing their further transport to cytochrome *a* and oxygen. Since oxygen was not consumed in these experiments, the data were reported as P:2e instead of P:O ratios. Under these conditions, succinate and NADH oxidation, the latter generated by oxidation of β-hydroxybutyrate, resulted in ATP formation. NADH added directly to the reaction mixture actually resulted in lower P:2e ratios than did NADH generated by β-hydroxybutyrate oxidation. This result was probably due to oxidation of NADH by the by-pass pathway enzymes without phosphorylation. Oxidative phosphorylation was also obtained with NADPH as a result of the reduction of NAD by NADPH catalyzed by the pyridine nucleotide transhydrogenase.

A phosphorylative site at the approximate level of K_9H was suggested by Asano and Brodie (1965b). The basis for this proposal is that substrates oxidized aerobically with NAD-linked dehydrogenases yielded P:O ratios higher than those obtained with substrates oxidized with FAD-linked dehydrogenases. Apparently, under aerobic conditions, the bypass pathways do not function as well as they do anaerobically where low P:2e ratios were obtained with the pyridine nucleotide dehydrogenases. Another reason for proposing a phosphorylative site in this region is that it would correspond to a site in the comparable region of the mammalian respiratory chain.

B. ATP Yield During Oxidative Phosphorylation

The yield of ATP per atom of oxygen reduced is firmly established for mitochondria as 3 for the span between NADH and oxygen and 2 for the span between succinate and oxygen. It would appear that these same P:O ratios hold for *M. phlei*, but it is not yet certain that this generalization can be made for all aerobic bacteria with a functional cytochrome system. The problem of low P:O ratios obtained with bacterial preparations which can phosphorylate ADP during aerobic oxidation and the probable causes were discussed in the preceding section. A unique experimental approach to the problem of P:O yields is possible with bacteria by the study of aerobic growth yields.

During aerobic growth, a large amount of substrate is assimilated and therefore cannot be neglected as it can in the case of anaerobic growth. Early studies of aerobic growth yields did not take assimilation into account and when yields of between 70–80 g/dry wt/mole of glucose used were reported, it appeared that only 7–8 moles of ATP/mole glucose were obtained, assuming Y_{ATP} for aerobic growth is about 10 g/mole ATP as it is for anaerobic growth. This was considerably less than the theoretical value of 38 (Table 9.3). Whitaker and Elsden (1963)

TABLE 9.3. Theoretical ATP yield for complete oxidation of one mole of glucose, assuming P:O ratios of 3 for the span between NADH and oxygen and of 2 for the span between succinic dehydrogenase and oxygen.

Reaction	Moles ADP produced by: Substrate phosphorylation	Oxidative phosphorylation
Glucose + ATP → glucose-6-phosphate + ADP	−1	
Fructose-6-phosphate + ATP → fructose-1,6-diphosphate + ADP	−1	
2 Triose phosphate + 2ADP + 2 NAD → 2phosphoglyceric acid + 2ATP + 2NADH	2	6
2 Phosphoenolpyruvate + 2 ADP → 2 pyruvate + 2 ATP	2	
2 Pyruvate + 2 CoA + 2 NAD → 2 acetyl-CoA + 2 NADH		6
2 Isocitrate + 2 NADP → 2 2-oxoglutarate + 2 NADPH		6
2 2-Oxoglutarate + 2 NAD → 2 succinyl-CoA + 2 NADH		6
2 Succinyl-CoA + 2P_i + 2ADP → 2 succinate + 2 ATP + 2 CoA	2	
2 Succinate → 2 fumarate		4
2 Malate + 2 NAD → 2 oxaloacetate + 2 NADH		6
	4	34

circumvented the problem of assimilation in their study of aerobic growth yields when they determined the amount of cell dry weight as a function of the oxygen consumption. These studies and later studies by Hadjipetrou *et al.* (1964) reported yields of 20–30 g/atom of oxygen consumed during growth on glucose and it now appears that the P:O ratio must be close to 3. The yields with other substrates such as succinate were not as high and the P:O ratios may be lower than 3 in these cases. It seems probable that the P:O ratio should always be somewhat less than 3 because of the bypass reactions and non-phosphorylative oxidations such as those catalyzed by the oxygenase enzymes.

References

Asano, A. and Brodie, A. F. (1964). *J. Biol. Chem.* **239**, 4280.

Asano, A. and Brodie, A. F. (1965a). *Biochem. Biophys. Res. Comm.* **19**, 121.

Asano, A. and Brodie, A. F. (1965b). *J. Biol. Chem.* **240**, 4002.

Asano, A., Kaneshiro, T. and Brodie, A. F. (1965). *J. Biol. Chem.* **240**, 895.

Bartsch, R. G. and Kamen, M. D. (1960). *J. Biol. Chem.* **235**, 825.

Deeb, S. S. and Hager, L. P. (1964). *J. Biol. Chem.* **239**, 1024.

Enzyme Nomenclature. Recommendation of the International Union of Biochemistry on the Nomenclature and Classification of Enzymes, Together with Their Units and the Symbols of Enzyme Kinetics. Elsevier Publishing Company, Amsterdam (1965).

Hadjipetrou, L. P., Gerrits, J. P., Teulings, F. A. G. and Stouthamer, A. H. (1964). *J. Gen. Microbiol.* **35**, 139.

Hori, K. (1961). *J. Biochem.* (*Tokyo*) **50**, 481.

Horio, T., Higashi, T., Sasagawa, M., Kusai, K., Nakai, M. and Okunuki, K. (1960). *Biochem. J.* **77**, 194.

Horio, T., Higashi, T., Yananaka, T., Matsubara, H. and Okunuki, K. (1961). *J. Biol. Chem.* **236**, 944.

Jackson, F. L. and Lawton, V. D. (1959). *Biochem. Biophys. Acta* **35**, 76.

Layne, E. C. and Mason, A. (1958). *J. Biol. Chem.* **231**, 889.

Morita, S. (1960). *J. Biochem.* (*Tokyo*) **48**, 870.

Murthy, P. S. and Brodie, A. F. (1964). *J. Biol. Chem.* **239**, 4292.

Orlando, J. A. (1962). *Biochim. Biophys. Acta* **57**, 373.

Orlando, J. A. and Horio, T. (1961). *Biochim. Biophys. Acta* **50**, 367.

Postgate, J. (1961). *Symp. Intern. Union Biochem. Canberra* **2**, 407.

Smith, L. (1961). *In* "The Bacteria" (I. C. Gunsalus and R. Y. Stanier, eds.), Academic Press, New York, Vol. II, p. 365.

Vernon, L. P. and White, F. G. (1957). *Biochim. Biophys. Acta* **25**, 321.

Whitaker, A. M. and Elsden, S. R. (1963). *J. Gen. Microbiol.* **31**, 22.

Yamanaka, T. and Okunuki, K. (1963a). *Biochim. Biophys. Acta* **67**, 379.

Yamanaka, T. and Okunuki, K. (1963b). *Biochim. Biophys. Acta* **67**, 407.

Yamanaka, T., Miki, K. and Okunuki, K. (1963). *Biochim. Biophys. Acta* **77**, 654.

10. OXIDATION OF HYDROCARBONS

I. Oxidation of Saturated Hydrocarbons

A. Terminal Oxidation

Oxidation of hydrocarbons by bacteria to the corresponding fatty acids during growth has been demonstrated enough times to suggest that this is the major pathway of hydrocarbon oxidation (Foster, 1962; McKenna and Kallio, 1965; Van der Linden and Thijsse, 1965). Formation of fatty acids from hydrocarbons probably involves an initial attack at the terminal methyl group. Organisms used in these studies were frequently soil bacteria isolated from enrichment cultures, especially *Pseudomonas aeruginosa* and other soil pseudomonads. The observation of Stewart *et al.* (1959) that cetyl palmitate accumulated in the culture medium of a Gram negative coccus grown in hexadecane medium supports the view that alcohols are intermediates in the oxidation of hydrocarbons, since cetyl alcohol and palmitic acid are the corresponding alcohol and acid expected from hexadecane. These investigators also found that $^{18}O_2$ of atmospheric oxygen was incorporated directly into cetyl palmitate. There is considerable evidence from studies with whole cells that β-oxidation follows terminal oxidation (Van der Linden and Thijsse, 1965).

Terminal oxidation of a hydrocarbon was demonstrated in a cell-free system prepared from *Pseudomonas oleovorans* by Baptist *et al.* (1963). These investigators identified octanol, octaldehyde and octanoate as products of octane oxidation by their cell-free systems. Oxidation of octanol to octaldehyde by the enzyme preparation required NAD and thus fits the pattern of other alcohol dehydrogenases. Oxidation of octane to octanol required oxygen, NADH, ferrous iron and two separate enzyme fractions from the pseudomonad (Gholson *et al.*, 1963). These requirements are typical of mixed function oxidases (Mason, 1957), enzyme systems which catalyze the reduction of molecular oxygen with a reduced pyridine nucleotide and the incorporation of one atom of oxygen into substrate and one atom into water. Oxidation of octaldehyde was catalyzed by an NAD-dependent aldehyde dehydrogenase and was presumably followed by β-oxidation of octanoate.

B. Other Pathways of Hydrocarbon Oxidation

Leadbetter and Foster (1959) and Lukins and Foster (1963) discovered the formation of methyl ketones from saturated hydrocarbons by whole cells of *Pseudomonas methanica* and of *Mycobacterium* species. Formation of methyl ketones must involve initial attack on the hydrocarbon at the methylene carbon adjacent to the terminal carbon.

Kester and Foster (1963) isolated a *Corynebacterium* species from soil which produced dicarboxylic acids from the hydrocarbon supplied. The first product detected in culture filtrates was the acid formed by terminal oxidation, next, the ω-hydroxy fatty acid appeared, and finally the dicarboxylic acid. Oxidation of the methyl carbon farthest removed from the carboxyl carbon is termed ω-oxidation. ω-Oxidation was studied with purified enzymes from a pseudomonad (Kusunose *et al.*, 1964a, b), and also appears to be due to the action of a mixed function oxidase since introduction of the ω-hydroxy group into lauric acid required oxygen, NADH and the combined action of two enzyme fractions. C_{10}, C_{14} and C_{16} fatty acids were also substrates for the hydroxylation system, but lauric acid was the preferred substrate. The ω-hydroxy acid was oxidized to the ω-aldehyde-acid by an NAD-linked dehydrogenase.

II. Oxidation of Aromatic Hydrocarbons

Catechol and protocatechuate are key intermediates in the oxidation of several aromatic hydrocarbons and tryptophane by soil pseudomonads (Fig. 10.1; Stanier, 1950). Enzymes of the route from mandelic acid to benzoic acid were studied with cell-free preparations made from *P. fluorescens*, (later reclassified as *P. putida*) by Gunsalus *et al.* (1953). L-Mandelate was oxidized by a particulate mandelate dehydrogenase to benzoylformate (Fig. 10.2). This organism also formed mandelate racemase which interconverted D- and L-mandelate. Benzoylformate decarboxylase (benzoylformate carboxy-lyase) catalyzed the decarboxylation of benzoylformate to benzaldehyde and carbon dioxide with diphosphothiamine as the cofactor. Two pyridine nucleotide-linked dehydrogenases for benzaldehyde were found, one NAD-linked and one NADP-linked, but only the latter (benzaldehyde:NADP oxidoreductase) is listed in "Enzyme Nomenclature."

Taniuchi *et al.* (1964) studied the conversion of anthranilic acid to catechol by an enzyme which they named anthranilic acid hydroxylase.

$$\text{Anthranilic acid} + O_2 + \text{NADH} \rightarrow \text{catechol} + CO_2 + NH_3 + \text{NAD}$$

This reaction is similar to that catalyzed by the mixed function oxygenases except that two atoms of oxygen are incorporated into catechol (Kobayashi *et al.*, 1964).

The metabolism of catechol is outlined in Fig. 10.3. Catechol oxygenase (catechol: oxygen 1,2-oxidoreductase), frequently referred to as

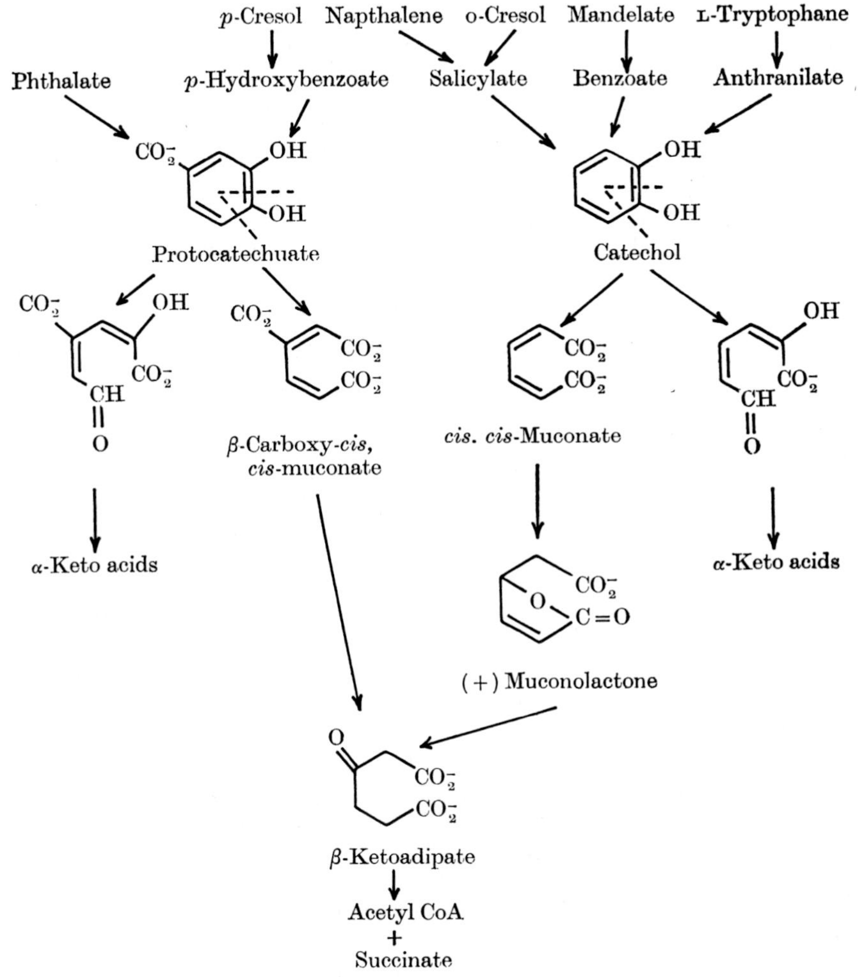

FIG. 10.1. Pathways of aromatic hydrocarbon metabolism showing the central roles of protocatechuate and catechol. (From Ornston and Stanier, 1964).

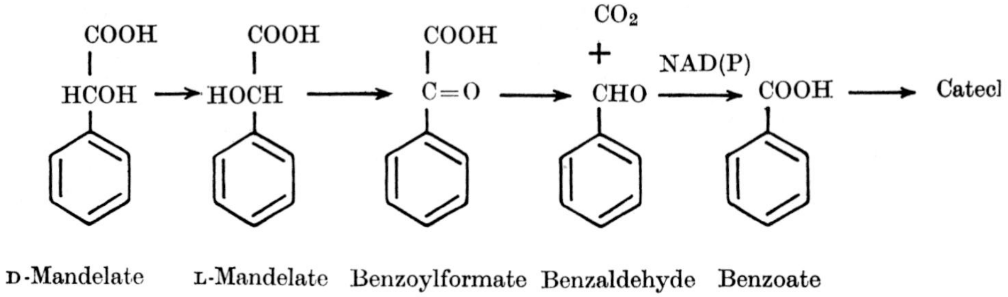

CO_2 + ; NAD(P); HCOH → HOCH → C=O → CHO → COOH → Catecl

D-Mandelate L-Mandelate Benzoylformate Benzaldehyde Benzoate

FIG. 10.2. Conversion of mandelate to benzoate by *Pseudomonas putida*.

pyrocatechase, catalyzes the oxidative cleavage of catechol to *cis, cis*-muconate (Hayaishi *et al.*, 1957) with the introduction of two atoms of oxygen into the molecule. Mason (1957) termed this type of enzyme an oxygen transferase. Sistrom and Stanier (1954) studied the enzymes involved in the conversion of *cis, cis*-muconate to β-ketoadipate with enzyme preparations of *P. fluorescens*. Muconate-lactonizing enzyme catalyzed the reversible formation of (+) muconolactone from *cis, cis*-muconate. Muconolactone isomerase was responsible for the migration of the double bond to form β-ketoadipate enol-lactone (Ornston and Stanier, 1964, 1966), and a hydrolase from the same organism converted this intermediate to β-ketoadipate. The metabolism of β-ketoadipate by *P. fluorescens* was studied by Katagiri and Hayaishi (1957). They found that β-ketoadipate was converted to acetyl-CoA and succinate, the first step being catalyzed by coenzyme A transferase and the second step by a thiolase-like enzyme.

β-Ketoadipate + succinyl-CoA → β-ketoadipoyl-CoA + succinate
β-Ketoadipoyl-CoA + CoA → acetyl-CoA + succinyl-CoA

Sum: β-Ketoadipate + CoA → acetyl-CoA + succinate

An alternate pathway of catechol metabolism occurs in other species of *Pseudomonas* (Fig. 10.1; Dagley *et al.*, 1964). Metapyrocatechase (catechol:oxygen 2,3-oxidoreductase) catalyzes a cleavage of catechol at the carbon adjacent to the bond split by catechol oxygenase to produce α-hydroxy muconicsemialdehyde. One strain of *Pseudomonas* metabolized α-hydroxymuconic semialdehyde to acetaldehyde and pyruvate (Dagley *et al.*, 1964) while another strain produced acetate and pyruvate (Nishizuka *et al.*, 1962).

Certain substituted aromatic compounds are converted to protocatechuate instead of catechol by bacteria which oxidize aromatic

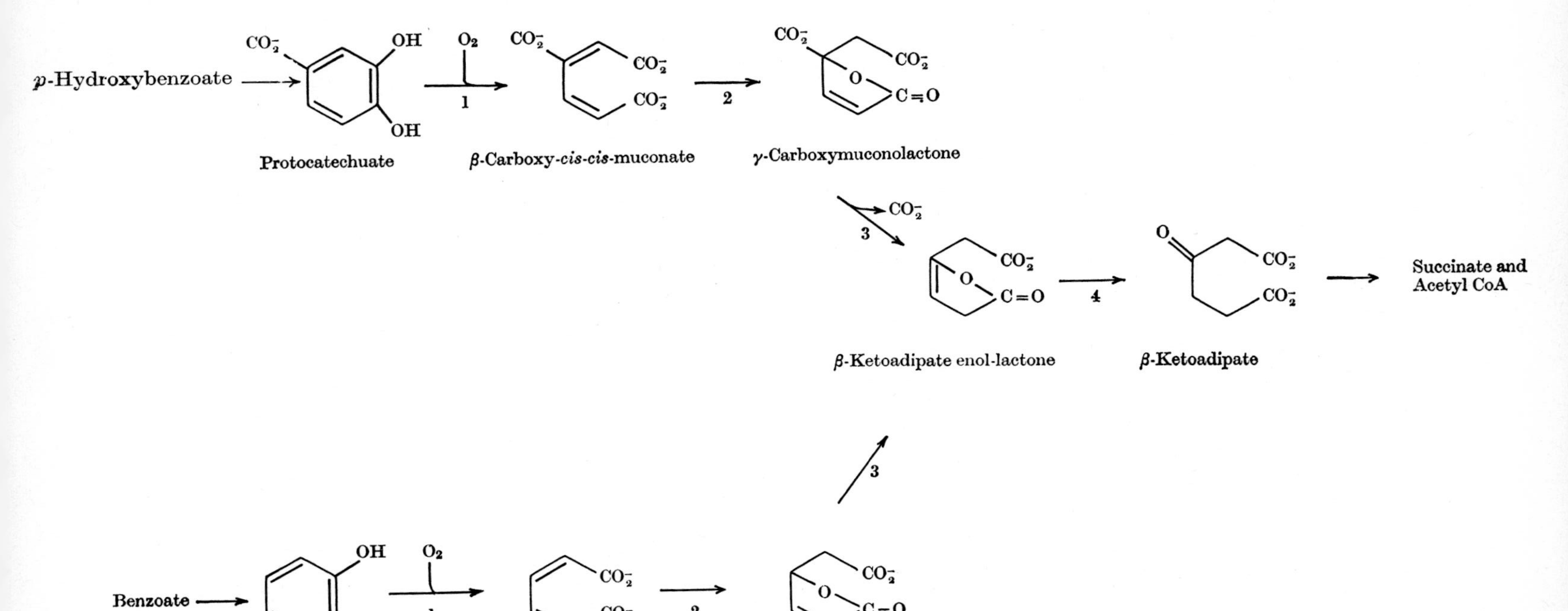

FIG. 10.3. Formation of β-ketoadipate from protocatechuate and catechol. (From Ornston and Stanier, 1966).

hydrocarbons (Fig. 10.1), but the metabolism of protocatechuate resembles that of catechol in many respects. Protocatechuate oxygenase (protocatechuate:oxygen 3,4-oxidoreductase) causes a fission of the aromatic ring between the adjacent hydroxyls (Stanier and Ingraham, 1954) resulting in β-carboxy-*cis, cis*-muconate (Fig. 10.3). A lactonizing enzyme from *P. fluorescens* catalyzes the formation of γ-carboxy-muconolactone and a decarboxylase from the same organism causes the direct formation of β-ketoadipate enol-lactone (Ornston and Stanier, 1964, 1966), the first common intermediate in the catechol and protocatechuate pathways.

An alternate pathway for protocatechuate metabolism exists as is the case for catechol (Fig. 10.1). An enzyme from a *Pseudomonas* species catalyzes a cleavage of protocatechuate to form α-hydroxy-γ-carboxy-muconic semialdehyde (Dagley *et al.*, 1964), a cleavage analogous to that resulting from the action of metapyrocatechase on catechol. α-Hydroxy-γ-carboxymuconic semialdehyde is further metabolized to two moles of pyruvate and one of formate.

Oxidation of phenathrene, anthracene (Evans *et al.*, 1965) and naphthalene (Davies and Evans, 1964) has also been studied with pseudomonads isolated from soil, and in all three instances the pathways are similar. The initial step is the introduction of 1,2-hydroxyl groups into the end ring, followed by ring fission between the hydroxyls. Eventually, three carbons of the ring attacked are removed as pyruvate, leaving an ortho-hydroxy aromatic aldehyde. Salicylaldehyde is obtained from naphthalene and is converted to salicyclic acid which is then oxidized by the catechol pathway. 1-Hydroxy-2-naphthaldehyde is obtained from phenanthrene and is oxidized by a similar pathway to 1,2-dihydroxynaphthalene, the first intermediate on the naphthalene pathway, and therefore the phenanthrene and naphthalene pathways converge at this point. Anthracene yields 2-hydroxy-3-naphthaldehyde and eventually 2,3-dihydroxynaphthalene which is metabolized by an unknown pathway.

III. Miscellaneous Oxidations

Marcus and Talalay (1956) used an organism originally isolated from soil with testosterone as the sole carbon source to prepare steroid dehydrogenases. They purified two enzymes, an α-hydroxysteroid dehydrogenase (3-α-hydroxysteroid:NAD(P) oxidoreductase) and a β-hydroxysteroid dehydrogenase (3 [or 17]-β-hydroxysteroid:NAD(P) oxidoreductase), which catalyzed the following reactions in the order named:

Androsterone + NAD(P) → androstane-3,17-dione + NAD(P)H

Testosterone + NAD(P) → 5-androstene-3,17-dione + NAD(P)H

In addition to their metabolic significance, these dehydrogenases are of interest because of their use in the determination of steroid blood levels.

The first step in the oxidation of D(+) camphor by *P. putida* is the insertion of an oxygen atom into the ring adjacent to the carbonyl functional group to produce a lactone (Fig. 10.4). A mixed function

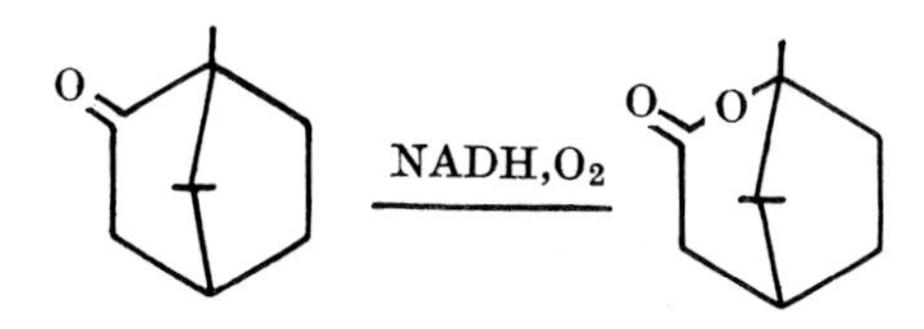

FIG. 10.4. Oxidation of D(+) camphor by a mixed function oxygenase of *Pseudomonas putida*.

oxygenase system which catalyzes this process has been purified and found to be composed of two proteins, termed E_1 and E_2 (Conrad *et al.*, 1965). E_1 contains FMN and is reduced by NADH; E_2 contains iron which is in turn reduced by FMNH of E_1. Reduced E_2 then reacts with oxygen and camphor to insert the oxygen atom into the ring.

References

Baptist, J. N., Gholson, R. K. and Coon, M. J. (1963). *Biochim. Biophys. Acta* **69**, 40.

Conrad, H. E., DuBus, R., Namvedt, M. J. and Gunsalus, I. C. (1965). *J. Biol. Chem.* **240**, 495.

Dagley, S., Chapman, P. J., Gibson, D. T. and Wood, J. M. (1964). *Nature* **202**, 775.

Davies, J. I. and Evans, W. C. (1964). *Biochem. J.* **91**, 251.

Evans, W. C., Fernley, H. N. and Griffiths, E. (1965). *Biochem. J.* **95**, 819.

Foster, J. W. (1962). *Antonie van Leeuwenhoek J. Microbiol. Serol.* **28**, 241.

Gholson, R. K., Baptist, J. N. and Coon, M. J. (1963). *Biochemistry* **2**, 1155.

Gunsalus, C. F., Stanier, R. Y. and Gunsalus, I. C. (1953). *J. Bacteriol.* **66**, 548.

Hayaishi, O., Katagiri, M. and Rothberg, S. (1957). *J. Biol. Chem.* **229**, 905.

Katagiri, M. and Hayaishi, O. (1957). *J. Biol. Chem.* **226**, 439.

Kester, A. S. and Foster, J. W. (1963). *J. Bacteriol.* **85**, 859.

Kobayashi, S., Kuno, S., Itada, N. and Hayaishi, O. (1964). *Biochem. Biophys. Res. Commun.* **16**, 556.

Kusunose, M., Kusunose, E. and Coon, M. J. (1964a). *J. Biol. Chem.* **239**, 1374.

Kusunose, M., Kusunose, E. and Coon, M. J. (1964b). *J. Biol. Chem.* **239**, 2135.

Leadbetter, E. R. and Foster, J. W. (1959). *Arch. Biochem. Biophys.* **82**, 491.

Lukins, H. B. and Foster, J. W. (1963). *J. Bacteriol.* **85**, 1074.
Marcus, P. I. and Talalay, P. (1956). *J. Biol. Chem.* **218**, 661.
Mason, H. S. (1957). *Advan. Enzymol.* **19**, 79.
McKenna, E. J. and Kallio, R. E. (1965). *Ann. Rev. Microbiol.* **19**, 183.
Nishizuka, Y., Ichiyama, A., Nakamura, S. and Hayaishi, O. (1962). *J. Biol. Chem.* **237**, 269.
Ornston, L. N. and Stanier, R. Y. (1964). *Nature* **204**, 1279.
Ornston, L. N. and Stanier, R. Y. (1966). *J. Biol. Chem.* **241**, 3776.
Shaw, R. (1966). *Nature* **209**, 1369.
Sistrom, W. R. and Stanier, R. Y. (1954). *J. Biol. Chem.* **210**, 821.
Stanier, R. Y. (1950). *Bacteriol. Rev.* **14**, 179.
Stanier, R. Y. and Ingraham, J. L. (1954). *J. Biol. Chem.* **210**, 799.
Stewart, J. E., Kallio, R. E., Stevenson, D. P., Jones, A. C. and Schissler, D. O. (1959). *J. Bacteriol.* **78**, 441.
Taniuchi, H., Hatanaka, M., Kuno, S., Hayaishi, O., Nakajima, M. and Kurihara, N. (1964). *J. Biol. Chem.* **239**, 2204.
Trippett, S., Dagley, S. and Stopher, D. A. (1960). *Biochem. J.* **76**, 9.
Van der Linden, A. C. and Thijsse, G. J. E. (1965). *Advan. Enzymol.* **27**, 469.

11. PROTEIN AND AMINO ACID CATABOLISM

I. Digestion of Proteins and Peptides

In the natural habitat of microorganisms, proteolysis is an important process which produces amino acids to be used for energy and for biosynthetic purposes. Many genera of bacteria are capable of forming proteolytic enzymes, but the Gram positive sporeformers are especially prone to produce these enzymes. Bacterial proteinases are frequently excreted into the environment (Pollock, 1962), and their formation is repressed by high levels of amino acids in the medium (Neumark and Citri, 1962). It is true of proteinases in general that they are more active against denatured proteins than against the native proteins. Several bacterial proteinases have been purified (Table 11.1) and many have been crystallized from the culture filtrate.

The alkaline proteinase of *Bacillus subtilis*, subtilisin, was originally crystallized by Güntelberg and Ottesen (1952) and has now been well characterized. A similar enzyme from *B. subtilis*, strain N′, named BPN′ (for *Bacillus* protease N′), was crystallized by Hagihara *et al.* (1958). *Bacillus subtilis* forms a neutral proteinase which has also been obtained in a high state of purity by McConn *et al.* (1964) and appears to be a zinc enzyme. Subtilisin is comparatively unspecific in its attack on peptide linkages and hydrolyzes both internal and terminal peptide bonds (Hagihara, 1960) with the release of free amino acids as well as peptides. This observation has been made with several bacterial proteinases and they do not seem to classify as typical endopeptidases or exopeptidases.

The active site of subtilisin was identified by Sanger and Shaw (1960) as a serine residue, as is the case with several proteinases of animal origin, as well as certain other hydrolytic enzymes. The method used to identify the active site takes advantage of the fact that certain hydrolytic enzymes including proteinases become inactivated after one equivalent of diisopropylphosphofluoridate (DFP) reacts with the enzyme. An investigation of this phenomenon with ^{32}P-labeled DFP uncovered the fact that in all cases DFP was bound to a serine residue.

TABLE 11.1. A partial list of organisms from which proteolytic and peptidolytic enzymes have been purified.

Organism	Reference
Proteinases	
Bacillus cereus	Wieland *et al.* (1960)
Bacillus subtilis (subtilisin)	Güntelberg and Ottensen (1952)
Bacillus subtilis, N′ (BPN′)	Hagihara *et al.* (1958)
Bacillus subtilis (neutral proteinase)	McConn *et al.* (1964)
Bacillus stearothermophilus	O'Brien and Campbell (1957)
Clostridium botulinum	Ashmarin and Vorontosov (1963)
Clostridium histolyticum (collagenase)	Seifter *et al.* (1959); Mandl *et al.* (1964)
Clostridium histolyticum (proteinase)	Labouesse and Gros (1960)
Proteus mirabilis	Hampson *et al.* (1963)
Proteus vulgaris	Mills and Wilkens (1958)
Pseudomonas aeruginosa	Morihara (1963)
Streptococcus faecalis var. *liquifaciens*	Shugart and Beck (1964); Bleiweis and Zimmerman (1964).
Streptococcus lactis	Williamson *et al.* (1964)
Group A streptococcus	Elliott (1950)
Peptidases	
Bacillus brevis	Erlanger (1957)
Clostridium botulinum	Millonig (1956)
Clostridium histolyticum	Mandl *et al.* (1957)
Lactobacillus casei	Brandsaeter (1957)
Vibrio comma	Saxena and Krishna Murti (1963)

There is considerable evidence that the unique reactivity of this particular serine is due to its location at the active center of the enzyme (Sanger, 1963).

Peptidase activity by bacteria has been reported numerous times (Table 11.1), but there are fewer studies with purified peptidases than with proteinases and the peptidase preparations are not as well characterized. Lists of organisms most frequently reported to have active peptide digesting enzymes correlate well with those having high proteolytic activity. Peptidases are located either extracellularly or intracellularly. Both *Lactobacillus casei* (Leach and Snell, 1960) and *E. coli* (Kessel and Lubin, 1963) contain intracellular peptidases which apparently have the function of hydrolyzing small peptides taken into the cell.

II. Transport of Peptides and Amino Acids

The observations that peptides are frequently better nutritional sources of amino acids for bacteria than are the free amino acids has been made many times. In some of these cases it is now certain that the peptides in question are transported into the cell more rapidly than are the amino acids and are subsequently hydrolyzed by intracellular peptidases mentioned in section I. Active transport of peptides and amino acids into the bacterial cell has been conclusively demonstrated and resembles sugar transport in all main points.

Levine and Simmonds (1960, 1962) isolated a mutant of *E. coli* K-12 which grew on glycine very slowly compared to the parent strain, although the mutant grew normally on glycylglycine. Both mutant and parent strain accumulated glycylglycine by a constituitive transport system, but the mutant possessed an impaired transport system for glycine.

Kessel and Lubin (1963) obtained two mutants of *E. coli*, one lacking glycylglycine peptidase and the other lacking the transport system for glycine. The first mutant was unable to accumulate glycylglycine from low levels of the peptide in the medium, but cell-free extracts contained the peptidase. The second mutant was able to accumulate glycylglycine, but cell-free extracts did not contain the peptidase. The existence of these mutants demonstrated that the transport system and the peptidase of *E. coli* are separate systems. The peptide transport system of *E. coli* may have a broad specificity since a number of peptides were transported.

Leach and Snell (1960) characterized the transport systems of *L. casei* for dipeptides of alanine and glycine and for the free amino acids. The accumulation of these substrates was proportional to the substrate concentration at lower concentrations, but the system became saturated at high substrate concentrations. It was possible to demonstrate a requirement for an energy source for transport of alanylglycine and glycine as well as inhibition of transport by dinitrophenol. Transport of either glycylalanine or alanylglycine was considerably faster than the transport of the free amino acids. L-Alanylglycine and glycyl-L-alanine competed with each other for accumulation and appeared to be concentrated by the same transport system. The dipeptide transport system was stereospecific, however, since glycyl-D-alanine did not compete with glycyl-L-alanine. It was not possible to demonstrate the presence of either of the dipeptides inside the cell due to the action of the dipeptidase. An interesting observation was that D-alanine and L-alanine were both concentrated by *L. casei* and apparently by the same transport system. D-alanine is of benefit to *L. casei* as a component of the cell wall.

One of the earliest investigators of active transport of amino acids by bacteria was E. F. Gale, who studied amino acid accumulation by Gram positive bacteria (Gale, 1953). Gale and his associates were the first to document the accumulation of amino acids against a concentration gradient, in particular the concentration of lysine and glutamate by *S. faecalis* and *S. aureus*; and the requirement for an energy source for this process. Since these early experiments of Gale, many other examples of active transport of amino acids by bacteria have been described.

The concentration of the branched chain amino acids by *E. coli* was reported by Cohen and Rickenberg (1956) at about the same time as the reports on galactoside permease from the same laboratory (see review by Kepes and Cohen, 1962). Concentration of valine required an energy source such as glucose or succinate and was inhibited by azide and dinitrophenol. The rate of accumulation was proportional to the concentration of valine and was saturated at low molarities of valine. L-leucine and L-isoleucine inhibited valine accumulation and apparently all three branched chain amino acids were accumulated by the same system. D-isomers of leucine, isoleucine and valine did not inhibit accumulation nor did several other amino acids and amino acid derivatives.

Holden and Holman (1959) found that *L. arabinosus* was able to accumulate L-glutamate to the extent of 10% of the cell dry weight and required glucose for the process. Halpern and Lupo (1965) studied the active transport of glutamate by a mutant of *E. coli* which was able to use glutamate as a carbon and energy source in contrast to the typical wild-type *E. coli*. Active transport of proline by *E. coli* was demonstrated by Britten *et al.* (1955) and Kessel and Lubin (1962). *Salmonella typhimurium* was found by Ames (1964) to contain a permease for histidine and apparently individual permeases for the three aromatic amino acids. A *Flavobacterium* isolated from soil by Durham and Martin (1966) accumulated both D- and L-tryptophane.

Mora and Snell (1963) studied transport of glycine, D-alanine and L-alanine by *S. faecalis* whole cells and protoplasts. All three amino acids appeared to be accumulated by the same transport system since they competed with each other. The transport system of both whole cells and protoplasts was saturated at low substrate concentrations and concentrated the amino acids in an unchanged form, thus possessing many characteristics of permeases. Whole cells and protoplasts behaved alike in most instances, but some interesting differences were found. The accumulation of all three amino acids by protoplasts, but not by whole cells, was stimulated by potassium ion and inhibited by sodium ion. Pyridoxal stimulated accumulation of glycine and D-alanine (but not L-alanine) by protoplasts but not intact cells. Another interesting finding was that D-cycloserine, but not L-cycloserine, inhibited uptake of both D- and L-alanine. Kessel and Lubin (1965) reported that the transport of glycine, D-alanine and D-cycloserine by *E. coli* was also accomplished by a single transport system.

Kaback and Stadtman (1966) used an interesting approach to study amino acid transport when they employed isolated cell membranes of *E. coli* W to investigate proline uptake. They were indeed able to demonstrate proline accumulation by the membranes which was specific for proline, was stimulated by glucose, and inhibited by azide and DNP. Cell membranes prepared from a mutant of *E. coli* W, which lacked the proline transport system, did not take up proline from the medium.

III. Metabolism of Amino Acids

A. Fermentation of Amino Acids

1. *Alanine*

Cardon and Barker (1947) isolated *C. propionicum* from an enrichment culture with alanine as the energy source and reported that *C. propionicum* fermented alanine according to the following equation:

$$3\ \text{Alanine} + 2H_2O \rightarrow 2\ \text{propionate} + \text{acetate} + CO_2 + 3NH_3.$$

Clostridium propionicum also fermented acrylate and lactate with the production of propionate and lactate, and when Leaver *et al.* (1955) found that lactate was fermented without randomization of carbons 2 and 3, it seemed likely that acrylate was an intermediate in the lactate fermentation and possibly the alanine fermentation as well. Vagelos *et al.* (1959) studied the metabolism of acrylate by *C. propionicum* and discovered acrylyl-CoA aminase (β-alanyl-CoA ammonia-lyase), the enzyme which catalyzes the addition of ammonia to acrylyl-CoA:

$$\text{Acrylyl-CoA} + NH_3 \rightarrow \beta\text{-alanyl-CoA}.$$

This enzyme apparently has nothing to do with the alanine fermentation, but its level is increased considerably as a result of growth on β-alanine and may be involved in the metabolism of that amino acid. The pathway of the alanine fermentation by *C. propionicum* is still unknown, but the acrylate pathway seems to be established in the fermentation of lactate by this organism as discussed in Chapter 6 (see Fig. 6.15).

2. *Arginine*

One of the early studies of amino acid metabolism by fermentative bacteria dealt with the decomposition of arginine to ornithine, carbon dioxide and ammonia (Fig. 11.1). The first step in the fermentation is the hydrolytic removal of ammonia from arginine by arginine desiminase (L-arginine iminohydrolase) to citrulline and ammonia. This enzyme has been reported in several species of fermentative bacteria as well as *P. aeruginosa* (Korzenovsky, 1955) and purified from *S. faecalis* (Petrack

$$\text{Arginine} + H_2O \rightarrow \text{citrulline} + NH_3$$
$$\text{Citrulline} + H_3PO_4 \rightarrow \text{ornithine} + \text{carbamoylphosphate}$$
$$\text{Carbamoylphosphate} + \text{ADP} \rightarrow \text{carbamic acid} + \text{ATP}$$

$$\text{Sum: Arginine} + H_2O + \text{ADP} + H_3PO_4 \rightarrow \text{ornithine} + NH_3 + \text{carbamic acid} + \text{ATP}$$

FIG. 11.1. Arginine fermentation by Gram positive cocci. Carbamic acid dissociates spontaneously to carbon dioxide and ammonia.

et al., 1957). The next step in the arginine fermentation is the phosphorolytic cleavage of citrulline to ornithine and carbamoylphosphate by ornithine transcarbamylase (carbamoylphosphate: L-ornithine carbamoyl-transferase, Fig. 11.1). This enzyme has been obtained in purified form from *Streptococcus lactis* (Ravel *et al.*, 1959). The decomposition of citrulline by several organisms was known to result in the formation of ATP (Slade, 1955) and this final step was shown to be the result of the action of carbamate kinase (ATP: carbamate phosphotransferase) by *S. faecalis* (Jones and Lipmann, 1960; Fig. 11.1). In comparison, carbamoylphosphate synthesis in animal tissues consumes 2 moles of ATP, requires acetylglutamate as a cofactor, and is practically nonreversible. These differences are interesting examples of comparative biochemistry since carbamoylphosphate is used by animal tissues for biosynthetic purposes and by *S. faecalis* for ATP formation. Bauchop and Elsden (1960) found that the $Y_{arginine}$ for *S. faecalis* was 10·5 grams cell dry weight per mole arginine, which agrees with the formation of one mole of ATP per mole of arginine fermented.

3. *Glutamate*

Woods and Clifton (1938) studied the fermentation of glutamate by *Clostridium tetanomorphum* and found that their fermentation balance data corresponded to the following equation:

$$5\ \text{Glutamate} + 6H_2O \rightarrow 6\ \text{acetate} + 2\ \text{butyrate} + 5CO_2 + 5NH_3 + H_2.$$

Later studies by Wachsman and Barker (1955) corroborated these findings and, in addition, demonstrated that carbons 1 and 2 of glutamate were the precursors of acetate, carbons 3 and 4 of glutamate were the precursors of butyrate, and carbon 5 was the precursor of carbon dioxide (Fig. 11.2). These data precluded the participation of the tricarboxylic acid cycle reactions in the decomposition of glutamate since in this case carbon dioxide would be formed from carbon 1 of glutamate.

Barker and his associates were mainly responsible for elucidating the reactions of the glutamate pathway of *C. tetanomorphum* (see review by

$$
\begin{array}{lcl}
\left.\begin{array}{l} {}^{1}COOH \\ | \\ H_2N{}^{2}CH \end{array}\right\} & \longrightarrow & \begin{array}{l} {}^{1}COOH \\ | \\ {}^{2}CH_3 \end{array} \\
| & & + \\
\left.\begin{array}{l} {}^{3}CH_2 \\ | \\ {}^{4}CH_2 \end{array}\right\} & \longrightarrow & {}^{3}CH_3{}^{4}CH_2{}^{3}CH_2{}^{4}COOH \\
| & & + \\
{}^{5}COOH & \longrightarrow & {}^{5}CO_2
\end{array}
$$

FIG. 11.2. Origin of carbon in products of glutamate-^{14}C fermentation by *C. tetanomorphum.*

Barker, 1961). The first reaction in the fermentation results in the formation of L-*threo*-β-methylaspartate from glutamate and is catalyzed by the glutamate mutase system (Barker *et al.*, 1958a; Fig. 11.3). The

$$
\begin{array}{c}
COOH \\ | \\ H_2NCH \\ | \\ CH_2 \\ | \\ CH_2 \\ | \\ COOH
\end{array}
\rightarrow
\begin{array}{c}
COOH \\ | \\ H_2NCH \\ | \\ HC\text{—}COOH \\ | \\ CH_3
\end{array}
\rightarrow
\begin{array}{c}
HOOC\text{—}CH + NH_3 \\ \| \\ C\text{—}COOH \\ | \\ CH_3
\end{array}
\xrightarrow{+H_2O}
\begin{array}{c}
COOH \\ | \\ CH_2 \\ | \\ HOC\text{—}COOH \\ | \\ CH_3
\end{array}
\rightarrow
\begin{array}{c}
COOH \\ | \\ CH_3 \\ + \\ CO\text{—}COOH \\ | \\ CH_3
\end{array}
$$

L-Glutamate | L-*threo*-β-Methyl-aspartate | Mesaconate | (+) Citra-malate | Acetate + Pyruvate

FIG. 11.3. Reactions in the decomposition of glutamate by *Clostridium tetanomorphum.*

glutamate mutase system is now known to consist of at least 2 enzymio components (Barker *et al.*, 1964) one of which contains a form of vitamin B_{12} as a cofactor (Barker *et al.*, 1958b). The second step in the glutamate fermentation is the removal of ammonia from β-methylasparate with the formation of mesaconate, an unsaturated dicarboxylic acid with the *trans* configuration (Wachsman, 1956). The enzyme which catalyzed the deamination was purified from *C. tetanomorphum* (Barker *et al.*, 1959) and was designated β-methylaspartase (L-*threo*-3-methylaspartate ammonia-lyase). The next step in the fermentation of glutamate is the addition of water to mesaconate to form (+) citramalic acid and is catalyzed by the mesaconase system. This enzyme system was purified from *C. tetanomorphum* by Blair and Barker (1966), and, like the glutamate mutase system, consists of two protein components which require cysteine for activity. The next step in the glutamate fermentation is the aldol cleavage of (+) citramalate to acetate and pyruvate, and was first studied with *C. tetanomorphum* enzyme preparations (Barker, 1961). A similar system exists in pseudomonads where citramalyl-CoA is an intermediate in the degradation of itaconic and citraconic acids. Citramalyl-CoA is formed by a coenzyme A transferase reaction between succinyl-CoA and citramalic acid and is then cleaved to acetyl-CoA and pyruvate (Cooper and Kornberg, 1962). Finally, butyrate formation would be expected to occur following oxidative decarboxylation of pyruvate by the pathway outlined for *C. kluyveri*. (Fig. 6.6).

4. *Histidine*

The products of the histidine fermentation by *C. tetanomorphum* are similar to those of the glutamate fermentation by the same organism except that a higher yield of ammonia is obtained and nearly one mole of formamide is produced (Wachsman and Barker, 1955; Barker, 1961). Glutamate is known to be an intermediate in histidine fermentation (Wickremasinghe and Fry, 1954) and is formed from histidine such that carbon 1 of histidine becomes carbon 5 of glutamate (Wachsman and Barker, 1955). Fermentation of histidine-2-^{14}C by *C. tetanomorphum* resulted in butyrate labeled in carbons 2 and 4 (Wachsman and Barker, 1955) as would be expected. The pathway of the histidine fermentation appears to be similar to that followed in the aerobic metabolism of histidine by *A. aerogenes*, described in section III.B.11 of this chapter. Urocanic acid was produced by *C. tetanomorphum* from histidine (Wickremasinghe and Fry, 1954) and enzyme extracts of this organism accumulated a compound with the properties of formininoglutamate.

5. *Glycine*

Cardon and Barker also reported in 1947 on the fermentation of glycine by an organism now known as *Peptococcus glycinophilus*, but originally assigned to the genus *Diplococcus*. According to their studies, glycine fermentation could be represented as follows:

$$4 \text{ Glycine} + 2H_2O \rightarrow 4NH_3 + 3 \text{ acetate} + 2CO_2.$$

Studies with glycine-1-^{14}C and -2-^{14}C by Barker *et al.* (1948) revealed an unexpected complexity to the fermentation. The methyl carbon of glycine appeared in both carbons of acetate but only the carboxyl carbon of glycine appeared in carbon dioxide. Added radioactive carbon dioxide was fixed mainly into the carboxyl carbon of acetate, although a significant amount appeared in the methyl carbon. The pathway of glycine fermentation by *P. glycinophilus* was worked out by Sagers and his associates (Fig. 11.4). The initial step in the fermentation is the labilization of the carboxyl group of glycine, which they demonstrated

$$2\ \text{Glycine} + 2\ \text{NAD} + 2\ \text{folate-H}_4 \rightarrow 2\ \text{folate-H}_4\text{-CH}_2\text{OH} + 2\ \text{NH}_3 + 2\ \text{CO}_2 + 2\ \text{NADH}$$

$$2\ \text{Folate-H}_4\text{-CH}_2\text{OH} + 2\ \text{glycine} \rightarrow 2\ \text{serine} + 2\ \text{folate-H}_4$$

$$2\ \text{Serine} \rightarrow 2\ \text{pyruvate} + 2\ \text{NH}_3$$

$$2\ \text{Pyruvate} + 2\ \text{NAD} + 2\ \text{CoA} \rightarrow 2\ \text{acetyl-CoA} + 2\ \text{CO}_2 + 2\ \text{NADH}$$

$$2\ \text{Acetyl-CoA} + 2\ \text{phosphate} \rightarrow 2\ \text{acetylphosphate} + 2\ \text{CoA}$$

$$2\ \text{Acetylphosphate} + 2\ \text{ADP} \rightarrow 2\ \text{acetate} + 2\ \text{ATP}$$

$$4\ \text{NADH} + 2\ \text{CO}_2 \rightarrow \text{acetate} + 4\ \text{NAD}$$

Sum:

$$4\ \text{Glycine} + 2\ \text{phosphate} + 2\ \text{ADP} \rightarrow 3\ \text{acetate} + 2\ \text{CO}_2 + 4\ \text{NH}_3 + 2\ \text{ATP}$$

FIG. 11.4. Glycine fermentation by *Peptococcus glycinophilus*.

by measuring the exchange of bicarbonate with the carboxyl carbon of glycine (Klein and Sagers, 1966). This reaction was somewhat unusual in that the combined action of two protein components was required; one protein was heat labile and contained pyridoxal phosphate, while the other protein was heat stable. Extracts of *P. glycinophilis* catalyzed the NAD-dependent cleavage of the labilized glycine with transfer of glycine carbon 2 to tetrahydrofolate with formation of hydroxymethyltetrahydrofolate, reduced NAD, carbon dioxide and ammonia. Klein and Sagers (1962) also verified the presence in their extracts of serine aldolase (L-serine:tetrahydrofolate 5,10 hydroxymethyl transferase) which catalyzed the formation of L-serine from glycine and hydroxymethyltetrahydrofolate and serine dehydrase (L-serine hydrolyase [deaminating]), which formed pyruvate and ammonia from serine. Other studies by Klein and Sagers (1962) outlined the conversion of pyruvate to acetyl-CoA by a pathway which required NAD and CoA. Acetylphosphate and acetate were formed from acetyl-CoA by the action of phosphotransacetylase and acetokinase (Chapter 6). It has been proposed that the 4 moles of reduced NAD formed in the fermentation of 4 moles of glycine fermented are utilized by reducing two moles of carbon dioxide to acetate (Sagers and Gunsalus, 1961), possibly by a pathway similar to that used by *C. thermoaceticum* (Chapter 6).

6. *Fermentation of pairs of amino acids; the Stickland reaction*

In 1934, L. H. Stickland published the first of a series of papers dealing with the fermentation of pairs of amino acids by members of the genus *Clostridium*, a process now known as the Stickland reaction. The review by Nisman (1954) contains a summary of much of the early work of Stickland and his colleagues. The Stickland reaction appears to be limited to certain proteolytic *Clostridium* species and is their most important energy yielding process. Stickland found that amino acids fermented by these organisms could be classified as either hydrogen donors or hydrogen acceptors, and that fermentation would not occur unless one member of each group was present. Stickland's test for hydrogen donors was to measure the reduction of a dye such as brilliant cresyl green by the bacterial suspension and the appropriate amino acid under anaerobic conditions. For hydrogen acceptors, the conditions were similar except that oxidation of a dye such as reduced benzyl viologen was measured. The best hydrogen donors were aliphatic amino acids, alanine, leucine, isoleucine, valine and norvaline, although several other amino acids reacted slowly (Kocholaty and Hoogerheide, 1938). The product in each case was the next lower fatty acid, carbon dioxide and ammonia (Nisman, 1954). The best hydrogen acceptors were glycine hydroxyproline, proline (Stickland, 1934) and ornithine (Woods, 1936), in which case the products of reduction were acetate from glycine and δ-aminovalerate from both proline and ornithine. Both isomers of proline and ornithine would serve as hydrogen acceptors (Woods, 1936). Experiments with glycine-^{14}C showed that reduction of glycine to acetate by *C. sticklandii* was a direct process (Stadtman *et al.*, 1958) in contrast to the pathway of acetate formation from glycine by *P. glycinophilus*.

Nisman (1954) and his associates reported that sonic extracts of *C. sporogenes* catalyzed the oxidation of L-alanine and L-valine with the reduction of NAD. They were also able to demonstrate the cleavage of pyruvate by these enzyme preparations which required NAD, thiamine pyrophosphate and coenzyme A, with the formation of acetylphosphate probably subsequent to acetyl-CoA formation. It seems reasonable to assume that the energy of the acetyl-CoA bond is obtained by the combined action of phosphotransacetylase and acetokinase as is the case with *C. kluyveri* (Fig. 11.5). Nisman and his associates presented similar data for the metabolism of the keto acids of valine and leucine.

Stadtman and Elliott (1957) found that at least two enzymes were involved in the reduction of L-proline by *C. sticklandii*, a proline racemase, and a proline reductase which was specific for the D-proline. The best reducing agent for D-proline reduction with the purified enzyme was 1,3-dimercaptopropanol, but a number of other dithiols including lipoic acid were also active.

The glycine reductase system of *C. sticklandii* was similar in certain respects to the proline reductase system but differed strikingly in that glycine reduction was accompanied by the formation of one mole of ATP per mole glycine reduced (Stadtman *et al.*, 1958). The precise nature of the catalytic process involved in glycine reduction is still a mystery, because the lability of the components of the system has

$$\text{Alanine} + \text{NAD} \rightarrow \text{pyruvate} + \text{NADH} + NH_3$$
$$\text{Pyruvate} + \text{NAD} + \text{CoA} \rightarrow \text{acetyl-CoA} + CO_2 + \text{NADH}$$
$$\text{Acetyl-CoA} + \text{phosphate} + \text{ADP} \rightarrow \text{acetate} + \text{ATP} + \text{CoA}$$
$$2\ \text{Glycine} + 2\ \text{NADH} + 2\ \text{ADP} \rightarrow 2\ \text{acetate} + 2\ \text{NAD} + 2\ \text{ATP} + 2\ NH_3$$

$$\text{Sum: Alanine} + 2\ \text{glycine} + 3\ \text{ADP} \rightarrow 3\ \text{acetate} + 3\ NH_3 + 3\ \text{ATP} + CO_2$$

FIG. 11.5. Reconstruction of events occuring in Stickland reaction involving alanine and glycine. With proline as the electron acceptor, 1 mole of ATP would be formed.

hindered their characterization. However, the electron transport chain which functions in glycine reduction has been partially characterized by Stadtman (1966). She isolated an acidic low molecular weight protein which she designated protein A and which was required for glycine reduction with a number of electron donors, including dimercaptans and NADH. Protein A was similar to ferredoxin, but distinct from it since ferredoxin was required in addition to protein A for reduction of glycine by NADH (but not by dimercaptans). FMN was also required for NADH oxidation and it appeared that there were still other proteins involved in the electron transport chain of *C. sticklandii* as well as protein A and ferredoxin. It seems likely that NADH is the physiological electron donor for glycine reduction and that dimercaptans function as artificial electron donors.

B. Oxidation of Amino Acids

1. *General amino acid oxidases*

Cell-free preparations of *Proteus vulgaris* (Stumpf and Green, 1944) catalyzed the consumption of oxygen with several L-amino acids and similar preparations of *Proteus morganii* (Stumpf and Green, 1946), and *P. aeruginosa* (Norton and Sokatch, 1966) catalyzed the oxidation of several D-amino acids. In all three cases, amino acids were oxidized to the corresponding keto acids with the consumption of one atom of oxygen per mole amino acid oxidized. This stoichiometry indicates that water is a product of the reaction rather than peroxide, as is the case with the animal amino acid oxidases, and suggests that the cytochrome system is involved in the reduction of oxygen by the bacterial preparations.

Tsukada (1966) purified two separate D-amino acid dehydrogenases from *P. fluorescens*; one enzyme was induced by growth on D-tryptophane and utilized methylene blue as an electron acceptor, the other enzyme was constituitive and used dichlorophenolindophenol as the electron acceptor. Neither enzyme used oxygen after purification. The substrate specificity of the two enzymes was similar and resembled D-amino acid oxidase of kidney, since aliphatic amino acids were among the best substrates and aspartate and glutamate were not oxidized. Both bacterial enzymes oxidized D-kynurenine as well. FAD was present in the purified preparations, but added FAD did not stimulate the rate of dye reduction. However, it is possible that FAD is bound too tightly to the enzyme to dissociate under ordinary conditions.

2. *Glycine*

Campbell (1955) studied glycine oxidation by a pseudomonad which he isolated from mud and which was grown in a glycine–yeast extract medium. Cell-free extracts of this organism catalyzed the oxidation of glycine to glyoxylate and ammonia and of glyxoylate to formate and carbon dioxide. No cofactors were required for glycine oxidation. Experiments with radioactive glycine showed that the carboxyl carbon of glycine was the precursor of carbon dioxide and that the α-carbon was the precursor of formate. Formate was also oxidized by the cell-free preparation to carbon dioxide.

Dagley *et al.* (1961) found a different pathway for glycine oxidation in their pseudomonad which was grown in a medium with glycine as the sole organic carbon compound. Glycine was again oxidized to glyoxyllate and ammonia by cell-free extracts, but glyoxylate itself was not further oxidized. Instead, glyoxylate was metabolized by the route outlined in Chapter 8 for the metabolism of glycollate *via* tartronic semialdehyde. Glyoxylate carboligase was present in extracts of cells grown on glycine as well as an enzyme which reduced tartronic semialdehyde to glycerate with NADH. Glycerate was readily converted to pyruvate by undialyzed enzyme extracts of the pseudomonad supplemented with ATP.

The explanation for the differences in glycine metabolism between the two pseudomonads is not certain; either two separate pathways exist in these organisms or, as Dagley *et al.* (1961) suggested, the difference may be attributed to the growth conditions used. The route described by Campbell may function exclusively for glycine catabolism, while the pathway described by Dagley and associates may be used both for energy production and for formation of biosynthetic intermediates.

3. *Serine and threonine*

Deamination of serine and threonine has been reported numerous times both in aerobically and anaerobically grown organisms. Wood and Gunsalus (1949) made the first studies with a purified enzyme using *E. coli* as the organism for study. The enzyme which they isolated was later shown to be specific for L-serine and L-threonine (Metzler and Snell, 1952) and catalyzed the deamination of serine and threonine as follows:

$$\text{L-Serine} \rightarrow \text{pyruvate} + NH_3$$
$$\text{L-Threonine} \rightarrow \text{2-oxobutyrate} + NH_3$$

L-Serine deaminase was activated by adenosine-5′-monophosphate and glutathione. Under aerobic conditions pyruvate and 2-oxobutyrate are metabolized by the usual oxo acid pathway to acetyl-CoA and propionyl-CoA. The metabolic fate of acetyl-CoA in bacteria is well known, but very little is known of the metabolism of propionyl-CoA by bacteria. It seems likely that one or more of the pathways of propionate metabolism known to occur in animal or plant cells (Kaziro and Ochoa, 1964) will be utilized by bacteria. Metzler and Snell (1952) separated a second serine and threonine deaminase from *E. coli* which was specific for the D-isomers of these two amino acids (D-serine hydro-lyase [deaminating]) and was distinct from the enzyme of Wood and Gunsalus. This enzyme likewise catalyzed the formation of pyruvate and 2-oxobutyrate from serine and threonine, but the prosthetic group was shown to be pyridoxal phosphate, and adenosine-5′-monophosphate and glutathione were without effect in this case. L-serine deaminase was later crystallized by two research groups working independently (Dupourgue *et al.*, 1966; Labow and Robinson, 1966).

A completely different route of threonine metabolism was uncovered by Elliott (1960) when he isolated aminoacetone as a product of the aerobic oxidation of L-threonine by *S. aureus*. Subsequently, Green and Elliott (1964) partially purified an L-threonine dehydrogenase from *S. aureus* which required NAD and which catalyzed the formation of aminoacetone, carbon dioxide and NADH from L-threonine. A similar enzyme from *Rhodopseudomonas sphaeroides* was described by Neuberger and Tait (1962). In both cases it seems likely that the immediate product of L-threonine dehydrogenase is aminoacetoacetate (Fig. 11.6) which then undergoes spontaneous decarboxylation to aminoacetone. Green and Elliott (1964) suggested that aminoacetone was metabolized by a circular pathway to methylglyoxal, lactate, pyruvate and acetyl-CoA, which they postulated could condense with glycine to form aminoacetoacetate to begin the cycle again.

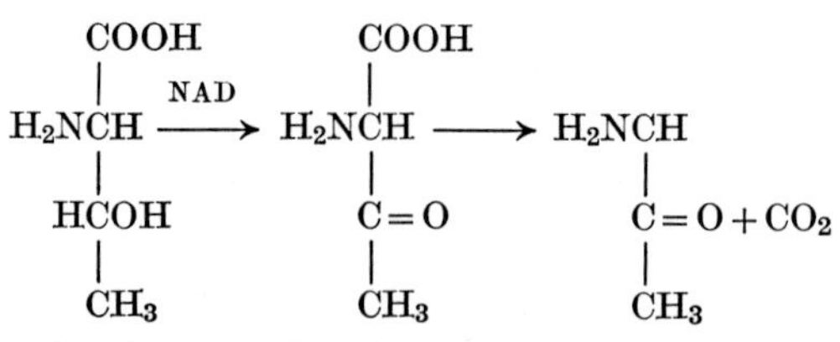

FIG. 11.6. Postulated route of aminoacetone formation from L-threonine.

4. *Aspartate*

Many aerobic, facultative and anaerobic bacteria contain the enzyme aspartase (L-aspartate ammonia-lyase) when grown in the presence of aspartic acid. Williams and McIntyre (1955) purified aspartase from *Bacterium cadaveris* and showed that the enzyme catalyzed the deamination of aspartic acid to fumaric acid and ammonia without any additional cofactors.

Several species of bacteria have been reported to catalyze β-decarboxylation of L-aspartate, including *Desulfovibrio desulfuricans* (Cattaneo-Lacombe *et al.*, 1958), *Alcaligenes faecalis* (Novogrodsky and Meister, 1964), *Chromobacter* species (Wilson, 1963) and *Clostridium perfringens* (Meister *et al.*, 1951). The crystalline enzyme from *A. faecalis* contains pyridoxal phosphate and is activated by oxoacids, pyruvate and 2-oxoglutarate being the best. This enzyme also slowly catalyzes a number of transamination reactions. The product of L-aspartate β-decarboxylation by L-aspartate 4-carboxy-lyase is L-alanine which is then metabolized by deamination to pyruvate. Markovetz *et al.* (1966) isolated a pseudomonad from soil enrichment culture which grew with D-aspartate as the carbon source and which contained aspartate racemase and β-decarboxylase. α-Decarboxylation of aspartate to β-alanine by L-aspartate 1-carboxy-lyase has also been reported in bacteria (David and Lichstein, 1950).

5. *Glutamate*

Bacterial glutamate dehydrogenase is NADP-linked as a rule, and probably functions in biosynthetic pathways to fix ammonia into the cell, although a catabolic role cannot be ruled out. Several years ago, Braunstein (1947) suggested a general route for deamination of L-amino acids by transamination with 2-oxoglutarate and oxidation of glutamate by glutamate dehydrogenase to regenerate 2-oxoglutarate.

L-Amino acid + 2-oxoglutarate → L-glutamate + 2-oxoacid
L-glutamate + NAD → 2-oxoglutarate + NADH + NH_3

Sum: L-amino acid + NAD → 2-oxoacid + NADH + NH_3

Nisman (1954) provided evidence that this pathway functions in clostridia which do not carry out the Stickland reaction. These organisms were unable to deaminate alanine unless 2-oxoglutarate was present, and cell-free extracts contained L-glutamate dehydrogenase specific for NAD as the electron acceptor (L-glutamate: NAD oxidoreductase [deaminating]). Evidence for this pathway of L-amino acid deamination in *P. aeruginosa* has also been reported (Norton and Sokatch, 1966).

An interesting pathway of glutamate metabolism was found by Vender *et al.* (1965) with a mutant of *E. coli* able to grow with L-glutamate as the sole organic carbon source. When grown on glutamate, the mutant contained a lowered amount of NADP-linked glutamate dehydrogenase as compared to the parent organism, but an increased level of aspartase. These investigators suggested that glutamate was metabolized by transamination with oxaloacetate to produce aspartate and 2-oxoglutarate. Aspartate was then deaminated by aspartase to produce fumarate which was converted to oxaloacetate by the enzymes of the tricarboxylic acid cycle. In this case, the inability of the parent to grow on glutamate seems to be due to the presence of glutamate decarboxylase, lacking in the mutant, which converts glutamate to γ-aminobutyrate, a nonmetabolizable compound.

6. *Alanine*

L-Alanine dehydrogenase (L-alanine:NAD oxidoreductase [deaminating]) has been reported in a number of organisms of the genus *Bacillus*, and NAD-linked enzymes have been purified from cell-free preparations of *B. subtilis* (Pierard and Wiame, 1960), *B. cereus* spores and vegetative cells (O'Connor and Halvorson, 1960; McCormick and Halvorson, 1964) and *M. tuberculosis* (Goldman, 1959). In all cases there is some suggestion that alanine dehydrogenase functions in the deamination of L-alanine, but this hypothesis is difficult to prove. An interesting situation exists with respect to germination of *Bacillus* spores where deamination of L-alanine to pyruvate is required for germination, and thus far L-alanine dehydrogenase is the only enzyme uncovered in these organisms capable of causing this reaction (O'Connor and Halvorson, 1960).

Another possible route of L-alanine deamination is by transamination with an appropriate oxoacid acceptor, an important route in animal tissues.

Pseudomonas aeruginosa contains an enzyme bound to the cell membrane which catalyzes the oxidation of D-alanine to pyruvate (Norton *et al.*, 1963). This enzyme is apparently cytochrome-linked since one atom of oxygen is consumed per mole alanine oxidized. The organism also contains an alanine racemase, thus it is possible that L-alanine is oxidized by conversion to D-alanine.

7. *Valine, leucine and isoleucine*

The available evidence suggests that the bacterial catabolism of the branched chain amino acids follows the outline of the pathways in animal tissues (Fig. 11.7). Growth of *P. aeruginosa* on valine resulted

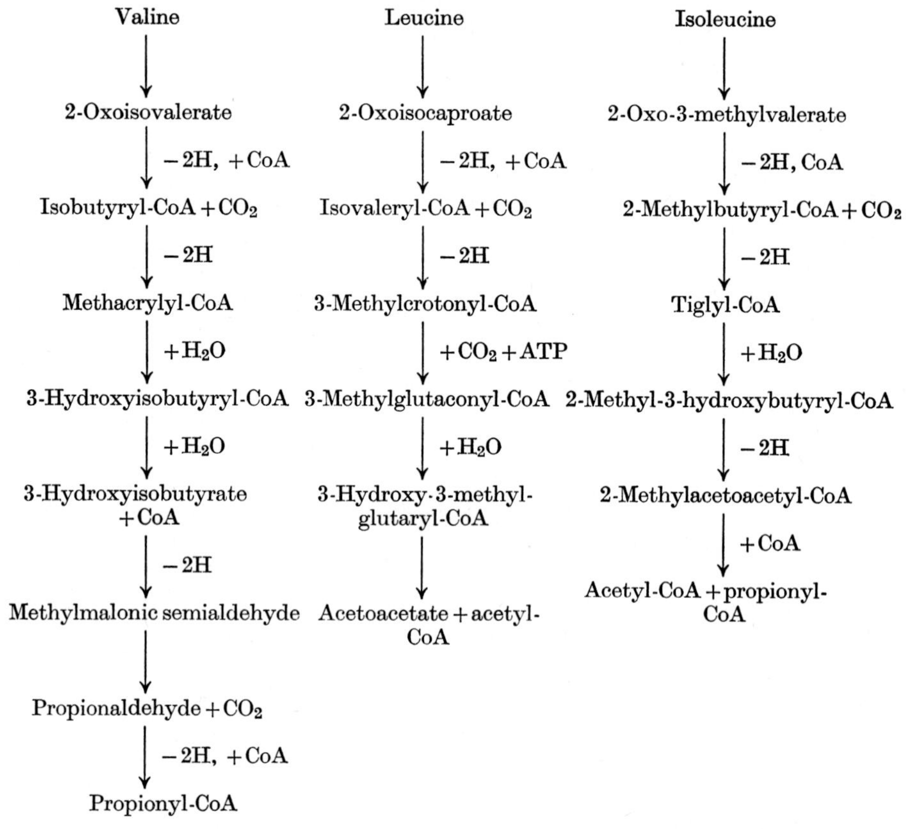

FIG. 11.7. Probable reactions in the oxidation of the branched-chain amino acids by bacteria, based on reactions known to occur in animal pathways and on detection of some of the enzymes in bacteria. Reviews by Greenberg (1961) and Meister (1965) should be consulted for details of these pathways.

in development of the ability of whole cells to oxidize isobutyrate and propionate as well as valine (Sokatch, 1966). When grown on DL-valine-4,4′^{14}C, *P. aeruginosa* synthesized alanine with isotope in carbons 1 and

3, which showed that the isopropyl carbons of valine were the precursors of pyruvate for alanine biosynthesis. The route of formation of propionate from valine and isobutyrate is not settled, but the isotope distribution of alanine suggests that decarboxylation of the methylmalonic semialdehyde to propionaldehyde and carbon dioxide is the preferred route (Fig. 11.7). Such a pathway would result in propionaldehyde and propionate labeled in positions 1 and 3, and pyruvate-1,3-^{14}C could be produced from such precursors by the acrylyl-CoA pathway (Fig. 6.15).

The initial step in the oxidative metabolism of all three branched chain amino acids is deamination to the corresponding 2-oxoacid, although the means by which this happens differs with the organism. One possible exception to this generalization may occur in the strain of *P. vulgaris* studied by Ekladius *et al.* (1957) which forms a decarboxylase active on all three branched chain amino acids and which converts them to the next lower amine. However, there was no evidence of further metabolism of the amines. Sanwal and Zink (1961) purified a leucine dehydrogenase from *B. cereus* which catalyzed the oxidative deamination of L-isomers of all three branched chain amino acids to the 2-oxoacids with NAD as the electron acceptor. Deamination of L-valine by enzyme extracts of *P. aeruginosa* occurred by transamination with 2-oxoglutarate to produce 2-oxoisovalerate and L-glutamate; the latter amino acid was then oxidized to regenerate 2-oxoglutarate (Norton and Sokatch, 1966). L-leucine and L-isoleucine were also active with the transaminase of *P. aeruginosa*. D-valine was oxidized by a particulate enzyme from *P. aeruginosa* to 2-oxoisovalerate with the consumption of one atom of oxygen per mole of 2-oxoacid produced. D-Isomers of leucine and isoleucine, as well as of several other amino acids, were oxidized by this enzyme preparation. Valine racemase was not present in this organism, hence pathways for the metabolism of D- and L-branched chain amino acids were separate. Little is known of the enzymes in bacteria involved in the further metabolism of valine although 3-hydroxyisobutyrate dehydrogenase has been detected in *A. aerogenes* (Robinson and Coon, 1957). The metabolism of propionate in aerobic bacteria is not well known, but lactyl-CoA dehydrase was first discovered in a pseudomonad isolated from a propionate enrichment culture by Vagelos *et al.* (1959).

Sasaki (1962) prepared 2-oxoisocaproate by the action of whole cells of *P. vulgaris* on leucine and partially purified an enzyme from extracts of this organism which catalyzed the decarboxylation of 2-oxoisocaproate to isovaleraldehyde and carbon dioxide. The possibility that isovaleraldehyde is then oxidized to isovaleryl-CoA exists, but the reaction has not yet been demonstrated.

Lynen *et al.* (1961) first discovered the carbon dioxide-fixing enzyme of the leucine pathway, 3-methylcrotonyl-CoA carboxylase (3-methylcrotonyl-CoA carbon dioxide ligase [ADP]) in species of *Mycobacterium* and *Achromobacter* isolated from soil, using a selective culture medium with isovaleric acid as the carbon source. 3-Methylcrotonyl-CoA carboxylase contained biotin and the enzyme from the *Mycobacterium* was unique in that it catalyzed the carboxylation of free biotin. The product of biotin carboxylation was identified as 1-N-carboxy(+)biotin by Lynen *et al.* (1961). 3-Methylcrotonyl-CoA carboxylase was also isolated from *Pseudomonas oleovorans* by Rilling and Coon (1960). Methylglutaconase (3-hydroxy-3-methylglutaryl-CoA hydro-lyase), the enzyme which catalyzes the hydration of 3-methylglutaconyl-CoA to 3-hydroxy-3-methylglutaryl-CoA, was first purified from extracts of sheep liver, but has been detected in extracts of the *Mycobacterium* species isolated from isovaleric acid enrichment culture (Hilz *et al.*, 1958).

8. *Tyrosine and phenylalanine*

Guroff and Ito (1963) found that extracts of a pseudomonad catalyzed the oxidation of phenylalanine to tyrosine when grown on either phenylalanine or tyrosine (but not on asparagine). The enzyme which catalyzed the reaction required oxygen, was stimulated by NADH and was probably the bacterial counterpart of mammalian phenylalanine hydroxylase (L-phenylalanine tetrahydropteridine: oxygen oxidoreductase [4-hydroxylating]) which functions with a reduced pteridine as the electron donor. Takashima *et al.* (1964) demonstrated the incorporation of ^{18}O from atmospheric oxygen into tyrosine with an enzyme preparation from *P. aeruginosa.* Suda and Takeda (1950) reported the presence of homogentisate oxygenase (homogentisate: oxygen oxidoreductase) in cell-free extracts of a pseudomonad grown on tyrosine, and Adachi *et al.* (1966) later crystallized and characterized the enzyme from *P. fluorescens.* The bacterial enzyme is very similar to the mammalian enzyme in that ferrous ions are required and the product of homogentisate oxidation is maleylacetoacetic acid.

9. *Tryptophane*

In their early studies of tryptophane metabolism, Stanier and Hayaishi (1951) uncovered evidence for the existence of two pathways of tryptophane metabolism in soil pseudomonads. All isolates oxidized tryptophane and kynurenine, but some oxidized anthranilate while others oxidized kynureninate. The initial reactions in tryptophane metabolism were common to both pathways, but a branch point occurred at the stage of kynurenine, the aromatic route leading to anthranilate and catechol, while the quinoline pathway led to anthranilate and other quinoline derivatives (Fig. 11.8).

Hayaishi and Stanier (1951) prepared cell-free extracts of their tryptophane-metabolizing pseudomonads which catalyzed several reactions of the aromatic pathway. The first reaction was catalyzed by tryptophane pyrrolase (L-tryptophane: oxygen oxidoreductase) and resulted in the production of formylkynurenine from tryptophane. Tryptophane pyrrolase was later purified from *P. fluorescens* by Tanaka and Knox (1959), who showed that it was an iron porphyrin oxygenase similar to the animal enzyme. Hayaishi *et al.* (1957) had earlier demonstrated the incorporation of ^{18}O of molecular oxygen during catalysis by an enzyme preparation which metabolized tryptophane to formate and kynurenine, and found the isotope in both formate and kynurenine. The enzyme which hydrolyzed formylkynurenine, kynurenine formylase (aryl-formylamine amidohydrolase), was demonstrated by Hayaishi and Stanier (1951) in cell-free extracts of their pseudomonad. The next reaction in the aromatic pathway was the cleavage of kynurenine to alanine and anthranilate by kynureninase (L-kynurenine hydrolase) which was purified from *P. fluorescens* by Hayaishi and Stanier (1952). Like the corresponding animal enzyme, bacterial kynureninase was activated by pyridoxal phosphate. Anthranilate is oxidized by pseudomonads by way of catechol, as outlined in Chapter 10.

The first step in the quinoline pathway is the deamination of kynurenine by transamination with 2-oxoglutarate to form kynureninate and L-glutamate (Miller *et al.*, 1953). The oxoacid expected from kynurenine does not accumulate but instead forms the quinoline derivative spontaneously. Subsequent metabolism of kynureninate is partially conjecture, but several groups of investigators have obtained good circumstantial evidence for the reactions suggested in Fig. 11.8. Hayaishi *et al.* (1961) prepared extracts of *Pseudomonas* species which metabolized kynureninate to glutamate, alanine and acetate. Horibata *et al.* (1961) used labeled kynureninate to show that glutamate originated from the benzene ring of kynureninate with carbon 9 the precursor of carbon 1 of glutamate, and that alanine and acetate originated from the heterocyclic ring along with the carboxyl carbon of kynureninate in the case of alanine. Taniuchi *et al.* (1963) found that the initial reaction in the metabolism of kynureninate by cell-free extracts of a pseudomonad required NADH to produce a compound which they tentatively identified as 7,8-dihydro-7,8-dihydroxykynureninate. The next reaction required NAD and they were able to identify the product as 7,8-dihydroxykynureninate. Kuno *et al.* (1961) first

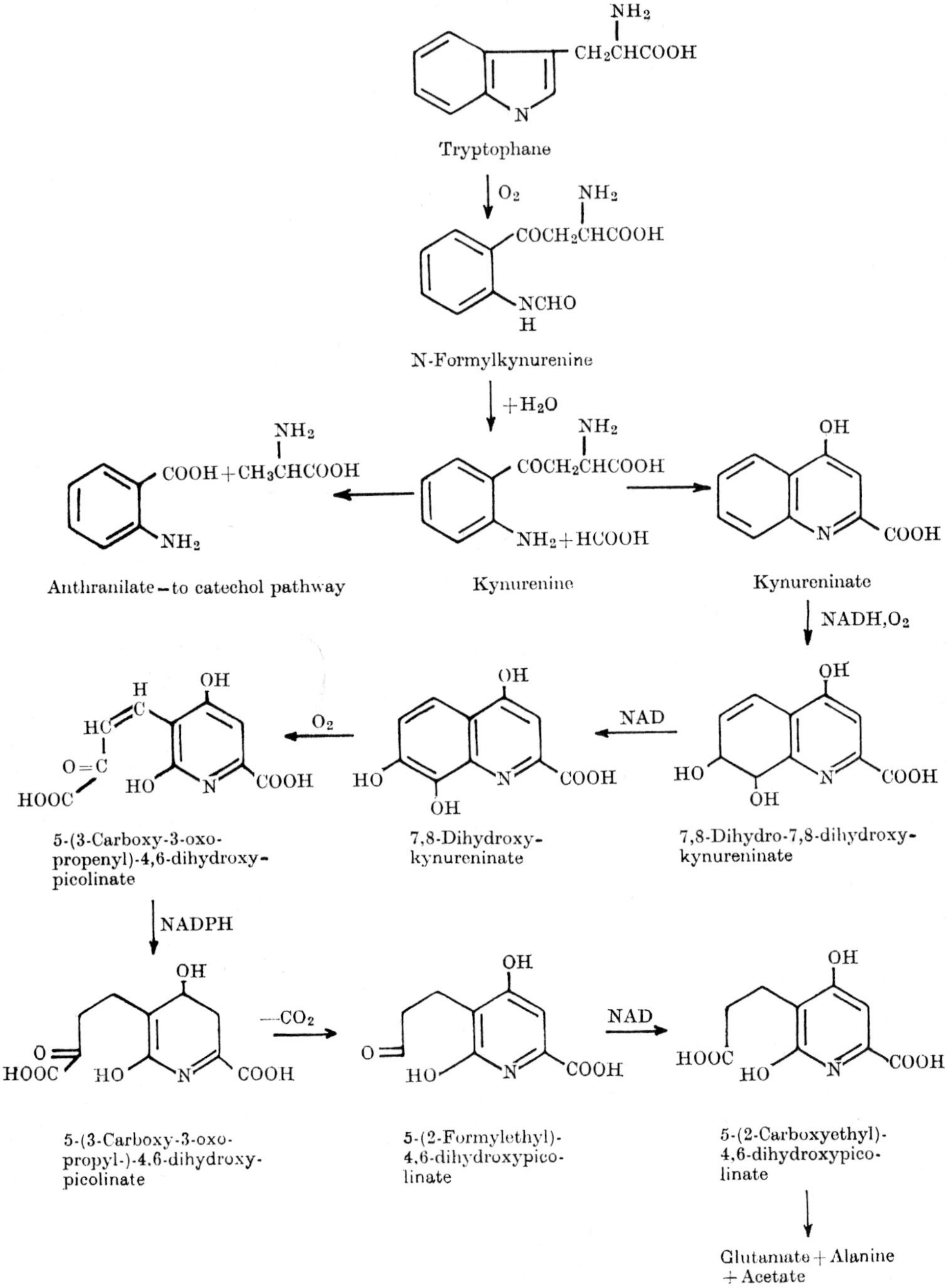

Tryptophane

$\downarrow O_2$

N-Formylkynurenine

$\downarrow +H_2O$

Anthranilate – to catechol pathway ← Kynurenine → Kynureninate

$COOH + CH_3CHCOOH$ (NH_2)

$NH_2 + HCOOH$

Kynureninate $\downarrow NADH, O_2$

7,8-Dihydro-7,8-dihydroxy-kynureninate $\xrightarrow{NAD}$ 7,8-Dihydroxy-kynureninate $\xrightarrow{O_2}$ 5-(3-Carboxy-3-oxo-propenyl)-4,6-dihydroxy-picolinate

$\downarrow NADPH$

5-(3-Carboxy-3-oxo-propyl-)-4,6-dihydroxy-picolinate $\xrightarrow{-CO_2}$ 5-(2-Formylethyl)-4,6-dihydroxypico-linate $\xrightarrow{NAD}$ 5-(2-Carboxyethyl)-4,6-dihydroxypico-linate

$\downarrow$

Glutamate + Alanine + Acetate

FIG. 11.8. Pathways of tryptophane metabolism in pseudomonads.

proposed the remaining steps in the metabolism of kynureninate and reported evidence for the NADPH- and NAD-requiring enzymes. Dagley and Johnson (1963) used cell-free extracts of an *Aerococcus* species to isolate two products of kynureninate metabolism which they partially characterized and which were probably 5-(3-carboxy-3-oxopropenyl)-4,6-dihydroxypicolinate and 5-(2-carboxyethyl)-4, 6-dihydroxypicolinate.

10. *Hydroxyproline*

The metabolism of hydroxyproline by *P. striata* was studied by Adams and his associates and is outlined in Fig. 11.9. The first step in

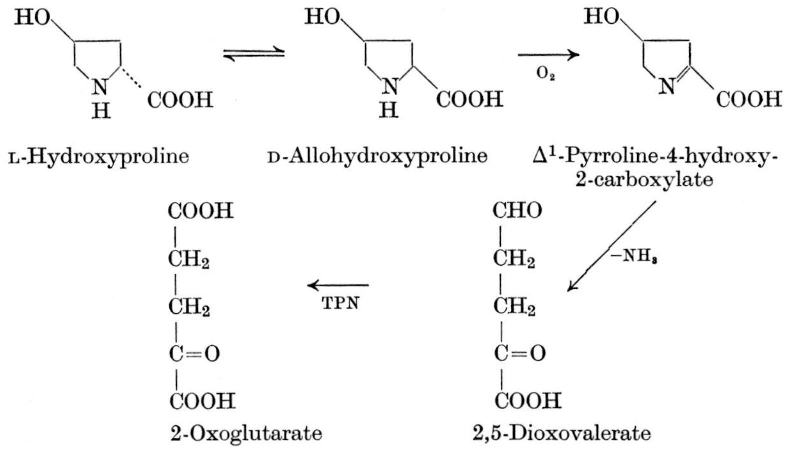

L-Hydroxyproline ⇌ D-Allohydroxyproline → (O_2) Δ^1-Pyrroline-4-hydroxy-2-carboxylate

→ ($-NH_3$) 2,5-Dioxovalerate (CHO–CH_2–CH_2–C=O–COOH) → (TPN) 2-Oxoglutarate (COOH–CH_2–CH_2–C=O–COOH)

FIG. 11.9. L-Hydroxyproline metabolism in *P. striata* (From Singh and Adams, 1965).

L-hydroxyproline metabolism was catalyzed by hydroxyproline-2-epimerase which Adams and Norton (1964) purified to a state of homogeneity. Hydroproline epimerase of *P. striata* catalyzed the following two reactions:

L-hydroxyproline → D-*allo*hydroxyproline

D-hydroxyproline → L-*allo*hydroxyproline

The enzyme was inducible by growth on L-hydroxyproline, had a molecular weight of about 18,000 g/mole, and contained no bound pyridoxal phosphate, pyridine nucleotides or flavine derivatives. The second enzyme of the hydroxyproline pathway is a rather specific D-*allo*hydroxyproline oxidase (Yoneya and Adams, 1961) which catalyzed the oxidation of D-*allo*hydroxyproline to Δ^1-pyrroline-4-hydroxy-2-carboxylate with the consumption of one atom of oxygen per mole of product formed. This enzyme was also induced by growth on L-hydroxyproline and was associated with the particulate fraction of the cell-free preparation. The operation of this enzyme in the hydroxyproline pathway provides a rare example of a proven function for a D-amino acid oxidase, since D-*allo*hydroxyproline is an obligatory intermediate in this pathway. Singh and Adams (1965) purified the third enzyme of the hydroxyproline pathway, Δ^1-pyrroline-4-hydroxy-2-carboxylate deaminase, also to a state close to homogeneity. This enzyme catalyzed the deamination of Δ^1-pyrroline-4-hydroxy-2-carboxylate to 2,5-dioxovalerate (2-oxoglutarate semialdehyde). Like the other enzymes of the pathway, the deaminase was induced by growth on L-hydroxyproline. The final reaction of the hydroxyproline pathway was the oxidation of 2,5-dioxovalerate to 2-oxoglutarate which was catalyzed by an NADP-linked dehydrogenase (Singh and Adams, 1965).

11. *Histidine*

The oxidative metabolism of histidine by *A. aerogenes* and *P. fluorescens* follows the same pathway in both organisms except for the final stages (Fig. 11.10). In both cases the first step is the deamination of histidine by the enzyme histidase (L-histidine ammonia-lyase) to produce urocanic acid (Magasanik and Bowser, 1955; Tabor *et al.*, 1952). Urocanic acid is converted to imidazolonepropionate by the enzyme urocanase (Revel and Magasanik, 1958) by a rather unusual enzymic reaction. Next the imidazolone ring is opened hydrolytically by the enzyme imidazolonepropionase (4-imidazolone-5-propionate hydrolase) with the production of N-formimino-L-glutamate (Magasanik and Bowser, 1955; Tabor and Mehler, 1954). At this point the *Aerobacter* and *Pseudomonas* pathways diverge. *Aerobacter aerogenes* contains N-formimino-L-glutamate formiminohydrolase which catalyzes the hydrolysis of formimino-L-glutamate to L-glutamate and formamide (Lund and Magasanik, 1965). *Aerobacter aerogenes* can oxidize L-glutamate but is unable to metabolize formamide which accumulates in the medium. *Pseudomonas fluorescens* contains an enzyme which hydrolyzes N-formimino-L-glutamate to ammonia and N-formyl-L-glutamate and another hydrolytic enzyme which converts N-formyl-L-glutamate to formate and L-glutamate (Tabor and Mehler, 1954).

12. *Arginine and proline*

De Hauwer *et al.* (1964) used *B. subtilis* grown in a medium with arginine as the sole nitrogen source to study arginine metabolism (Fig.

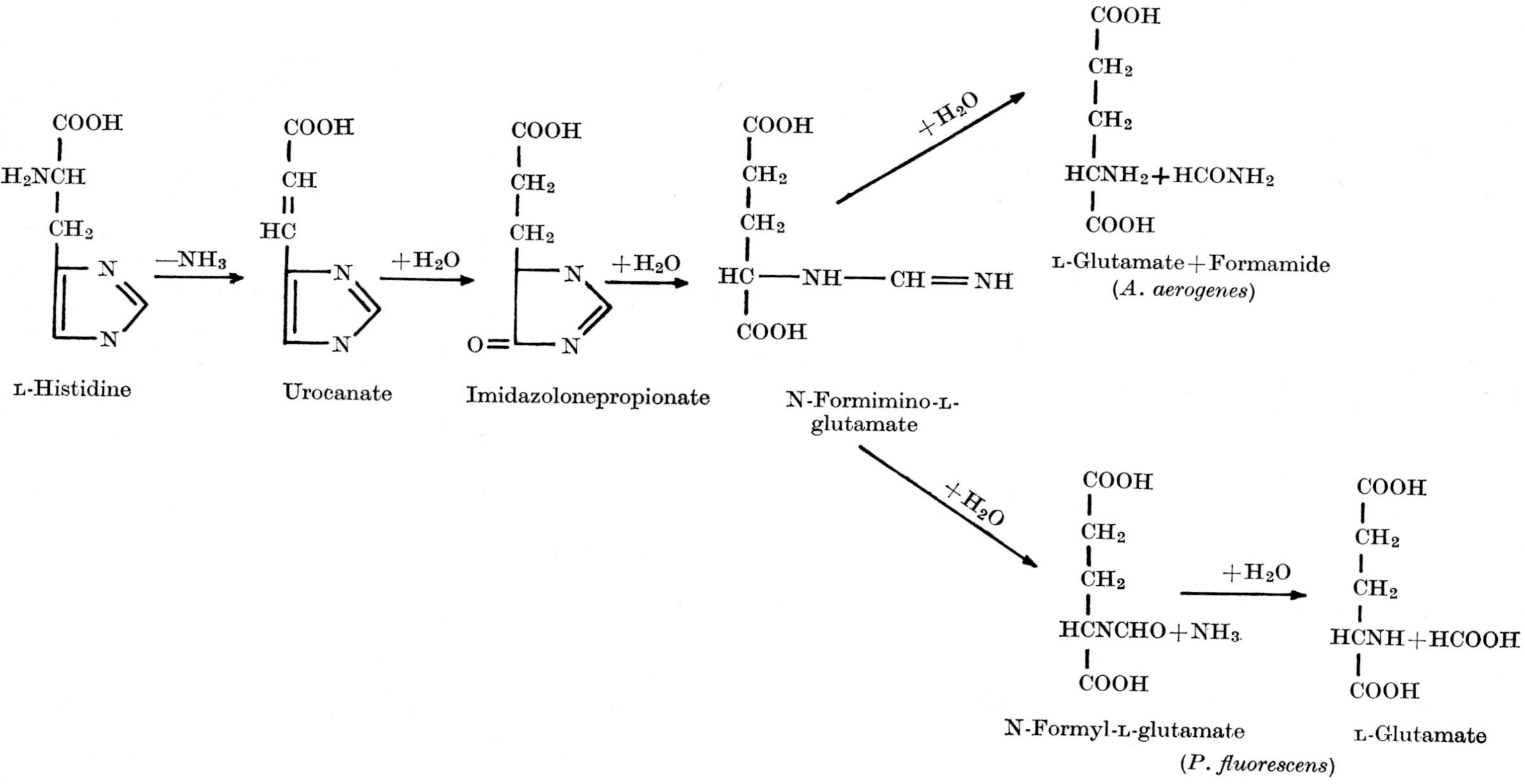

FIG. 11.10. Oxidative metabolism of histidine in *Aerobacter* and *Pseudomonas*.

11.11). Arginase (L-arginine amidinohydrolase) and ornithine transaminase (L-ornithine:2-oxoacid aminotransferase) were constituitive enzymes in their strain of *B. subtilis*, but an NADP-linked Δ^1-pyrroline-5-carboxylate dehydrogenase was formed in response to growth in the

$$\text{Arginine} + H_2O \rightarrow \text{ornithine} + \text{urea}$$
$$\text{Ornithine} + \text{2-oxoglutarate} \rightarrow \Delta^1\text{-pyrroline-5-carboxylate}$$
$$\Delta^1\text{-pyrroline-5-carboxylate} + \text{NADP} \rightarrow \text{glutamate} + \text{NADPH}$$

FIG. 11.11. Arginine metabolism by *B. subtilis*.

presence of arginine as the sole nitrogen source. When proline was supplied as the nitrogen source, *B. subtilis* formed an L-proline oxidase and NADP-linked Δ^1-pyrroline-5-carboxylate dehydrogenase; hence, the arginine and proline pathways converge at the point of Δ^1-pyrroline-5-carboxylate. The dehydrogenases for Δ^1-pyrroline-5-carboxylate were different as judged by column chromatography although they catalyzed the same reaction.

13. *Lysine*

Itada *et al.* (1961) reported the discovery of an L-lysine oxygenase in extracts of a *Pseudomonas* species which catalyzed the oxidative decarboxylation of lysine to 5-aminovalerate and carbon dioxide. They also showed that atmospheric oxygen was incorporated into the product. The enzyme was recently crystallized by Takeda and Hayaishi (1966) who also showed that the immediate product was actually 5-aminovaleramide.

The primary pathway of lysine degradation in animal tissues goes through pipecolate (Fig. 11.12), and this is probably the case with

$$\text{Lysine} \xrightarrow{-NH_3} \text{2-Oxo-6-aminocaproate} \xrightarrow{-2H} \Delta^1\text{-Piperideine-2-carboxylate}$$
$$\xrightarrow{+2H} \text{Pipecolate} \xrightarrow{-2H} \Delta^1\text{-Piperideine-6-carboxylate} \longrightarrow \text{2-Amino-6-oxocaproate}$$
$$\xrightarrow{-2H} \text{2-Aminoadipate} \xrightarrow{-NH_3} \text{2-Oxoadipate} \xrightarrow{-2H,\ -CO_2,\ +\ CoA} \text{Glutaryl-CoA}$$
$$\xrightarrow{-2H} \text{Glutaconyl-CoA} \xrightarrow{-CO_2} \text{Crotonyl-CoA} \xrightarrow{+H_2O} \text{3-Hydroxybutyryl-CoA}$$
$$\xrightarrow{+H_2O} \text{Acetoacetyl-CoA} \xrightarrow{+CoA} \text{2 Acetyl-CoA}$$

FIG. 11.12. Probable reactions in the metabolism of lysine by *Pseudomonas*.

bacteria, although there is very little evidence in existence on the early reactions in bacteria. However, pipecolate metabolism by *Pseudomonas* was studied by Basso *et al.* (1962) and they used cell suspensions of their organism to oxidize pipecolate to an equilibrium mixture of Δ^1-piperideine-6-carboxylate and 2-amino-6-oxocaproate. Calvert and Rodwell (1966) purified an NAD-linked dehydrogenase from this same organism which acted on L-2-amino-6-oxocaproate to produce 2-aminoadipate. Deamination of 2-aminoadipate and oxidative decarboxylation would result in glutaryl-CoA. Numa *et al.* (1964) prepared enzyme extracts of *P. fluorescens* grown on glutarate which catalyzed the oxidation of glutaryl-CoA to carbon dioxide and a mixture of crotonyl-CoA and 3-hydroxybutyryl-CoA. They were able to resolve the enzyme for FAD and to show that FAD was required for catalysis. Glutaconyl-CoA was considered to be an intermediate since it was converted to crotonyl-CoA and 3-hydroxybuturyl-CoA by their enzyme preparations. Presumably, bacterial crotonase was responsible for the hydration of crotonyl-CoA.

14. *Cysteine, homocysteine and methionine*

The decomposition of cysteine and cystine by cell suspensions of *Proteus* species and *E. coli* to pyruvate, ammonia and hydrogen sulfide and of homocysteine to 2-oxobutyrate, ammonia and hydrogen sulfide has been reported frequently (Tarr, 1933; Desnuelle and Fromageot, 1939; Kallio and Porter, 1950; Kallio, 1951). The process is not understood, but it seems to require pyridoxal phosphate, and it is possible that the initial reaction is transamination to yield 3-mercaptopyruvate. Both *E. coli* and *A. aerogenes* have been found to contain an enzyme which causes the decomposition of 3-mercaptopyruvate to pyruvate and elemental sulfur (Meister *et al.*, 1951). When excess thiol is present, sulfur is reduced chemically to hydrogen sulfide with oxidation of the thiol to disulfide.

Another possible pathway of cysteine metabolism based on the assumption that cysteine was first oxidized to cysteine sulfinic acid was studied by Kearney and Singer (1953). In support of this idea they found that cell suspensions of *P. vulgaris* grown on beef infusion agar readily oxidized cysteine sulfinic acid. Cell-free preparations of this organism catalyzed the transamination of L-cysteine sulfinate with 2-oxoglutarate to 3-sulfinylpyruvate and glutamate. Manganese ion caused the non-enzymic decomposition of 3-sulfinylpyruvate to sulfite and pyruvate and, aerobically, of sulfite to sulfate.

Kallio and Larson (1955) used a strain of *Pseudomonas* isolated from a methionine enrichment culture to study methionine degradation. This organism was grown on DL-methionine and cell-free preparations contained a racemase activated by pyridoxal phosphate. One enzyme fraction catalyzed the oxidation of L-methionine to the corresponding 2-oxoacid with the consumption of one atom of oxygen per mole substrate utilized, and apparently without the necessity for added cofactors. However, no further metabolism of "ketomethionine" was demonstrated. Kallio and Larson prepared another enzyme fraction which catalyzed the formation of methylmercaptan, ammonia and 2-oxobutyrate from L-methionine. They demonstrated a requirement for pyridoxal phosphate for this reaction and named the enzyme L-methionine dethiomethylase. 2-Oxobutyrate would be expected to yield propionyl-CoA by the usual route of oxidative decarboxylation.

References

Adachi, K., Iwayama, Y., Tanioka, H. and Takeda, Y. (1966). *Biochem. Biophys. Acta* **118**, 88.

Adams, E. and Norton, I. L. (1964). *J. Biol. Chem.* **239**, 1525.

Ames, G. F. (1964). *Arch. Biochem. Biophys.* **104**, 1.

Ashmarin, I. P. and Vorontsov, I. V. (1963). *Biokhimiya* **28**, 676.

Barker, H. A. (1961). *In* "The Bacteria" (I. C. Gunsalus and R. Y. Stanier, eds.), Academic Press, New York, Vol. II, p. 151.

Barker, H. A., Volcani, B. E. and Cardon, B. P. (1948). *J. Biol. Chem.* **173**, 803.

Barker, H. A., Smyth, R. D., Wawszkiewicz, E. J., Lee, M. N. and Wilson, R. M. (1958a). *Arch. Biochem. Biophys.* **78**, 468.

Barker, H. A., Weissbach, H. and Smyth, R. D. (1958b). *Proc. Natl Acad. U.S.* **44**, 1093.

Barker, H. A., Smyth, R. D., Wilson, R. M. and Weissbach, H. (1959). *J. Biol. Chem.* **234**, 320.

Barker, H. A., Rooze, V., Suzuki, F. and Iodice, A. A. (1964). *J. Biol. Chem.* **239**, 3260.

Basso, L. V., Rao, D. R. and Rodwell, V. W. (1962). *J. Biol. Chem.* **237**, 2239.

Bauchop, T. and Elsden, S. R. (1960). *J. Gen. Microbiol.* **23**, 457.

Blair, A. H. and Barker, H. A. (1966). *J. Biol. Chem.* **241**, 400.

Bleiweis, A. S. and Zimmerman, L. N. (1964). *J. Bacteriol.* **88**, 653.

Brandsaeter, E. (1957). *Iowa State Coll. J. Sci.* **31**, 366.

Braunstein, A. E. (1947). *Advan. Protein Chem.* **3**, 1.

Britten, R. J., Roberts, R. B. and French, E. F. (1955). *Proc. Natl Acad. Sci. U.S.* **41**, 863.

Calvert, A. F. and Rodwell, V. W. (1966). *J. Biol. Chem.* **241**, 409.

Campbell, L. L. (1955). *J. Biol. Chem.* **217**, 669.

Cardon, B. P. and Barker, H. A. (1947). *Arch. Biochem. Biophys.* **12**, 165.

Cattaneo-Lacombe, J., Senez, J. C. and Beaumont, P. (1958). *Biochim. Biophys. Acta* **30**, 458.

Cohen, G. N. and Rickenberg, H. V. (1956). *Ann. Inst. Pasteur* **91**, 693.

Cooper, R. A. and Kornberg, H. L. (1962). *Biochim. Biophys. Acta* **62**, 438.

Dagley, S. and Johnson, P. A. (1963). *Biochim. Biophys. Acta* **78**, 577.

Dagley, S., Trudgill, P. W. and Callely, A. G. (1961). *Biochem. J.* **81**, 623.

David, W. E. and Lichstein, H. C. (1950). *Proc. Soc. Exptl Biol. Med.* **73**, 216.

De Hauwer, G., Lavalle, R. and Wiame, J. M. (1964). *Biochim. Biophys. Acta* **81**, 257.
Desnuelle, P. and Fromageot, C. (1939). *Enzymologia* **6**, 80.
Dupourge, D., Newton, W. A. and Snell, E. E. (1966). *J. Biol. Chem.* **241**, 1233.
Durham, N. N. and Martin, J. (1966). *Biochim. Biophys. Acta* **115**, 260.
Ekladius, L., King, H. K. and Sutton, C. R. (1957). *J. Gen. Microbiol.* **17**, 602.
Elliott, S. D. (1950). *J. Exptl Med.* **92**, 201.
Elliott, W. H. (1960). *Biochem. J.* **74**, 478.
Erlanger, B. F. (1957). *J. Biol. Chem.* **224**, 1073.
Gale, E. F. (1953). *Advan. Protein Chem.* **8**, 287.
Goldman, D. S. (1959). *Biochim. Biophys. Acta* **34**, 527.
Green, M. L. and Elliott, W. H. (1964). *Biochem. J.* **92**, 537.
Greenberg, D. M. (1961). *In* "Metabolic Pathways" (D. M. Greenberg, ed.), Academic Press, New York, Vol. II, p. 80.
Güntelberg, A. V. and Ottesen, M. (1952). *Nature* **170**, 802.
Guroff, G. and Ito, T. (1963). *Biochim. Biophys. Acta* **77**, 159.
Hagihara, B., Matsubara, H., Nakai, M. and Okunuki, K. (1958). *J. Biochem.* (*Tokyo*) **45**, 185.
Hagihara, B. (1960). *In* "The Enzymes" (P. D. Boyer, H. Lardy and K. Myrback, eds.), Academic Press, New York, Vol. 4, p. 193.
Halpern, Y. S. and Lupo, M. (1965). *J. Bacteriol.* **90**, 1288.
Hampson, S. E., Mills, G. L. and Spencer, T. (1963). *Biochim. Biophys. Acta* **73**, 476.
Hayaishi, O. and Stanier, R. Y. (1951). *J. Bacteriol.* **62**, 691.
Hayaishi, O. and Stanier, R. Y. (1952). *J. Biol. Chem.* **195**, 735.
Hayaishi, O., Rothberg, S., Mehler, A. H. and Saito, Y. (1957). *J. Biol. Chem.* **229**, 889.
Hayaishi, O., Taniuchi, H., Tashiro, M. and Kuno, S. (1961). *J. Biol. Chem.* **236**, 2492.
Hilz, H., Knappe, J., Ringelman, E. and Lynen, F. (1958). *Biochem. Z.* **329**, 476.
Holden, J. T. and Holman, J. (1959). *J. Biol. Chem.* **234**, 865.
Horibata, K., Taniuchi, H., Tashiro, M., Kuno, S. and Hayaishi, O. (1961). *J. Biol. Chem.* **236**, 2991.
Itada, N., Ichihara, A., Makita, T., Hayaishi, O., Suda, M. and Sasaki, N. (1961). *J. Biochem.* (*Tokyo*) **50**, 118.
Jones, M. E. and Lipmann, F. (1960). *Proc. Natl Acad. Sci. U.S.* **46**, 1194.
Kaback, H. R. and Stadtman, E. R. (1966). *Proc. Natl Acad. Sci. U.S.* **55**, 920.
Kallio, R. E. (1951). *J. Biol. Chem.* **192**, 371.
Kallio, R. E. and Larson, A. D. (1955). *In* "Amino Acid Metabolism" (W. D. McElroy and H. B. Glass, eds.), The Johns Hopkins Press, Baltimore, p. 616.
Kallio, R. E. and Porter, S. R. (1950). *J. Bacteriol.* **60**, 607.
Kaziro, Y. and Ochoa, S. (1964). *Advan. Enzymol.* **26**, 283.
Kearney, E. B. and Singer, T. P. (1953). *Biochim. Biophys. Acta* **11**, 276.
Kepes, A. and Cohen, G. (1962). *In* "The Bacteria" (I. C. Gunsalus and R. Y. Stanier, eds.), Academic Press, New York, Vol. IV, p. 179.
Kessel, D. and Lubin, M. (1962). *Biochim. Biophys. Acta* **57**, 32.
Kessel, D. and Lubin, M. (1963). *Biochim. Biophys. Acta* **71**, 656.
Kessel, D. and Lubin, M. (1965). *Biochemistry* **4**, 561.
Klein, S. M. and Sagers, R. D. (1962). *J. Bacteriol.* **83**, 121.
Klein, S. M. and Sagers, R. D. (1966). *J. Biol. Chem.* **241**, 197.

Kocholaty, W. and Hoogerheide, J. C. (1938). *Biochem. J.* **32**, 437.
Korzenovsky, M. (1955). *In* "Amino Acid Metabolism" (W. D. McElroy and H. B. Glass, eds.), The Johns Hopkins Press, Baltimore, p. 309.
Kuno, S., Tashiro, M., Taniuchi, H., Horibata, K., Hayaishi, O., Seno, S., Tokuyama, T. and Sakan, T. (1961). *Federation Proc.* **20**, 3.
Labouesse, B. and Gros, P. (1960). *Bull. Soc. Chim. Biol.* **42**, 543.
Labow, R. and Robinson, W. G. (1966). *J. Biol. Chem.* **241**, 1239.
Leach, F. R. and Snell, E. E. (1960). *J. Biol. Chem.* **235**, 3523.
Leaver, F. W., Wood, H. G. and Stjernholm, R. (1955). *J. Bacteriol.* **70**, 521.
Levine, E. M. and Simmonds, S. (1960). *J. Biol. Chem.* **235**, 2902.
Levine, E. M. and Simmonds, S. (1962). *J. Biol. Chem.* **237**, 3718.
Lund, P. and Magasanik, B. (1965). *J. Biol. Chem.* **240**, 4316.
Lynen, F., Knappe, J., Lorch, E., Jütting, G., Ringelmann, E. and Lachance, J. P. (1961). *Biochem. Z.* **335**, 123.
Magasanik, B. and Bowser, H. R. (1955). *J. Biol. Chem.* **213**, 571.
Meister, A., Fraser, P. E. and Tice, S. V. (1954). *J. Biol. Chem.* **206**, 561.
Mandl, I., Ferguson, L. T. and Zaffuto, S. F. (1957). *Arch. Biochem. Biophys.* **69**, 564.
Mandl, I., Keller, S. and Manahan, J. (1964). *Biochemistry* **3**, 1737.
Markovetz, A. J., Cook, W. J. and Larson, A. D. (1966). *Can. J. Microbiol.* **12**, 745.
McConn, J. D., Tsuru, D. and Yasunobu, K. T. (1964). *J. Biol. Chem.* **239**, 3706.
McCormick, N. G. and Halvorson, H. O. (1964). *J. Bacteriol.* **87**, 68.
Meister, A. M. (1965). *In* "Biochemistry of the Amino Acids", Academic Press, New York, Vol. II, p. 729.
Meister, A. M., Sober, H. A. and Trice, S. V. (1951). *J. Biol. Chem.* **189**, 577.
Metzler, D. E. and Snell, E. E. (1952). *J. Biol. Chem.* **198**, 353.
Miller, I. L., Tsuchida, M. and Adelberg, E. A. (1953). *J. Biol. Chem.* **203**, 205.
Millonig, R. C. (1956). *J. Bacteriol.* **72**, 301.
Mills, G. L. and Wilkens, J. M. (1958). *Biochim. Biophys. Acta* **30**, 63.
Mora, J. and Snell, E. E. (1963). *Biochemistry* **2**, 136.
Morihara, K. (1963). *Biochim. Biophys. Acta* **73**, 113.
Neuberger, A. and Tait, G. H. (1962). *Biochem. J.* **84**, 317.
Neumark, R. and Citri, N. (1962). *Biochim. Biophys. Acta* **59**, 749.
Nisman, B. (1954). *Bacteriol. Rev.* **18**, 16.
Norton, J. E. and Sokatch, J. R. (1966). *J. Bacteriol.* **92**, 116.
Norton, J. E., Bulmer, G. S. and Sokatch, J. R. (1963). *Biochim. Biophys. Acta* **78**, 136.
Novogrodsky, A. and Meister, A. M., (1964). *J. Biol. Chem.* **239**, 879.
Numa, S., Ishimura, Y., Nakazawa, T., Okazaki, T. and Hayaishi, O. (1964). *J. Biol. Chem.* **239**, 3915.
O'Brien, R. T. and Campbell, L. L. (1957). *Arch. Biochem. Biophys.* **70**, 432.
O'Connor, R. J. and Halvorson, H. (1960). *Arch. Biochem. Biophys.* **91**, 290.
Petrack, B., Sullivan, L. and Ratner, S. (1957). *Arch. Biochem. Biophys.* **69**, 186.
Pierard, A. and Wiame, J. M. (1960). *Biochim. Biophys. Acta* **37**, 490.
Pollock, M. R. (1962). *In* "The Bacteria" (I. C. Gunsalus and R. Y. Stanier, eds.), Academic Press, New York, Vol. IV, p. 121.
Ravel, J. M., Grona, M. L., Humphreys, J. S. and Shive, W. (1959). *J. Biol. Chem.* **234**, 1452.
Revel, H. R. B. and Magasanik, B. (1958). *J. Biol. Chem.* **233**, 930.

Rilling, H. C. and Coon, M. J. (1960). *J. Biol. Chem.* **235**, 3087.
Robinson, W. G. and Coon, M. J. (1957). *J. Biol. Chem.* **225**, 511.
Sagers, R. D. and Gunsalus, I. C. (1961). *J. Bacteriol.* **81**, 541.
Sanger, F. (1963). *Proc. Chem. Soc.* 76.
Sanger, F. and Shaw, D. C. (1960). *Nature* **187**, 872.
Sanwal, B. D. and Zink, M. W. (1961). *Arch. Biochem. Biophys.* **94**, 430.
Sasaki, S. (1962). *J. Biochem. (Tokyo)* **51**, 335.
Saxena, K. C. and Krishna Murti, C. R. (1963). *Indian J. Chem.* **1**, 530.
Seifter, S., Gallop, P. M., Klein, L. and Mailman, E. (1959). *J. Biol. Chem.* **234**, 285.
Shugart, L. R. and Beck, R. W. (1964). *J. Bacteriol.* **88**, 586.
Singh, R. M. M. and Adams, E. (1965). *J. Biol. Chem.* **240**, 4344.
Slade, H. D. (1955). *In* "Amino Acid Metabolism" (W. D. McElroy and H. B. Glass, eds.), The Johns Hopkins Press, Baltimore, p. 321.
Sokatch, J. R. (1966). *J. Bacteriol.* **92**, 72.
Stadtman, T. C. (1966). *Arch. Biochem. Biophys.* **113**, 9.
Stadtman, T. C. and Elliott, P. (1957). *J. Biol. Chem.* **228**, 983.
Stadtman, T. C., Elliott, P. and Tiemann, L. (1958). *J. Biol. Chem.* **231**, 961.
Stanier, R. Y. and Hayaishi, O. (1951). *Science* **114**, 326.
Stickland, L. H. (1934). *Biochem. J.* **28**, 1746.
Stumpf, P. K. and Green, D. E. (1944). *J. Biol. Chem.* **153**, 387.
Stumpf, P. K. and Green, D. E. (1946). *Federation Proc.* **5**, 157.
Suda, M. and Takeda, Y. (1950). *J. Biochem. (Tokyo)* **37**, 375.
Tabor, H., Mehler, A. H., Hayaishi, O. and White, J. (1952). *J. Biol. Chem.* **196**, 121.
Tabor, H. and Mehler, A. H. (1954). *J. Biol. Chem.* **210**, 559.
Takashima, K., Fujimoto, D. and Tamiya, N. (1964). *J. Biochem. (Tokyo)* **55**, 122.
Takeda, H. and Hayaishi, O. (1966). *J. Biol. Chem.* **241**, 2733.
Tanaka, T. and Knox, W. E. (1959). *J. Biol. Chem.* **234**, 1162.
Taniuchi, H. and Hayaishi, O. (1963). *J. Biol. Chem.* **238**, 283.
Tarr, H. L. A. (1933). *Biochem. J.* **27**, 759.
Tsukada, K. (1966). *J. Biol. Chem.* **241**, 4522.
Vagelos, P. R., Earl, J. M. and Stadtman, E. R. (1959). *J. Biol. Chem.* **234**, 765.
Vender, J., Jayaraman, K. and Rickenberg, H. V. (1965). *J. Bacteriol.* **90**, 1304.
Wachsman, J. T. (1956). *J. Biol. Chem.* **223**, 19.
Wachsman, J. T. and Barker, H. A. (1955). *J. Biol. Chem.* **217**, 695.
Wickremasinghe, R. L. and Fry, B. A. (1954). *Biochem. J.* **58**, 268.
Wieland, T., Griss, G., Haider, K. and Haccius, B. (1960). *Arch. Mikrobiol.* **35**, 415.
Williams, V. R. and McIntyre, R. T. (1955). *J. Biol. Chem.* **217**, 467.
Williamson, W. T., Tove, S. B. and Speck, M. L. (1964). *J. Bacteriol.* **87**, 49.
Wilson, E. M. (1963). *Biochem. Biophys. Acta* **67**, 345.
Wood, W. A. and Gunsalus, I. C. (1949). *J. Biol. Chem.* **181**, 171.
Woods, D. D. (1936). *Biochem. J.* **30**, 1934.
Woods, D. D. and Clifton, C. E. (1938). *Biochem. J.* **32**, 345.
Yoneya, T. and Adams, E. (1961). *J. Biol. Chem.* **236**, 3273.

12. METABOLISM OF INORGANIC COMPOUNDS

I. Hydrogen Oxidation

Hydrogenomonas species are able to grow autotrophically by means of their ability to oxidize hydrogen gas for energy and to reduce carbon dioxide to cell material. The oxidation of hydrogen is accomplished by the hydrogenase system which catalyzes the reaction

$$H_2 \rightarrow 2H^+ + 2e$$

which is the same reaction as that of the hydrogen electrode ($E_0' = -0{\cdot}420$ volt). Providing that electron transport occurs via the cytochrome system with the reduction of oxygen to water, 3 moles of ATP would be expected per mole hydrogen oxidized. Packer (1958) studied the reduction of electron transport enzymes in intact cells of *Hydrogenomonas ruhlandii* by hydrogen and observed that flavoproteins, cytochromes *b* and *c* as well as two carbon monoxide binding pigments were reduced. Thus it appears that the conventional aerobic electron transport system functions in hydrogen oxidation by *Hydrogenomonas*.

The hydrogenase system has been reported in a large number of aerobic and anaerobic bacteria and is usually assayed by measuring the rate of reduction of dyes with hydrogen or by measuring the evolution of hydrogen gas from a chemically reduced dye, in particular, methyl viologen (see review of Fromageot and Senez, 1960). It may also be assayed by measurement of the exchange of H from water with deuterium gas (Hoberman and Rittenberg, 1943) or by the conversion of *para*-hydrogen to *ortho*-hydrogen, a reaction also catalyzed by hydrogenase (Krasna and Rittenberg, 1954). There is considerable circumstantial evidence that hydrogenase is a metallo-flavoprotein.

Hydrogenase of *Hydrogenomonas* species is somewhat different from that of other hydrogenase preparations in that NAD is the specific electron acceptor, a point first established by Packer and Vishniac (1955), with a purified enzyme from *H. ruhlandii*. The enzyme studied by Packer and Vishniac was stimulated by manganese ion, phosphate, cysteine and FMN. Repaske (1962) was able to demonstrate a specific requirement for FMN by hydrogenase of *H. eutropha* and later described a NAD-menadione reductase in *H. eutropha* (Repaske and Lizotte, 1965). Bone (1963) used a purified hydrogenase from *H. ruhlandii* to study the mechanism of catalysis, and on the basis of inhibitor and kinetic studies, came to the conclusion that oxidized enzyme reacted with hydrogen to produce reduced enzyme which subsequently reacted with NAD to produce oxidized enzyme and reduced NAD. The reaction is reversible, and he found it possible to produce hydrogen gas from NADH as did Packer and Vishniac (1955). Bone also proved that the β-isomer of NADH was the product of hydrogenase catalysis.

II. Oxidation of Iron

Temple and Colmer (1951) settled a controversy of long standing when they isolated an iron-oxidizing bacterium, *Thiobacillus ferrooxidans*, which was able to grow in a completely inorganic medium with ferrous iron as the energy source at pH 2·5, which ruled out spontaneous iron oxidation known to occur rapidly at higher pH values. *Thiobacillus ferrooxidans* was able to grow with thiosulfate as the energy source as well, hence the assignment to the genus *Thiobacillus*. Leathen *et al.* (1956) isolated a similar iron-oxidizing bacterium which they designated *Ferrobacillus ferrooxidans*; it was unable to use thiosulfate as an energy source, but was able to use sulfur. Later, Silverman and Lundgren (1959a, b) improved the growth medium for both these organisms in order to obtain larger cell yields and demonstrated that intact cells of *F. ferrooxidans* oxidized iron according to the following equation:

$$4FeSO_4 + O_2 + 2H_2SO_4 \rightarrow 2Fe_2(SO_4)_3 + 2H_2O.$$

The electrode potential of the ferrous-ferric couple, 0·77 V, is very close to that of the oxygen–water couple, 0·81 V, and the energy yield appears to be very small. However, Michaelis and Friedheim (1931) found that the potential of the ferrous–ferric couple could be lowered and even made electronegative by the use of anions such as oxalate and pyrophosphate which form complexes with iron, and it is possible that such complexes might be formed in or near the cell.

On the basis of polarographic evidence, Dugan and Lundgren (1965) postulated that the first step in iron oxidation was the formation of an oxygenated iron complex. Blaylock and Nason (1963) observed that cell-free extracts of *F. ferrooxidans* treated with dithionite exhibited absorption bands characteristic of cytochromes *c*, *b* and *a*; but when ferrous iron was used, only reduced bands of cytochromes *c* and *a* appeared. Blaylock and Nason purified an iron oxidase particle from these preparations which catalyzed the oxidation of ferrous salts with oxygen as the electron acceptor with a pH optimum of 4·9. They were also able to solubilize and purify extensively a component of the particle which catalyzed the reduction of horse heart cytochrome *c* by iron but which was exceedingly heat stable, casting some doubt on the enzymic nature of this component. The iron oxidase also contained cytochrome oxidase which was probably the terminal oxidase of the particle. Again, it seems reasonable to assume that electron transport during iron oxidation is mediated by the cytochrome system and that ATP is produced by oxidative phosphorylation.

III. Oxidation of Ammonia to Nitrate

Nitrification is the process of ammonia oxidation to nitrate by soil bacteria and is performed by organisms of the family Nitrobacteraceae. Hofman and Lees (1953) were the first to provide evidence for hydroxylamine, the only established intermediate in the oxidation of ammonia to nitrite. They showed that whole cells of *Nitrosomonas* species produced hydroxylamine from ammonium ion when poisoned with hydrazine, an inhibitor of hydroxylamine oxidation. Whole cells of *Nitrosomonas* were able to oxidize hydroxylamine to nitrite. Anderson (1964, 1965) studied the further oxidation of hydroxylamine by whole cells and cell-free extracts of *Nitrosomonas* and was able to rule out hydrazine and hyponitrite as possible intermediates. Anaerobically, the cell-free extracts converted hydroxylamine to nitrous oxide and nitric oxide. Nitric oxide was oxidized by the extracts in the presence of ferricyanide and cytochrome *c*, presumably to nitrite. It now seems likely that the observed formation of nitrous oxide is a spontaneous side reaction (Lees, 1960; see also Fig. 12.1).

Falcone *et al.* (1963) prepared a particulate enzyme from *Nitrosomonas europaea* which catalyzed the oxidation of hydroxylamine with various electron acceptors and which produced nitrite aerobically. Spectroscopic studies of their enzyme preparations revealed the appearance of bands of reduced cytochromes *c*, *b* and *a* after the addition of hydroxylamine. Hooper and Nason (1965) purified hydroxylamine-cytochrome *c* reductase from both *Nitrosomonas europaea* and *Nitrocystis oceanus* and found the preparations from both sources to be similar.

Aleem and Alexander (1958) prepared sonic extracts of *Nitrobacter agilis* which catalyzed the oxidation of nitrite to nitrate. Aleem and Nason (1959) showed that the enzyme was particulate and required iron for catalysis, and that nitrite oxidation was accompanied by a reduction of cytochromes *c* and *a* of the particle.

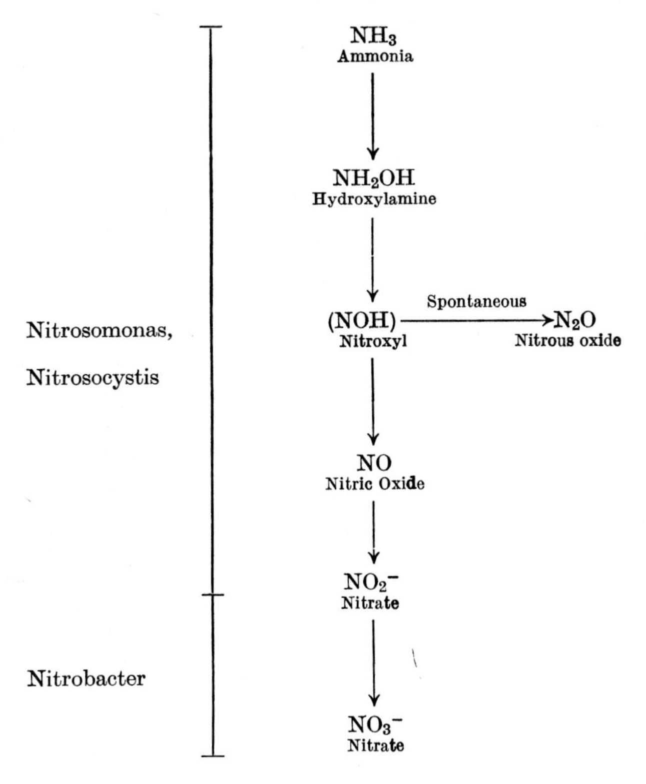

FIG. 12.1. Probable pathway of ammonia oxidation by nitrifying bacteria.

Oxidative phosphorylation has been demonstrated with cell-free preparations of nitrifying bacteria. Burge *et al.* (1963) detected the incorporation of ^{32}P into ATP catalyzed by cell-free preparations of *Nitrosomonas europaea* which were able to oxidize hydroxylamine to nitrite. They were unable to show a dependence of phosphorylation on hydroxylamine oxidation with particulate enzyme preparations, although this was possible with soluble enzyme preparations. Later, Ramaiah and Nicholas (1964) were able to show a clear-cut dependence of phosphorylation on hydroxylamine oxidation with a washed particle preparation from *Nitrosomonas europaea*. Glucose, hexokinase and glucose-6-phosphate dehydrogenase were added in order to trap phosphate incorporated into ADP, and P:2e ratios of 0·16 were obtained by this method.

Aleem and Nason (1960) used a preparation of the particulate nitrite oxidizing system described by them in an earlier publication (Aleem and

Nason, 1959) to demonstrate oxidative phosphorylation by *Nitrobacter agilis*. These investigators used a conventional glucose–hexokinase trap in order to detect ATP formation and were able to demonstrate a dependence of oxidative phosphorylation on nitrite oxidation with P:O values of the order of 0·1–0·2. Malavolta *et al.* (1960) also demonstrated the incorporation of ^{32}P into ATP by cell-free preparations of *Nitrobacter agilis*.

IV. Nitrate Respiration

A. Reduction of Nitrate, Nitrite and Nitric Oxide

Nitrate respiration is typically an anaerobic process and is defined as electron transport with nitrate as the terminal electron acceptor (Nason, 1962). The product of nitrate respiration is usually nitrite, although certain bacteria can reduce nitrite further to produce nitrogen, nitrous oxide (N_2O) and nitric oxide (NO). When nitrogen is reduced to the stage of nitrogen gas or further, the process is referred to as denitrification since these compounds are gases and are lost from the soil. Nitrate respiration is distinct from nitrate assimilation, which results in the reduction of nitrate to ammonia to be used for biosynthesis of amino acids (Chapter 16) and is typically an aerobic process. In any event nitrate reduction seems to follow a pathway the reverse of that illustrated in Fig. 12.1.

Taniguchi and Itagaki (1960) obtained the respiratory nitrate reductase of *E. coli* in a highly purified state by solubilizing the enzyme from a respiratory particle. The level of the respiratory nitrate reductase was considerably elevated as a result of anaerobic growth in the presence of nitrate. The best natural electron donors for nitrate reduction by the intact particle were FMNH, FADH, formate and NADH, while the best artificial electron donors were reduced methyl and benzyl viologen. Addition of formate to the particle under anaerobic conditions resulted in the appearance of bands of reduced cytochrome b_1 which disappeared following the addition of nitrate. Taniguchi and Itagaki further found that nitrate reduction was inhibited by 2-heptyl-4-hydroxyquinoline-N-oxide, which suggested that cytochrome *b* participated in nitrate reduction, and by amytal, which suggested that a flavin enzyme participated in the electron transport chain. Nitrate reductase was eluted from the particle by a brief heat treatment at pH 8·3 followed by a cold incubation for 15–20 hours. The solubilized nitrate reductase did not contain flavin or cytochromes but did contain one mole of molybdenum and about 40 moles of iron per mole enzyme. The solubilized enzyme still utilized reduced methyl viologen as an electron donor, but now formate and NADH were no longer effective.

Cytochrome-linked nitrate reductases have been purified from several other organisms including *A. fisheri* (Sadana and McElroy, 1957), *P. aeruginosa* (Fewson and Nicholas, 1961), *B. cereus* (Hackenthal and Hackenthal, 1965), *N. agilis* (Straat and Nason, 1965) and *B. stearothermophilus* (Downey, 1966).

The second step in nitrate respiration is reduction of nitrite to nitric oxide. Walker and Nicholas (1961) purified the nitrite reductase of *P. aeruginosa* some 600-fold and found that reduced forms of 1,4-napthoquinone, methylene blue, pyocyanine, FAD and FMN were the best electron donors. Bands of reduced cytochrome appeared in the enzyme preparations following reduction. Yamanaka (1964) later discovered that his preparations of crystalline cytochrome oxidase prepared from *P. aeruginosa* rapidly reduced nitrite to nitric oxide with reduced cytochrome c_{551} as the electron donor, and he suggested a dual function for this enzyme in both nitrate and oxygen respiration.

Najjar and Chung (1956) studied both nitrite and nitric oxide reduction by ammonium sulfate fractions of *Pseudomonas stutzeri* and *B. subtilis*. Reduction of either substrate was effected by reduced pyridine nucleotides and stimulated by FAD and FMN. The products of nitrite reduction were nitric oxide and small amounts of nitrogen gas while reduction of nitric oxide resulted in nitrogen gas only. Fewson and Nicholas purified the nitric oxide reductase of *P. aeruginosa* using reduced pyocyanine as the electron donor. Reduced pyridine nucleotides did not serve as electron donors for the purified enzyme, but either FAD or FMN was required for reduction of nitric oxide with reduced pyocyanine.

B. Phosphorylation Coupled to Nitrate Respiration

Takahashi *et al.* (1963) estimated that the amount of energy available in the total oxidation of glucose with nitrate as the electron acceptor is 422 kcal/mole glucose, which compares favorably with 674 kcal/mole glucose oxidized aerobically.

$$C_6H_{12}O_6 + 12KNO_3 \rightarrow 6CO_2 + 6H_2O + 12KNO_2 + 422 \text{ kcal}$$

Yamanaka *et al.* (1962) were the first to report evidence for oxidative phosphorylation coupled to nitrate respiration by cell-free extracts of bacteria. They observed that enzyme preparations of *P. aeruginosa* catalyzed incorporation of ^{32}P into organic phosphate which was dependent on reduction of nitrate to nitrite. With lactate as the electron donor, $P:NO_3^-$ ratios of about 0·3 were obtained. A short time later Ohnishi (1963) reported similar findings with cell-free preparations of

P. denitrificans. Naik and Nicholas (1966) observed oxidative phosphorylation coupled with reduction of nitrate and nitrite by enzyme preparations of *Micrococcus denitrificans*, but not with nitric oxide, nitrous oxide or hydroxylamine. With extracts of *P. denitrificans* only nitrate reduction was accompanied by phosphorylation. Ota (1965) separated three protein fractions from the soluble portion of the cell-free extracts of *E. coli* which were required for oxidative phosphorylation coupled to nitrate respiration. Two of these coupling factors also functioned in phosphorylation coupled to oxygen respiration, and thus it is possible that some enzymic machinery is shared by phosphorylative processes associated with nitrate and oxygen respiration.

V. Oxidation of Sulfur Compounds

Members of the genus *Thiobacillus* grow aerobically with inorganic sulfur compounds as the energy source (see Vishniac and Santer, 1957, for the physiology of these organisms). All members of the genus with the exception of *T. novellus* are strict autotrophs. Most species of *Thiobacillus* grow best with thiosulfate as the energy source, but sulfide, sulfur and thiocyanate can also be used by certain species for growth. *Thiobacillus denitrificans* grows anaerobically with thiosulfate as the energy source and nitrate as the terminal electron acceptor. Photosynthetic bacteria of the families Thiorhodaceae and Chlorobacteriaceae oxidize reduced sulfur compounds to provide electrons for the reduction of carbon dioxide to cell material, and several heterotrophic bacteria are also able to oxidize sulfur compounds. Ironically, *Beggiatoa alba*, the organism used by Winogradsky to formulate the concept of chemolithotrophic nutrition, has never been grown under strictly autotrophic conditions. Much work has been done on the pathway of sulfur oxidation in thiobacilli, but these studies have been complicated by the occurrence of non-enzymic reactions which may or may not have physiological significance.

London and Rittenberg (1964) prepared enzyme extracts of *T. thioparus* and *T. thiooxidans* which catalyzed the oxidation of sulfide, thiosulfate, tetrathionate and trithionate to sulfate. When the reaction was halted before complete oxidation, London and Rittenberg were able to show the transient formation of tetrathionate and trithionate. No cofactors were required for the oxidation; in fact, best results were obtained with extracts treated with charcoal before dialysis. On the basis of these results, they proposed that sulfur oxidation by thiobacilli proceeded as follows:

$$4S^{=} \rightarrow 2S_2O_3^{=} \rightarrow S_4O_6^{=} \rightarrow SO_3^{=} + S_3O_6^{=} \rightarrow 4SO_3^{=} \rightarrow 4SO_4^{=}.$$

The inclusion of the polythionates as intermediates in sulfur oxidation is in accord with earlier observations that these compounds accumulated in culture fluids during growth (Vishniac and Santer, 1957).

On the other hand, Peck (1960) proposed a different pathway of thiosulfate oxidation employing, in part, reactions known to occur in the reduction of sulfate (see Section VI of this chapter), and was able to demonstrate the occurance of the postulated reactions known to occur in the enzyme extracts of *T. thioparus*. Peck suggested that the first step in thiosulfate metabolism was the reduction of thiosulfate to sulfide and sulfite (Fig. 12.2), and was able to demonstrate such a reac-

$$4H^+ + 4e + 2S_2O_3^{2-} \rightarrow 2SO_3^{2-} + 2H_2S$$

$$2H_2S + O_2 \rightarrow 2S + 2H_2O$$

$$2SO_3^{2-} + 2\text{ Adenosine-5'-monophosphate} \rightarrow 2\text{ adenosine-5'-phosphosulfate} + 4e$$

$$2\text{ Adenosine-5'-phosphosulfate} + 2PO_4^{3-} \rightarrow 2\text{ adenosine-5'-diphosphate} + 2SO_4^{2-}$$

$$2\text{ Adenosine-5'-diphosphate} \rightarrow \text{adenosine-5'-monophosphate} + \text{adenosine-5'-triphosphate}$$

$$\text{Sum: } 2S_2O_3^{2-} + O_2 + \text{adenosine-5'-monophosphate} + 2PO_4^{3-} + 4H^+ \rightarrow 2S + 2SO_4^{2-} + \text{adenosine-5'- triphosphate} + 2H_2O$$

FIG. 12.2. Peck's proposal for the metabolism of thiosulfate by thiobacilli.

tion using reduced glutathione as the electron donor. Presumably the more reduced sulfur atom of thiosulfite, the "outer sulfur", becomes sulfide and the more oxidized sulfur becomes sulfite. His enzyme preparations also contained sulfide oxidase which catalyzed the oxidation of hydrogen sulfide to sulfur with atmospheric oxygen. The next postulated step in the process was the oxidation of sulfite in the presence of adenosine-5'-monophosphate to produce adenosine-5'-phosphosulfate (APS). This reaction was actually studied in the reverse direction, i.e. by following sulfite formation using reduced methyl viologen as an electron donor. Sulfate was released from APS by the action of ADP-sulfurylase (ADP: sulfate adenyltransferase) which produced ADP and sulfate from APS and inorganic phosphate. The phosphosulfate linkage of APS is a high energy bond, and the energy can ultimately be obtained in the form of ATP by the action of myokinase (ATP: AMP phosphotransferase) on ADP with the formation of AMP and ATP.

At this time it does not seem possible to reconcile the proposals of London and Rittenberg and of Peck, particularly since the former investigators used cell-free preparations apparently freed of cofactors which are essential to Peck's hypothesis. Research from other laboratories does not help to clarify the situation. Experiments with thiosulfate labeled specifically in either the reduced or oxidized sulfur atom have shown that the oxidized sulfur appears as sulfate much earlier than does the reduced sulfur atom, which tends to favor Peck's hypothesis but does not really exclude the tetrathionate pathway of thiosulfate oxidation (Kelly and Syrett, 1966).

Trudinger (1958) described an enzyme from a *Thiobacillus* species which catalyzed the oxidation of thiosulfate to tetrathionate. This enzyme was also present in *Chromatium* D, a sulfur green photosynthetic organism grown on thiosulfate as the electron donor (Smith, 1966), as well as an enzyme which catalyzed the cleavage of thiosulfate to sulfide and sulfite with lipoic acid as the thiol in an analogous fashion to Peck's first reaction (Smith and Lascelles, 1966). Suzuki and Silver (1966) isolated an iron oxidase from *T. thioparus* which catalyzed the incorporation of atmospheric oxygen into elemental sulfur to produce thiosulfate. The enzyme required catalytic amounts of glutathione and Suzuki and Silver suggested that the initial product of oxidation was actually sulfite which then reacted spontaneously with sulfur to form thiosulfate.

Oxidations used by thiobacilli for growth with their respective yields of free energy are listed in Table 12.1. It is assumed that the aerobic

TABLE 12.1. Change in Free Energy ($\Delta G_0'$) associated with Oxidation or Reduction of Inorganic Compounds by Chemolithotrophic and Heterotrophic Bacteria

Organism	Reaction	$\Delta G_0'$	Ref.
Hydrogenomonas	$H_2 + \frac{1}{2}O_2 \rightarrow H_2O$	−56 kcal	a
Nitrosomonas	$NH_3 + \frac{1}{2}O_2 \rightarrow NH_2OH$	+3·7	b
	$NH_2OH + O_2 \rightarrow HNO_2 + H_2O$	−69·1	b
Nitrobacter	$NO_2^- + \frac{1}{2}O_2 \rightarrow NO_3^-$	−17·5	a
Thiobacillus	$HS^- + 2O_2 \rightarrow SO_4^{2-}$	−171	c
	$S + 3/2O_2 \rightarrow SO_4^{2-} + 2H^+$	−119·9	c
	$S_2O_3^{2-} + 5/2O_2 \rightarrow 2SO_4^{2-}$	−237·6	c
	$3S_2O_3^{2-} + 5NO_3^- \rightarrow 5/2N_2 + 6SO_4^{2-}$	—549	c
Desulfovibrio	$\text{Lactate} + \frac{1}{2}SO_4^{2-} \rightarrow \text{acetate} + CO_2 + \frac{1}{2}S^{2-}$	−24·08	d

References: a = Baas-Becking and Parks (1927); b = Anderson (1965); c = Fromageot and Senez (1960); d = Senez (1962).

thiobacilli obtain energy for growth by oxidative phosphorylation and that *T. denitrificans* uses phosphorylation associated with nitrate respiration. Although neither process has been demonstrated in cell-free extracts of thiobacilli, the presence of cytochromes has been confirmed. Trudinger (1961) observed the appearance of absorption bands of reduced cytochromes with cell suspensions of *Thiobacillus* X and isolated three *c*-type cytochromes from the organism. Cook and Umbreit (1963) demonstrated the occurance of a *c*-type cytochrome in extracts of *T. thiooxidans*, and Aubert *et al.* (1958) isolated a *c*-type cytochrome from *T. denitrificans*. Peck (1962) has also pointed out that one mole of ATP is formed by means of substrate level phosphorylation during thiosulfate oxidation (Fig. 12.2).

VI. Reduction of Sulfate; Sulfate Respiration

Sulfate is used as a terminal electron acceptor by certain anaerobic bacteria in a fashion comparable to the utilization of nitrate. As with nitrate, sulfate is also reduced by bacteria for the purpose of forming sulfur-containing amino acids, a process referred to as assimilatory sulfate reduction and considered in the discussion of amino acid biosynthesis. Sulfate respiration is apparently limited in nature, so far having been reported only in organisms of the genus *Desulfovibrio* and certain Gram negative spore-formers typified by *Clostridium nigrificans*. The sulfate-reducing spore-formers are not typical *Clostridium* species, however, and Campbell and Postgate (1965) proposed the creation of a new genus *Desulfatomaculum* to accommodate these organisms.

Desulfovibrio species are obligate anaerobes and grow with a limited number of organic carbon and energy sources, lactate being the substrate most frequently used. In most cases, sulfate is required as the electron acceptor in the medium and is reduced to sulfide, but this is not so with pyruvate as the energy source (Table 12.1; and see reviews of Postgate, 1959, 1965, for the physiology of these organisms). Lactate and pyruvate are metabolized to acetylphosphate, thus providing some substrate phosphorylation (Sadana, 1954; Millet, 1954). Earlier reports that *Desulfovibrio* was able to grow autotrophically with hydrogen gas as the energy source have been disputed by Postgate (1965) and Mechalas and Rittenberg (1960). The latter authors showed that hydrogen is able to serve as an energy source for *Desulfovibrio* providing suitable carbon compounds are present in the medium for biosynthetic reactions. The existence of hydrogenase in these organisms is well established and the enzyme has been highly purified from cell-free preparations of *D. desulfuricans* (Sadana and Morey, 1961). *Desulfovibrio desulfuricans* contains a single *c*-type cytochrome designated c_3 which has a low E_0' of $-0{\cdot}204$ mV and which contains two moles of haematohaem per mole enzyme. Cytochrome c_3 has been shown to function as an electron donor in sulfate and sulfite reduction and is involved in oxidative phosphorylation by *Desulfovibrio*. Peck (1966) recently succeeded in demonstrating phosphorylation coupled with sulfite reduction in extracts of *Desulfovibrio gigas* with P:2H ratios of 0·1–0·2. *Desulfovibrio* also contains a porphyro-protein, desulfoviridin, whose function has not yet been discovered.

Considerable progress has been made towards an understanding of the pathway of sulfate reduction by *Desulfovibrio* (see review of Peck, 1962). Ishimoto (1959) and Peck (1959) independently discovered that reduction of sulfate to sulfite with hydrogen by cell-free extracts of *Desulfovibrio* required ATP. Sulfate activation by *D. desulfuricans* (Peck, 1959) as well as by yeast during assimilatory sulfate reduction (Robbins and Lipmann, 1958) occurs as a result of the action of ATP-sulfurylase (ATP:sulfate adenyltransferase) which catalyzes the conversion of ATP and sulfate to APS and pyrophosphate (Fig. 12.3).

Lactate oxidation:

$$2\ \text{Lactate} \rightarrow 2\ \text{pyruvate} + 4e$$
$$2\ \text{Pyruvate} + 2PO_4^{3-} \rightarrow 2\ \text{acetylphosphate} + 2CO_2 + 4e$$
$$2\ \text{Acetylphosphate} + AMP \rightarrow 2\ \text{acetate} + ATP$$

$$2\ \text{Lactate} + 2PO_4^{3-} + AMP \rightarrow 2\ \text{acetate} + 2CO_2 + 8e + ATP$$

Sulfate reduction:

$$SO_4^{2-} + ATP \rightarrow \text{adenosine-5'-phosphosulfate} + P_2O_7^{4-}$$
$$P_2O_7^{4-} + H_2O \rightarrow 2PO_4^{3-} + 2H^+$$

$$\text{Adenosine-5'-phosphosulfate} + 2e \rightarrow SO_3^{2-} + AMP$$
$$SO_3^{2-} + 6e + 6H^+ \rightarrow S^{2-} + 3H_2O$$

$$SO_4^{2-} + 8e + ATP + 4H^+ \rightarrow 2PO_4^{3-} + S^{2-} + 2H_2O + AMP$$

Overall reaction:

$$2\ \text{Lactate} + SO_4^{2-} + 4H^+ \rightarrow 2\ \text{acetate} + 2CO_2 + S^{2-} + 2H_2O$$

FIG. 12.3. Balanced reaction for the reduction of sulfate to sulfide by *Desulfovibrio desulfuricans* with lactate as the electron donor.

Akagi and Campbell (1962) purified and characterized ATP-sulfurylase of *D. desulfuricans* and *C. nigrificans*. These same investigators also characterized a pyrophosphatase purified from *D. sulfuricans* (Akagi and Campbell, 1963) which serves to "pull" the ATP-sulfurylase reaction in favor of APS formation. Peck (1959) demonstrated the reduction of APS to sulfite and AMP by enzyme preparations of *Desulfovibrio* and showed that cytochrome c_3 was required for reduction of APS with hydrogen as the electron donor. Ishimoto and Fujimoto (1961) demonstrated the reduction of sulfite to sulfide with hydrogen by extracts of *Desulfovibrio* and proved that cytochrome c_3 also participated in this reaction.

References

Akagi, J. M. and Campbell, L. L. (1962). *J. Bacteriol.* **84**, 1194.

Akagi, J. M. and Campbell, L. L. (1963). *J. Bacteriol.* **86**, 563.

Aleem, M. I. H. and Alexander, M. (1958). *J. Bacteriol.* **76**, 510.

Aleem, M. I. H. and Nason, A. (1959). *Biochem. Biophys. Res. Commun.* **1**, 323.

Aleem, M. I. H. and Nason, A. (1960). *Proc. Natl Acad. Sci. U.S.* **46**, 763.

Anderson, J. H. (1964). *Biochem. J.* **91**, 8.

Anderson, J. H. (1965). *Biochem. J.* **95**, 688.

Aubert, J. P., Milhaud, G., Moncel, C. and Millet, J. (1958). *Compt. Rend.* **246**, 1616.

Baas-Becking, L. G. M. and Parks, G. S. (1927). *Physiol. Rev.* **7**, 85.

Blaylock, B. A. and Nason, A. (1963). *J. Biol. Chem.* **238**, 3453.

Bone, D. H. (1963). *Biochim. Biophys. Acta* **67**, 589.

Burge, W. D., Malavolta, E. and Delwiche, C. C. (1963). *J. Bacteriol.* **85**, 106.

Campbell, L. L. and Postgate, J. R. (1965). *Bacteriol. Rev.* **29**, 359.

Cook, T. M. and Umbreit, W. W. (1963). *Biochemistry* **2**, 194.

Downey, R. J. (1966). *J. Bacteriol.* **91**, 634.

Dugan, P. R. and Lundgren, D. G. (1965). *J. Bacteriol.* **89**, 825.

Falcone, A. B., Shug, A. L. and Nicholas, D. J. D. (1963). *Biochim. Biophys. Acta* **77**, 199.

Fewson, C. A. and Nicholas, D. J. D. (1961). *Biochim. Biophys. Acta* **49**, 335.

Fromageot, C. and Senez, J. C. (1960). *In* "Comparative Biochemistry" (M. Florkin and H. S. Mason, eds.), Academic Press, New York, Vol. I, p. 347.

Hackenthal, E. and Hackenthal, R. (1965). *Biochim. Biophys. Acta* **107**, 189.

Hoberman, H. D. and Rittenberg, D. (1943). *J. Biol. Chem.* **147**, 211.

Hofman, T. and Lees, H. (1953). *Biochem. J.* **54**, 579.

Hooper, A. B. and Nason, A. (1965). *J. Biol. Chem.* **240**, 4044.

Ishimoto, M. (1959). *J. Biochem.* (*Tokyo*) **46**, 105.

Ishimoto, M. and Fujimoto, D. (1961). *J. Biochem.* (*Tokyo*) **50**, 299.

Kelly, D. P. and Syrett, P. J. (1966). *Biochem. J.* **98**, 537.

Krasna, A. I. and Rittenberg, D. (1954). *J. Am. Chem. Soc.* **76**, 3015.

Leathen, W. W., Kinsel, N. A. and Braley, S. A. (1956). *J. Bacteriol.* **72**, 700.

Lees, H. (1960). *Ann. Rev. Microbiol.* **14**, 83.

London, J. and Rittenberg, S. C. (1964). *Proc. Natl Acad. Sci. U.S.* **52**, 1183.

Malavolta, E., Delwiche, C. C. and Burge, W. D. (1960). *Biochem. Biophys. Res. Commun.* **2**, 445.

Mechalas, B. J. and Rittenberg, S. C. (1960). *J. Bacteriol.* **80**, 501.

Michaelis, L. and Friedheim, E. (1931). *J. Biol. Chem.* **91**, 343.

Millet, J. (1954). *Compt. Rend.* **238**, 408.

Naik, M. S. and Nicholas, D. J. D. (1966). *Biochim. Biophys. Acta* **113**, 490.

Najjar, V. A. and Chung, C. W. (1956). *In* "A Symposium on Inorganic Nitrogen Metabolism" (W. D. McElroy and B. Glass, eds.), The Johns Hopkins Press, Baltimore, p. 260.

Nason, A. (1962). *Bacteriol. Rev.* **26**, 16.

Ohnishi, T. (1963). *J. Biochem. (Tokyo)* **53**, 71.

Ota, A. (1965). *J. Biochem. (Tokyo)* **58**, 137.

Packer, L. (1958). *Arch. Biochem. Biophys.* **78**, 54.

Packer, L. and Vishniac, W. (1955). *Biochem. Biophys. Acta* **17**, 153.

Peck, H. D. (1959). *Proc. Natl Acad. Sci. U.S.* **45**, 701.

Peck, H. D. (1960). *Proc. Natl Acad. Sci. U.S.* **46**, 1053.

Peck, H. D. (1962). *Bacteriol. Rev.* **26**, 67.

Peck, H. D. (1966). *Biochem. Biophys. Res. Commun.* **22**, 112.

Postgate, J. R. (1959). *Ann. Rev. Microbiol.* **13**, 505.

Postgate, J. R. (1965). *Bacteriol. Rev.* **29**, 425.

Ramaiah, A. and Nicholas, D. J. D. (1964). *Biochim. Biophys. Acta* **86**, 459.

Repaske, R. (1962). *J. Biol. Chem.* **237**, 1351.

Repaske, R. and Lizotte, C. L. (1965). *J. Biol. Chem.* **240**, 4774.

Robbins, P. W. and Lipmann, F. (1958). *J. Biol. Chem.* **233**, 686.

Sadana, J. C. (1954). *J. Bacteriol.* **67**, 547.

Sadana, J. C. and McElroy, W. D. (1957). *Arch. Biochem. Biophys.* **67**, 16.

Sadana, J. C. and Morey, A. V. (1961). *Biochim. Biophys. Acta* **50**, 153.

Senez, J. C. (1962). *Bacteriol. Rev.* **26**, 95.

Silverman, M. P. and Lundgren, D. G. (1959a). *J. Bacteriol.* **77**, 642.

Silverman, M. P. and Lundgren, D. G. (1959b). *J. Bacteriol.* **78**, 326.

Smith, A. J. (1966). *J. Gen. Microbiol.* **42**, 371.

Smith, A. J. and Lascelles, J. (1966). *J. Gen. Microbiol.* **42**, 357.

Takahashi, H., Taniguchi, S. and Egami, F. (1963). *In* "Comparative Biochemistry" (M. Florkin and H. S. Mason, eds.), Academic Press, New York, Vol. V, p. 91.

Taniguchi, S. and Itagaki, E. (1960). *Biochim. Biophys. Acta* **44**, 263.

Temple, K. L. and Colmer, A. R. (1951). *J. Bacteriol.* **62**, 605.

Straat, P. A. and Nason, A. (1965). *J. Biol. Chem.* **240**, 1412.

Suzuki, I. and Silver, M. (1966). *Biochim. Biophys. Acta* **122**, 22.

Trudinger, P. A. (1958). *Biochim. Biophys. Acta* **30**, 211.

Trudinger, P. A. (1961). *Biochem. J.* **78**, 673.

Vishniac, W. and Santer, M. (1957). *Bacteriol. Rev.* **21**, 195.

Walker, G. C. and Nicholas, D. J. D. (1961). *Biochim. Biophys. Acta* **49**, 350.

Yamanaka, T. (1964). *Nature* **204**, 253.

Yamanaka, T., Ota, A. and Okunuki, K. (1962). *J. Biochem. (Tokyo)* **51**, 253.

13. PHOTOSYNTHETIC ENERGY METABOLISM

There are three important physiological differences between photosynthesis of green plants and bacteria (see also Chapter 1):

1. Bacterial photosynthesis is anaerobic.
2. The reducing agent used by photoautotrophic bacteria is an inorganic compound other than water, usually hydrogen or a sulfur compound, and oxygen is not involved.
3. Although sulfur purple bacteria and sulfur green bacteria use carbon dioxide as the sole carbon source, the non-sulfur purple bacteria typically use an organic compound as the carbon source and are thus photosynthetic heterotrophs.

I. Energy Considerations

Einstein's law of photochemical equivalence states that atoms or molecules which absorb light absorb one quantum of light per molecule. The amount of light energy absorbed by one mole of absorbant is defined as an einstein, and the amount of energy per einstein is inversely proportional to the wavelength. Bacterial photosynthetic pigments absorb light in the region of 800 to 900 mμ (Fig. 13.1), and the energy available from light of this wavelength is of the order of 30–35 kcal/einstein. Green plant photosynthetic pigments absorb light in the region of 600 to 650 mμ where about 40 kcal/einstein are obtained. The energy available to bacteria and plants for photosynthesis is enough to cause excitation of electrons but not ionization. On the other hand, approximately 118 kcal/mole are required for the reduction of carbon dioxide to the oxidation level of formaldehyde, although the total energy demand during photosynthetic carbohydrate formation may be higher, depending on the route. The quantum number, the number of quanta required to provide enough energy for the reduction of one molecule of carbon dioxide to the stage of formaldehyde, has been measured numerous times for both bacterial and plant photosynthesis and is of the order of 8 to 10 quanta in both cases, although values as low as 3 to 4 quanta

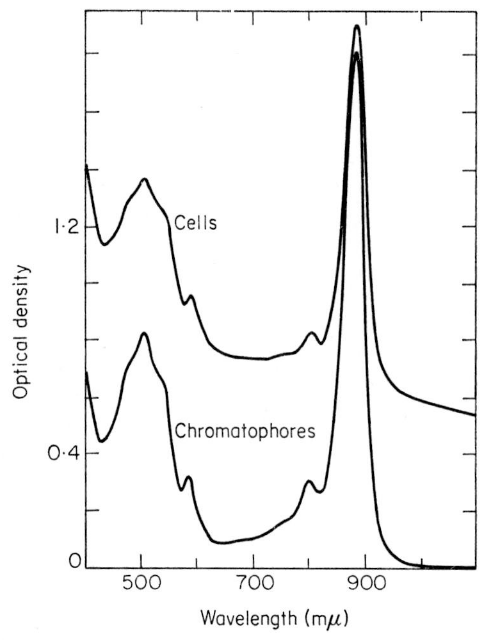

FIG. 13.1. Spectra of a cell suspension and a preparation of chromatophores from *Rhodospirillum rubrum* (from Schachman *et al.*, 1952).

have been reported from Warburg's laboratory (see review of Gaffron, 1960).

II. The Bacterial Photosynthetic Apparatus

Bacteriochlorophyll is the main photosynthetic pigment of the *Thiorhodaceae* and *Athiorhodaceae* and the structure of this tetrapyrrole is shown in Fig. 13.2 (Golden *et al.*, 1958). Two photosynthetic pigments have been extracted from the green sulfur bacterium, *Chlorobium*, and partially characterized and are designated chlorobium chlorophyll (650) and chlorophyll (660) on the basis of the absorption spectrum in ether. Holt and Hughes (1961) showed that the 650 pigment was derived from a homolog of 2-desvinyl-α-hydroxyethylpyropheophorbide *a* and that the 660 pigment was a derivative of δ-methyl-2-desvinyl-2-α-hydroxyethylpyropheophorbide *a* (Holt, *et al.* 1962). Photosynthetic bacteria contain large amounts of carotenoids, lycopene (Fig. 13.3) being the most commonly encountered, although in excess of 30 different carotenoids have been isolated (Jensen, 1963). The main function of carotenoids is to protect the cell against photooxidative damage (Stanier, 1959), with a secondary function as auxilliary light absorbers.

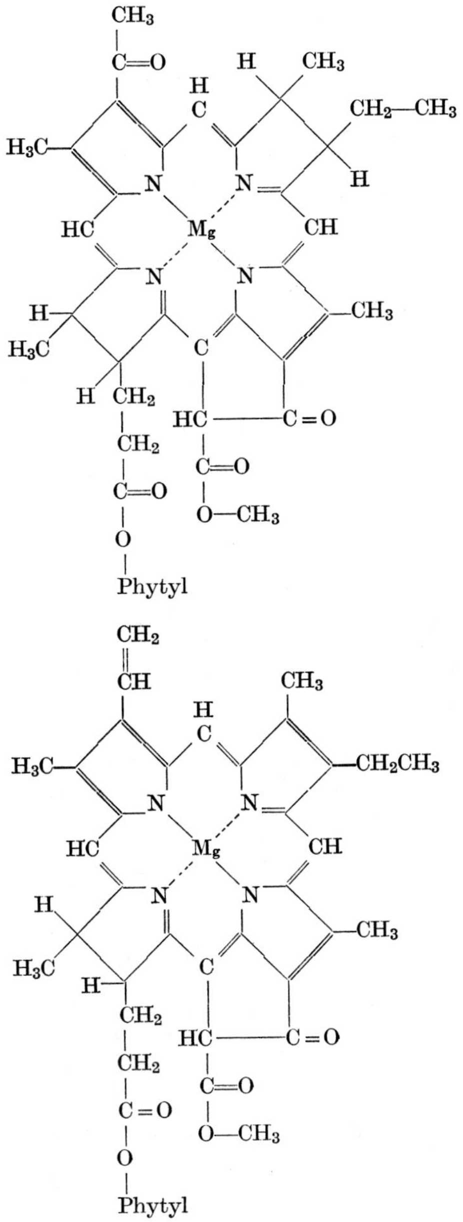

FIG. 13.2. Structure of bacteriochlorophyll (top) and chlorophyll *a* (bottom) (from Lascelles, 1963).

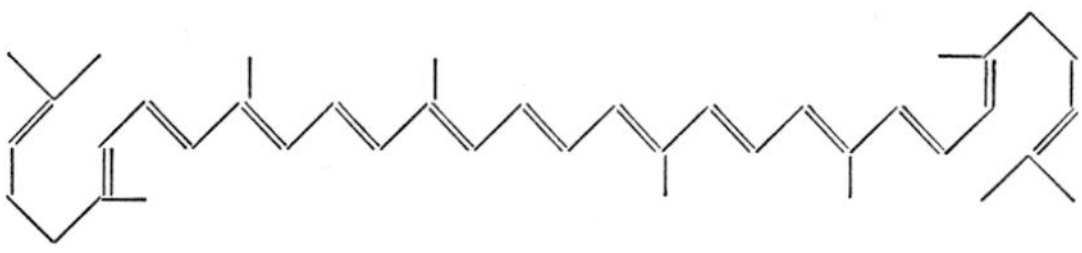

FIG. 13.3. Structure of lycopene.

Several investigators have used photosynthetic bacteria to prepare subcellular particles (termed chromatophores) which are functionally similar to chloroplasts of higher plants (Fuller *et al.*, 1963; Cohen-Bazire, 1963). Chromatophores contain both chlorophyll and carotenoids, but they do not have the high degree of organization of chloroplasts.

III. Photosynthesis and Generation of ATP

In spite of a tremendous amount of effort on the part of several laboratories, the exact details of the conversion of light energy into ATP during photosynthesis are still unknown, although the general outlines of the process have taken shape. Most of the work has been done with higher plants, and we must use this information to supplement our understanding of bacterial photosynthesis. For example, in green plant photosynthesis, the lowest quantum numbers are obtained when red light is supplemented with a small amount of green light, a phenomenon called the "enhancement effect" (see review of Vernon and Avron, 1965). A partial explanation of this phenomenon is that plant cells contain two pigment systems for the absorption of light; the first system absorbs light in the far red region of the spectrum while the second system absorbs in the green region and appears to be involved in the splitting of water to oxygen and reducing power. The enhancement effect does not occur in bacterial photosynthesis where the reducing power is supplied by an electron donor other than water (Blinks and van Niel, 1963). Thus it appears that photosynthetic pigments of bacteria fulfil a function similar to that of the first pigment system of higher plants, namely photosynthetic phosphorylation.

In 1959, Arnon first proposed a general theory for the conversion of light energy into ATP, based on experiments from his own and several other laboratories. A simplified version of his theory which has been modified to accommodate subsequent studies and which excludes the reactions associated with the pigment system involved in the splitting of water appears in Fig. 13.4. The initial event in photosynthetic energy generation by plants and bacteria is the absorption of light energy by the chloroplast or chromatophore with the production of excited

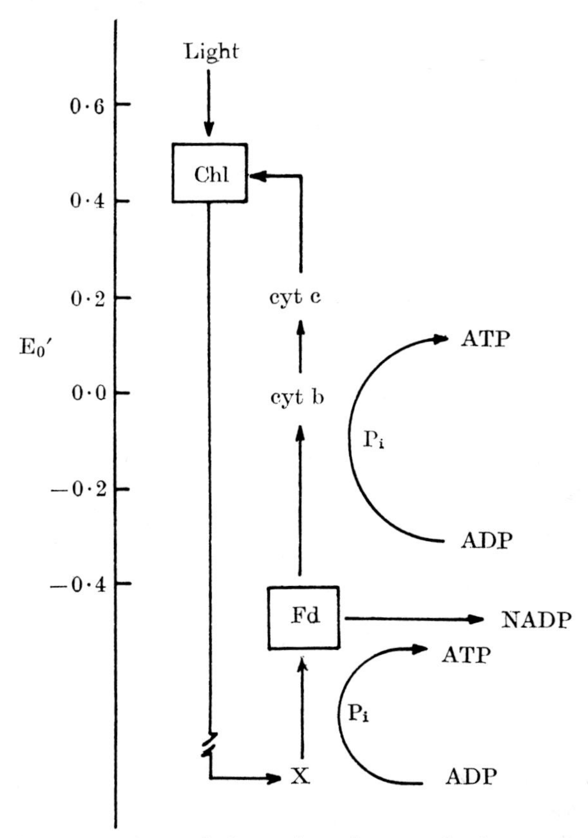

Fig. 13.4. Electron transport in bacteria and green plants as a consequence of light absorption by the primary photosynthetic pigment system. The reactions involved in the splitting of water are not shown in order to emphasize the relationship to bacterial photosynthesis. Abbreviations used are: Chl = chloroplast, Fd = ferredoxin.

electrons. Chloroplasts contain very little water and much of their behavior resembles that of solid semiconductors rather than chemicals in solution (see discussion of Clayton, 1962). Excited electrons, or excitons, can move about freely in a semiconductor creating an area of electron deficiency, or "hole", which is also free to migrate.

Arnon (1965) showed that ferredoxin can act as an electron acceptor for excitons by demonstrating the light-dependent reduction of ferredoxin by plant chloroplasts and enzymes. All species of ferredoxin have a low oxidation–reduction potential, that from *Chromatium* being −0·490 V (Arnon, 1965). Since the potential for chlorophyll, including bacteriochlorophyll, is of the order of +0·45 V, it is apparent that reduction of ferredoxin by chlorophyll is thermodynamically unfavorable and requires energy which is supplied by light.

Arnon discovered that light-dependent reduction of ferredoxin was associated with ATP production (Fig. 13.4), one mole of ATP being produced for every two moles of ferredoxin reduced. The formation of ATP as a consequence of ferredoxin reduction suggests that an electron acceptor of even lower potential than ferredoxin exists. Such an acceptor has been predicted on the basis of thermodynamic considerations and would be expected to have a potential of the order of −0·9 V (Bassham, 1963).

In plant preparations a flavoprotein, ferredoxin:NADP reductase, catalyzes the transfer of electrons from reduced ferredoxin to NADP with the formation of NADPH which is then used for the reduction of carbon dioxide (Arnon, 1965). This process results in an electron deficiency in the photosynthesizing system which plants make up by the photolysis of water to form a reducing agent and an oxidizing agent, the latter being removed as oxygen. In contrast, photoautotrophic bacteria do not depend on photolysis of water for the generation of reducing power, but instead use an exogenous electron donor such as sulfide. Nevertheless, ferredoxin is involved in the generation of reducing power in photosynthetic bacteria. Evans and Buchanan (1965) showed that ferredoxin was required for the light-dependent reduction of carbon dioxide to pyruvate by extracts of *Chlorobium thiosulfatophilum* with sulfide as the electron donor (Chapter 14). In addition, enzymes which catalyzed the transfer of electrons from reduced ferredoxin to NAD and NADP have been demonstrated in both photoautotrophic and photoheterotrophic bacteria (Weaver *et al.*, 1965; Yamanaka and Kamen, 1965). It thus appears that the fundamental difference between green plant and bacterial photosynthesis is the source of the reducing agent; it is derived from the photolysis of water in green plants and is supplied as an exogenous electron donor in bacteria.

Electrons not used for the photoreduction of NADP or NAD are returned to the chloroplast or chromatophore which has been made electron-deficient by loss of excitons. The electron transport system involved is cytochrome-linked and electron transport by this route is accompanied by ATP production, termed cyclic photophosphorylation because of the closed pathway followed by the electrons. Several cytochromes of the *c*-type have been purified from photosynthetic bacteria, cytochromes of the *b*-type have been reported, and a pigmented protein termed Rhodospirillum heme protein (RHP) has been isolated from *R. rubrum* (Newton and Kamen, 1961). Cyclic photophosphorylation was first demonstrated in plant preparations by Arnon *et al.* (1954) and in bacterial preparations by Frenkel (1954).

Non-cyclic photophosphorylation is phosphorylation associated with electron transport which terminates with reduction of ferredoxin or NADP. In the cell, the ultimate electron acceptor would be carbon dioxide. This type of phosphorylation is useful to green plants where large amounts of reducing power are generated by the photochemical process, but not to photoautotrophic bacteria where an inorganic electron donor is used. It appears that the main function of the light reaction in bacteria is to provide ATP by the process of cyclic photophosphorylation.

References

Arnon, D. I. (1959). *Nature* **184**, 10.

Arnon, D. I. (1965). *Science* **149**, 1460.

Arnon, D. I., Allen, M. B. and Whatley, F. B. (1954). *Nature* **174**, 394.

Bassham, J. A. (1963). *Advan. Enzymol.* **25**, 39.

Blinks, L. R. and van Niel, C. B. (1963). *In* "Studies on Microalgae and Photosynthetic Bacteria" (Japanese Society of Plant Physiologists, eds.), University of Tokyo Press, Tokyo, p. 297.

Clayton, R. K. (1962). *Bacteriol. Rev.* **26**, 151.

Cohen-Bazire, G. (1963). *In* "Bacterial Photosynthesis" (H. Gest, A. San Pietro and L. P. Vernon, eds.), The Antioch Press, Yellow Springs, Ohio, p. 89.

Evans, M. C. W. and Buchanan, B. B. (1965). *Proc. Natl Acad. Sci. U.S* **53**, 1420.

Frenkel, A. W. (1954). *J. Am. Chem. Soc.* **76**, 5568.

Fuller, R. C., Conti, S. F. and Mellin, D. B. (1963). *In* "Bacterial Photosynthesis" (H. Gest, A. San Pietro and L. P. Vernon, eds.), The Antioch Press, Yellow Springs, Ohio, p. 71.

Gaffron, H. (1960). *In* "Plant Physiology" (F. C. Steward, ed.), Academic Press, New York, Vol. IB, p. 3.

Golden, J. H., Linstead, R. B. and Whitham, G. H. (1958). *J. Chem. Soc.* 1725.

Holt, A. S. and Hughes, D. W. (1961). *J. Am. Chem. Soc.* **83**, 499.

Holt, A. S., Hughes, D. W., Kende, H. J. and Purdie, J. W. (1962). *J. Am. Chem. Soc.* **84**, 2835.

Jensen, S. L. (1963). *In* "Bacterial Photosynthesis" (H. Gest, A. San Pietro and L. P. Vernon, eds.), The Antioch Press, Yellow Springs, Ohio, p. 19.

Lascelles, J. (1963). *In* "Bacterial Photosynthesis" (H. Gest, A. San Pietro and L. P. Vernon, eds.), The Antioch Press, Yellow Springs, Ohio, p. 35.

Newton, J. W. and Kamen, M. D. (1961). *In* "The Bacteria" (I. C. Gunsalus and R. Y. Stanier, eds.), Academic Press, New York, Vol. II, p. 397.

Schachman, H. K., Pardee, A. B. and Stanier, R. Y. (1952). *Arch. Biochem. Biophys.* **38**, 245.

Stanier, R. Y. (1959). "The Harvey Lectures, 1958–1959", Academic Press, New York, p. 219.

Vernon, L. P. and Avron, M. (1965). *Ann. Rev. Biochem.* **34**, 269.

Weaver, P., Tinker, K. and Valentine, R. C. (1965). *Biochem. Biophys. Res. Commun.* **21**, 195.

Yamanaka, T. and Kamen, M. D. (1965). *Biochem. Biophys. Res. Comm.* **18**, 611.

Part Three

BIOSYNTHETIC METABOLISM

14. AUTOTROPHIC CARBON DIOXIDE FIXATION

I. Production of Reducing Power

A. In Chemoautotrophs

Reduction of one mole of carbon dioxide to cellular carbohydrate requires approximately 118 kcal and the equivalent of two moles of a reduced two-electron carrier, probably of low oxidation-reduction potential such as pyridine nucleotide. The most logical source of reducing power for chemolithoautotrophs would be electrons generated by oxidation of the inorganic energy source. However, reduction of pyridine nucleotide by many of the compounds which serve as energy sources for chemolithotrophs is frequently an endergonic reaction (see the discussion of Gibbs and Schiff, 1960). The necessary energy could be supplied by ATP, and, indeed, Aleem *et al.* (1963) were able to demonstrate that cell-free extracts of *Nitrobacter agilis* catalyzed the reduction of NAD by reduced horse cytochrome c when ATP was added. Aleem made similar observations using enzyme preparations of *Nitrosomonas europaea* (1966a) and *Thiobacillus novellus* (1966b) and was able to couple the reduction of pyridine nucleotide by extracts of *T. novellus* to the oxidation of thiosulfate. The ratio of ATP utilized to pyridine nucleotide reduced was about 5:1 with cell-free preparations of *N. europaea* (Aleem, 1966a). Little is known of the electron carriers involved in this process, but a flavoprotein enzyme was implicated in the reduction of pyridine nucleotide.

B. In Photoautotrophs

There is no evidence for the formation of reducing power by photosynthetic bacteria as a consequence of the photochemical cleavage of water; in fact, the lack of an enhancement effect (Chapter 13) suggests that cleavage of water does not occur in bacterial photosynthesis. For this and other reasons outlined in Chapter 13, we are forced to the conclusion that photoautotrophic bacteria use exogenous electron donors such as hydrogen and sulfur compounds as the ultimate source of electrons for reduction of carbon dioxide. The need for reducing power by photoheterotrophic bacteria is not nearly so great since these organisms assimilate organic compounds at or near the oxidation state of cell material. Evans and Buchanan (1965) were able to demonstrate the reduction of carbon dioxide with sulfide by enzyme preparations of *Chlorobium thiosulfatophilum*. Their reaction mixture contained a chlorophyll-rich particulate fraction prepared from extracts of Chlorobium cells, ferredoxin, sodium sulfide, acetyl-CoA and a purified preparation of pyruvate synthetase, an enzyme which catalyzes the reductive carboxylation of acetyl-CoA:

$$2\ \text{Ferredoxin}_{\text{red}} + \text{carbon dioxide} + \text{acetyl-CoA} \rightarrow 2\ \text{ferredoxin}_{\text{ox}} + \text{pyruvate} + \text{CoA}$$

Pyruvate formation required all the components of the system and was absolutely dependent on light. Under these conditions sodium sulfide acted as the electron donor, ferredoxin as the electron carrier and light energy was used to reduce ferredoxin.

An enzyme similar to plant ferredoxin:NADP reductase which catalyzes the transfer of electrons from reduced ferredoxin to NAD and NADP has been demonstrated in preparations of both photoautotrophic and photoheterotrophic bacteria (Chapter 13, Section 3). Therefore it is reasonable to conclude that electron transport from ferredoxin to pyridine nucleotide and eventually carbon dioxide occurs in bacteria as well as in green plants, the only difference being the ultimate source of electrons (see Fig. 13.4).

II. The Reductive Pentose Cycle

A. In Algae

M. Calvin and his co-workers were primarily responsible for discovery of the reactions involved in, and subsequent to, the fixation of carbon dioxide in green plants. The procedures which these investigators developed are discussed in the monograph of Bassham and Calvin (1957). They exposed living algae very briefly to highly radioactive $^{14}CO_2$ and then killed the cells by squirting them into boiling methanol. Next, the methanol extracts were chromatographed in two dimensions and the finished chromatogram placed in contact with X-ray film. Emission of the β-particles from radioactive intermediates on the paper exposed the film, and the darkened areas were compared with location of intermediates on the paper and identified by comparison with a map prepared by chromatrography of known compounds.

Calvin and his associates identified phosphoglyceric acid as the intermediate which first became radioactive. Other compounds which became labeled quickly were sugar phosphates, principally ribulose-1,5-diphosphate, hexosemonophosphate and sedoheptulose phosphate. Studies of the isotope distribution of sugar phosphates eluted from chromatograms uncovered a distinct relationship to the oxidative pentose cycle. Phosphoglyceric acid was labeled in carbon 1, fructose

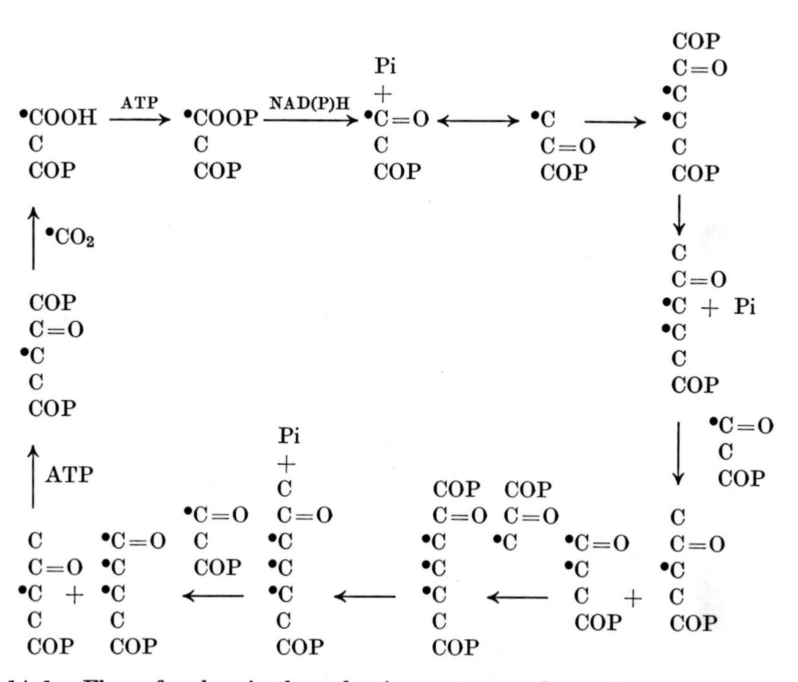

FIG. 14.1. Flow of carbon in the reductive pentose cycle starting with isotopic carbon dioxide. Compare with Fig. 7.1, "Flow of carbon in the oxidative pentose cycle", Fig. 6.11, "Hexose phosphate synthesis from pentose phosphate" and Fig. 6.14, "Hexose phosphate synthesis by *S. faecalis*".

equally in carbons 3 and 4, sedoheptulose equally in carbons 3, 4 and 5 and ribulose equally in carbons 1 and 2, but much more heavily in carbon 3. Calvin proposed a cyclic pathway for regeneration of the acceptor of carbon dioxide which would account for these findings now known as the reductive pentose cycle (Fig. 14.1). In the key reaction, ribulose-1,5-diphosphate acted as the acceptor for carbon dioxide resulting in the formation of two moles of phosphoglyceric acid (Fig. 14.2). Quayle *et al.* (1954) were the first to obtain evidence for this reaction with a cell free system and later Weisbach *et al.* (1956) purified the

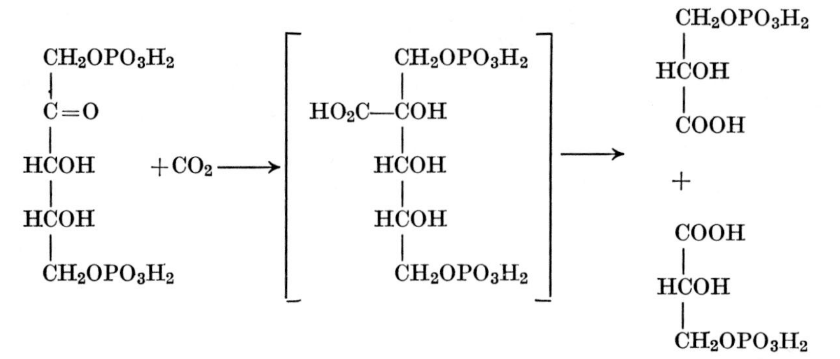

$$\begin{array}{c} CH_2OPO_3H_2 \\ | \\ C{=}O \\ | \\ HCOH \\ | \\ HCOH \\ | \\ CH_2OPO_3H_2 \end{array} + CO_2 \longrightarrow \left[\begin{array}{c} CH_2OPO_3H_2 \\ | \\ HO_2C{-}COH \\ | \\ HCOH \\ | \\ HCOH \\ | \\ CH_2OPO_3H_2 \end{array} \right] \longrightarrow \begin{array}{c} CH_2OPO_3H_2 \\ | \\ HCOH \\ | \\ COOH \\ + \\ COOH \\ | \\ HCOH \\ | \\ CH_2OPO_3H_2 \end{array}$$

D-Ribulose-1,5-diphosphate   3-Phospho-D-glycerate

FIG. 14.2. Carboxylation of ribulose-1,5-diphosphate. The intermediate in brackets has been postulated but is too unstable to be isolated if it exists.

enzyme from spinach leaves and named it ribulose diphosphate carboxylase (3-phospho-D-glycerate carboxy-lyase). The second reaction unique to the reductive pentose cycle is catalyzed by phosphoribulosekinase (ATP:D-ribulose-5-phosphate 1-phosphotransferase) and is the phosphorylation of ribulose-5-phosphate at position 1 to form ribulose 1,5-diphosphate (Hurwitz *et al.*, 1956).

The other reactions of the reductive pentose cycle were catalyzed by enzymes already known. Phosphoglyceric acid formed by the carboxylation of ribulose-1,5-diphosphate was converted to triose phosphate by glycolytic enzymes. Fructose-1,6-diphosphate was formed from triose phosphate by the action of aldolase and was converted to fructose-6-phosphate by hydrolysis of phosphate at carbon 1 catalyzed by hexosediphosphatase (D-fructose-1,6-diphosphate 1-phosphohydrolase). Hexosediphosphatase was originally identified in organ extracts by Gomori (1943). Racker and Schroeder (1958) purified a phosphatase from spinach leaves which was specific for fructose-1,6-diphosphate although another protein fraction contained phosphatases which acted on both sedoheptulose-1,7-diphosphate and fructose-1,6-diphosphate. Carbons 1 and 2 of fructose-6-phosphate were transfered to glyceraldehyde-3-phosphate to yield xylulose-5-phosphate and erythrose-4-phosphate in a reaction catalyzed by transketolase. Bassham and Calvin (1957) proposed that aldolase was responsible for the formation of heptulose phosphate in a manner analogous to the formation of fructose-6-phosphate, by the condensation of erythrose-4-phosphate and dihydroxyacetonephosphate to form sedoheptulose-1,7-diphosphate followed by hydrolysis of the phosphate at carbon 1. A similar result would be obtained by the condensation of carbons 1 to 3 of fructose-6-phosphate with erythrose-4-phosphate catalyzed by transaldolase to yield sedoheptulose-7-phosphate directly as well as glyceraldehyde-3-phosphate. Either sequence would account for the observed labeling patterns of sugar phosphates. In any case, sedoheptulose-7-phosphate and glyceraldehyde-3-phosphate were converted to two molecules of pentose phosphate by a second transketolase reaction, pentose phosphates were converted to ribulose-5-phosphate and finally to ribulose-1,5-diphosphate by the action of phosphoribulokinase.

Balanced reactions for the formation of one mole of hexosemonophosphate from the fixation of 6 moles of carbon dioxide are given in Table 14.1. Although only 18 moles of ATP are consumed, considerable amounts of energy are required for the reduction of pyridine nucleotide, either by light or by ATP in the case of chemolithotrophic organisms.

B. In Chemolithotrophs

When the significance of the reductive pentose cycle in algae was appreciated, several microbiologists turned their attention to chemolithotrophic bacteria to see if this pathway would account for the ability of these organisms to grow autotrophically. Santer and Vishniac (1955) tested cell free extracts of *Thiobacillus thioparus* for their ability to fix $^{14}CO_2$ with ribulose-1,5-diphosphate as the acceptor and found that 3-phosphoglycerate-1-^{14}C was produced. Suzuki and Werkman (1958) reported similar findings with cell-free preparations of *Thiobacillus thiooxidans*. Aubert *et al.* (1957) provided some of the most convincing evidence for the operation of the reductive pentose cycle in thiobacilli. They found that labeling patterns of phosphoglycerate, fructose-6-phosphate, sedoheptulose-7-phosphate and ribulose-1,5-diphosphate formed during brief exposures of *Thiobacillus denitrificans* to radioactive carbon dioxide were virtually identical to those observed by Calvin and his colleagues in their study of algae. Bergmann *et al.* (1958) showed that autotrophically grown *Hydrogenomonas facilis* fixed $^{14}CO_2$ into phosphoglyceric acid, and McFadden and Tu (1965) demonstrated the presence of ribulosediphosphate carboxylase in extracts of this organism. Aleem (1965) verified the existence of ribulosediphosphate carboxylase in *Nitrobacter agilis* and Campbell *et al.* (1966) measured the level of each enzyme of the reductive pentose cycle in *Nitrocystis oceanus*. It is apparent from these studies that the reductive pentose cycle is wide-spread among the chemolithoautrophic bacteria, at least as judged by the distribution of ribulosediphosphate carboxylase (Table 14.2).

TABLE 14.1. Balanced reactions for the synthesis of one mole of hexosemonophosphate from six moles of carbon dioxide by the reductive pentose cycle according to Bassham and Calvin (1957).

Reaction	Enzyme
6 Ribulose-1,5-diphosphate + 6 carbon dioxide → 12 3-phosphoglycerate	Ribulosediphosphate carboxylase
12 3-Phosphoglycerate + 12 ATP → 12 1,3-diphosphoglycerate + 12 ADP	Phosphoglycerate kinase
12 1,3-Diphosphoglycerate + 12 NAD(P)H → 12 glyceraldehyde-3-phosphate + 12P_i + 12 NAD(P)	Glyceraldehydephosphate dehydrogenase
5 Glyceraldehyde-3-phosphate → 5 dihydroxyacetonephosphate	Triosephosphate isomerase
3 Glyceraldehyde-3-phosphate + 3 dihydroxyacetonephosphate → 3 fructose-1, 6-diphosphate	Aldolase
3 Fructose-1, 6-diphosphate → 3 fructose-6-phosphate + 3 P_i	Hexosediphosphatase
2 Fructose-6-phosphate + 2 glyceraldehyde-3-phosphate → 2 xylulose-5-phosphate + 2 erythrose-4-phosphate	Transketolase
2 Erythrose-4-phosphate + 2 dihydroxyacetonephosphate → 2 sedoheptulose-1, 7-diphosphate	Aldolase
2 Sedoheptulose-1, 7-diphosphate → 2 sedoheptulose-7-phosphate + 2P_i	Hexosediphosphatase
2 Sedoheptulose-7-phosphate + 2 glyceraldehyde-3-phosphate → 2 xylulose-5-phosphate + 2 ribose-5-phosphate	Transketolase
2 Ribose-5-phosphate → 2 ribulose-5-phosphate	Phosphoribose isomerase
4 Xylulose-5-phosphate → 4 ribulose-5-phosphate	Ribulose-5-phosphate-3-epimerase
6 Ribulose-5-phosphate + 6 ATP → 6 ribulose-1,5-diphosphate + 6 ADP	Phosphoribulokinase

Sum: 6 Carbon dioxide + 18 ATP + 12 NAD(P)H → fructose-6-phosphate + 17 P_i + 18 ADP + 12 NAD(P)

TABLE 14.2. A partial list of organisms which have been shown to carboxylate ribulose-1,5-diphosphate with the formation of 3-phosphoglycerate or which have been shown to contain ribulosediphosphate carboxylase by enzymic procedures.

Organism	Reference
Chemolithoautotrophs	
Thiobacillus denitrificans	Trudinger (1956)
Thiobacillus thioparus	Santer and Vishniac (1955)
Thiobacillus thiooxidans	Suzuki and Werkman (1958)
Hydrogenomonas facilis	Bergmann *et al.* (1958), McFadden and Tu (1965)
Nitrobacter agilis	Aleem (1965)
Nitrocystis oceanus	Campbell *et al.* (1966)
Photoautotrophs	
Chromatium D	Smillie *et al.* (1962).
Chlorobium thiosulfatophilum	Smillie *et al.* (1962).
Photoheterotrophs	
Rhodopseudomonas sphaeroides	Lascelles (1960)
*Rhodospirillum rubrum**	Glover *et al.* (1952)
*Rhodopseudomonas capsulatus**	Stoppani *et al.* (1955)

Photoheterotrophs designated with * were examined under photoautotrophic conditions.

C. IN *PSEUDOMONAS OXALATICUS* GROWN ON FORMATE

The only known case of operation of the reductive pentose cycle under conditions of what would ordinarily be considered heterotrophic growth was studied by Quayle and Keech (1959a, b) in *Pseudomonas oxalaticus* grown on formate. Using chromatographic techniques similar to those of Calvin, they found that isotope from formate-^{14}C appeared most rapidly in phosphoglyceric acid and other sugar phosphates and that ribulosediphosphate carboxylase was present in the organism when grown on formate, but not on oxalate (see Chapter 8). The organism also contained a particulate formate dehydrogenase which was responsible for the oxidation of formate to carbon dioxide.

D. IN PHOTOSYNTHETIC BACTERIA

There is fair evidence for the operation of the reductive carbon cycle in Athiorhodaceae which are able to grow photoautotrophically. Glover *et al.* (1952) showed that $^{14}CO_2$ fixation by *R. rubrum* with hydrogen as the electron donor resulted in formation of 3-phosphoglycerate. However, when acetate was added, most of the fixed carbon came from acetate and was present in the lipid fraction. Stoppani *et al.* (1955)

studied $^{14}CO_2$ fixation in *Rhodopseudomonas capsulatus* under photoautotrophic conditions and found that labeled sugar phosphates were formed as expected from the operation of the reductive pentose cycle. Lascelles (1960) found that ribulosediphosphate carboxylase was induced in *Rhodopseudomonas sphaeroides* when both light and anaerobic conditions were provided. Smillie *et al.* (1962) detected all the enzymes of the reductive pentose cycle in cell-free extracts of Chromatium D and *Chlorobium thiosulfatophilum* although ribulosediphosphate carboxylase activity was low in the latter organism.

Observations such as that provided by Glover *et al.* (1952) suggested that growth under photoheterotrophic conditions resulted in photoassimilation of the organic substrate in preference to autotrophic carbon dioxide fixation. The role of the organic substrate in metabolism of photoheterotrophs was clarified by Stanier *et al.* (1959; see also review of Stanier, 1961) when they showed that organic substrates were photoassimilated by *R. rubrum* to form storage materials in opposition to the commonly held belief that they served as reducing agents for the reduction of carbon dioxide. Even numbered fatty acids such as acetate and butyrate appeared inside the cell as reserve material in the form of *poly*-β-hydroxybutyrate while propionate, succinate and pyruvate appeared as carbohydrate, probably glycogen or a similar material. It was apparent that the energy for the formation of polymeric reserve materials was provided by photophosphorylation and that the reserve materials could then be used to provide energy or carbon as required.

III. The Reductive Carboxylic Acid Cycle

Evans *et al.* (1966) found that the earliest detectable product of $^{14}CO_2$ fixation by *C. thiosulfatophilum* was glutamate which accounted for about 75% of the fixed carbon after 30 sec. with small amounts of succinate, aspartate and sugar phosphates. They proposed that carbon dioxide fixation in this organism occured by a pathway which was in principle a reversal of the tricarboxylic acid cycle (Fig. 14.3). Since the tricarboxylic acid cycle is an exergonic pathway, energy must be supplied in order to drive the reductive carboxylic acid cycle. The necessary energy is provided in their scheme indirectly in the form of reduced ferredoxin which is formed in a reaction driven by light energy (Section I. B.) and which is of low enough potential that the reduction of carbon dioxide with ferredoxin as the electron donor is thermodynamically feasible.

The first step of the reductive carboxylic cycle is the reductive

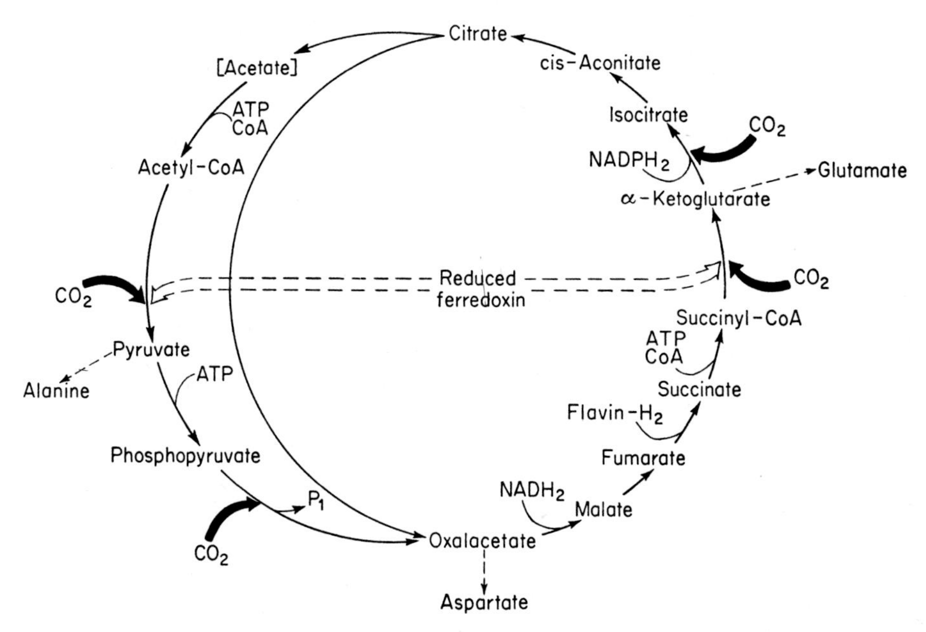

FIG. 14.3. The reductive carboxylic acid cycle (Evans *et al.*, 1966).

carboxylation of acetyl-CoA with reduced ferredoxin as the electron donor and is catalyzed by pyruvate synthetase (Table 14.3). Pyruvate is converted to phosphoenolpyruvate by phosphoenolpyruvate synthetase (Cooper and Kornberg, 1965) with the consumption of two high energy phosphate bonds. The second carbon dioxide fixation step is catalyzed by phosphoenolpyruvate carboxylase (orthophosphate: oxaloacetate carboxy-lyase [phosphorylating]; Bandurski and Greiner, 1953) and results in the formation of oxaloacetate. The conversion of oxaloacetate to succinyl-CoA is catalyzed by standard enzymes of the tricarboxylic acid cycle. The reductive carboxylation of succinyl-CoA to 2-oxoglutarate is a reaction analogous to the pyruvate synthetase reaction and is catalyzed by 2-oxoglutarate synthetase (Buchanan and Evans, 1965). Here again, ability of the enzyme to use reduced ferredoxin overcomes a thermodynamic problem since the corresponding reaction of the tricarboxylic acid pathway is the oxidative decarboxylation of 2-oxoglutarate which is non-reversible for all practical purposes. The pathway from 2-oxoglutarate to citrate again is by conventional tricarboxylic acid cycle reactions. Citritase catalyzes the cleavage of citrate to acetate and oxaloacetate, and the formation of the latter compound completes the inner circle of the reductive car-

TABLE 14.3. Balanced reactions for the synthesis of one mole of oxaloacetate from four moles of carbon dioxide by the reductive carboxylic acid cycle.

Reaction	Enzyme
Acetyl-CoA + carbon dioxide + 2 ferredoxin$_{red}$ → pyruvate + CoA + 2 ferredoxin$_{ox}$	Pyruvate synthetase
Pyruvate + ATP → phosphoenolpyruvate + AMP + P_i	Phosphoenolpyruvate synthetase
Phosphoenolpyruvate + carbon dioxide → oxaloacetate + P_i	Phosphoenolpyruvate carboxylase
Oxaloacetate + NADH → malate + NAD	Malate dehydrogenase
Malate → fumarate + H_2O	Fumarase
Fumarate + flavin$_{red}$ → succinate + flavin$_{ox}$	Succinate dehydrogenase
Succinate + ATP + CoA → succinyl-CoA + ADP + P_i	Succinate thiokinase
Succinyl-CoA + carbon dioxide + 2 ferredoxin$_{red}$ → 2-oxoglutarate + 2 ferredoxin$_{ox}$ + CoA	2-Oxoglutarate synthetase
2-Oxoglutarate + NADPH + carbon dioxide → isocitrate + NADP	Isocitrate dehydrogenase
Isocitrate → citrate	Aconitase
Citrate → acetate + oxaloacetate	Citritase
Acetate + ATP + CoA → acetyl-CoA + AMP + PP_i	Acetyl-CoA synthetase

Sum: 4 Carbon dioxide + 3 ATP + NADH + NADPH + flavin$_{red}$ + 4 ferredoxin$_{red}$ → oxaloacetate + 2 AMP + ADP + NAD + NADP + flavin$_{ox}$ + 4 ferredoxin$_{ox}$ + 3 P_i + PP_i

It is assumed that ferredoxin acts as a one-electron carrier.

boxylic acid cycle. Acetate is activated to acetyl-CoA by acetyl-CoA synthetase (acetate:CoA ligase [AMP]):

$$\text{Acetate} + \text{CoA} + \text{ATP} \rightarrow \text{acetyl-CoA} + \text{AMP} + \text{pyrophosphate}$$

which completes the outer circle of the cycle. Evans *et al.* (1966) detected all of the required enzymes in extracts of *C. thiosulfatophilum* most of which were present in amounts high enough to account for the overall rate of carbon dioxide fixation by whole cells.

There is a distinct possibility that this pathway occurs in other photosynthetic bacteria either as the sole pathway or in conjunction with the reductive pentose cycle. Radioactive glutamate has frequently been observed as one of the early products of carbon dioxide fixation and pyruvate synthetase has been found in anaerobic sporeformers (Chapter 16).

References

Aleem, M. I. H. (1965). *Biochim. Biophys. Acta.* **107**, 14.

Aleem, M. I. H. (1966a). *Biochim. Biophys. Acta.* **113**, 216.

Aleem, M. I. H. (1966b). *J. Bacteriol.* **91**, 729.

Aleem, M. I. H., Lees, H. and Nicholas, D. J. D. (1963). *Nature* **200**, 759.

Aubert, J. P., Milhaud, G. and Millet, J. (1957). *Ann. inst. Pasteur* **92**, 515.

Bandurski, R. S. and Greiner, C. M. (1953). *J. Biol. Chem.* **204**, 781.

Bassham, J. A. and Calvin, M. (1957). "The Path of Carbon in Photosynthesis." Prentice-Hall, Englewood Cliffs, N.J.

Bergmann, F. H., Towne, J. C. and Burris, R. H. (1958). *J. Biol. Chem.* **230**, 13.

Buchanan, B. B. and Evans, M. C. W. (1965). *Proc. Natl. Acad. Sci. U.S.* **54**, 1212.

Campbell, A. E., Hellebust, J. A. and Watson, S. W. (1966). *J. Bacteriol.* **91**, 1178

Cooper, R. A. and Kornberg, H. L. (1965). *Biochim. Biophys. Acta.* **104**, 618.

Evans, M. C. W. and Buchanan, B. B. (1965). *Proc. Natl. Acad. Sci. U.S.* **53**, 1420.

Evans, M. C. W., Buchanan, B. B. and Arnon, D. I. (1966). *Proc. Natl. Acad. Sci. U.S.* **55**, 928.

Gibbs, M. and Schiff, J. A. (1960). *In* "Plant Physiology" (F. C. Steward, ed.). Academic Press, New York, Vol. IB., p. 279.

Glover, J., Kamen, M. D. and Van Genderen, H. (1952). *Arch. Biochem. Biophys.* **35**, 384.

Gomori, G. (1943). *J. Biol. Chem.* **148**, 139.

Hurwitz, J., Weissbach, A., Horecker, B. L. and Smyrniotis, P. Z. (1956). *J. Biol. Chem.* **218**, 769.

Lascelles, J. (1960). *J. Gen. Microbiol.* **23**, 499.

McFadden, B. A. and Tu, Chang-chu L., (1965). *Biochem. Biophys. Res. Commun.* **19**, 728.

Quayle, J. R., Fuller, R. C., Benson, A. A. and Calvin, M. (1954). *J. Am. Chem. Soc.* **76**, 3610.

Quayle, J. R. and Keech, D. B. (1959a). *Biochem. J.* **72**, 623.

Quayle, D. B. and Keech, D. B. (1959b). *Biochem. J.* **72**, 631.

Racker, E. and Schroeder, E. A. R. (1958). *Arch. Biochem. Biophys.* **74**, 326.

Santer, M. and Vishniac, W. (1955). *Biochim. Biophys. Acta.* **18**, 157.
Smillie, R. M., Rigopoulos, N. and Kelly, H. (1962). *Biochim. Biophys. Acta.* **56**, 612.
Stanier, R. Y. (1961). *Bacteriol. Rev.* **25**, 1.
Stanier, R. Y., Doudoroff, M., Kunisawa, R. and Contopoulov, R. (1959). *Proc. Natl. Acad. Sci. U.S.* **45**, 1246.
Stoppani, A. O. M., Fuller, R. C. and Calvin, M. (1955). *J. Bacteriol.* **69**, 491.
Suzuki, I. and Werkman, C. H. (1958). *Arch. Biochem. Biophys.* **77**, 112.
Trudinger, P. A. (1956). *Biochem. J.* **64**, 274.
Weissbach, A., Horecker, B. L. and Hurwitz, J. (1956). *J. Biol. Chem.* **218**, 795.

15. CARBOHYDRATE BIOSYNTHESIS

I. From Acetate

Many bacteria have the ability to grow in media with acetate as the sole carbon source, and it follows that carbohydrate as well as other cellular components are synthesized from acetate. Assimilation of acetate has been studied mainly in the enteric bacteria and pseudomonads, and these studies have been summarized by Kornberg and Elsden (1961) as well as related studies dealing with the assimilation of other two-carbon compounds. Bagatell *et al.* (1959) found that *E. coli* formed glucose-3,4-^{14}C when grown with acetate-1-^{14}C; and it seems probable that bacteria utilize a pathway similar to that used by animals in gluconeogenesis (Krebs, 1964).

The scheme for the biosynthesis of hexose from acetate as it occurs in bacteria is shown in Fig. 15.1. The first step in the assimilation of acetate is the formation of acetyl-CoA either by the acetokinase-phosphotransacetylase pathway (Chapter 8) or by the action of acetyl-CoA synthetase which has been reported in a few species of bacteria. Acetate is incorporated into citrate and isocitrate by tricarboxylic acid cycle reactions. The key reaction in the assimilation of acetate is catalyzed by isocitritase (*threo*-D_S-isocitrate glyoxylate lyase), an enzyme first reported by Campbell *et al.* (1953), which effects an aldolase-type cleavage of isocitrate to glyoxylate and succinate (Fig. 15.2). Smith and Gunsalus (1957) characterized the enzyme and the reaction and showed that the true substrate was *threo*-D_S-isocitrate, the same isomer utilized by isocitrate dehydrogenase. Isocitritase has been found in many species of bacteria, yeast and fungi, but not in higher animals (Olson, 1961). Another reaction unique to assimilation of acetate by microorganisms is the condensation of glyoxylate and acetyl-CoA to malate catalyzed by malate synthetase (L-malate glyoxylate-lyase [CoA-acetylating]) which was discovered by Wong and Ajl (1956) using extracts of acetate-grown *E. coli*. The significance of isocitritase and malate synthetase to bacteria during growth on acetate was first appreciated by Kornberg and Krebs (1957) who suggested that the

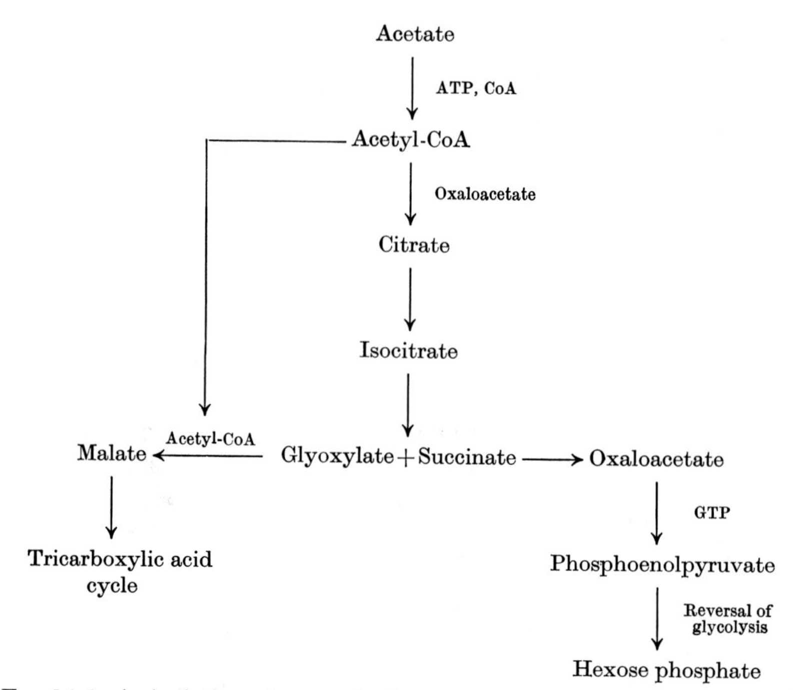

FIG. 15.1. Assimilation of acetate by bacteria leading to the formation of hexose.

"glyoxylate shunt" of the tricarboxylic acid cycle in bacteria served the purpose of replenishing the supply of four-carbon compounds, in particular oxaloacetate, drained away to provide intermediates for carbohydrate and amino acid biosynthesis. It is apparent from Fig. 15.1 that one mole of glyoxylate will remain for every mole of succinate withdrawn and that condensation of glyoxylate with acetyl-CoA will form another four-carbon compound to take the place of the one assimilated. Such a pathway has little value to higher animals, because they obtain amino acids from their diet and, therefore, has been lost during evolution

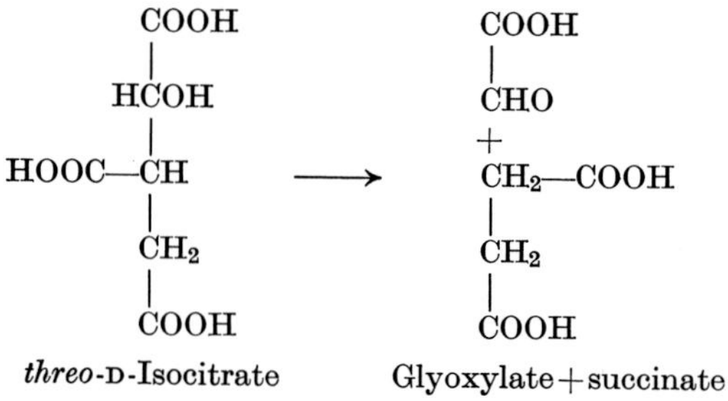

$$\begin{array}{c} COOH \\ | \\ HCOH \\ | \\ HOOC\text{—}CH \\ | \\ CH_2 \\ | \\ COOH \end{array} \longrightarrow \begin{array}{l} COOH \\ | \\ CHO \\ + \\ CH_2\text{—}COOH \\ | \\ CH_2 \\ | \\ COOH \end{array}$$

threo-D-Isocitrate Glyoxylate + succinate

FIG. 15.2. Cleavage of *threo*-D-isocitrate by isocitritase.

Succinate formed in the isocitritase reaction can be converted to phosphoenolpyruvate for hexose synthesis by way of oxaloacetate. Utter and Kurahashi (1954) described an enzyme from chicken liver, phosphoenolpyruvate carboxykinase (GTP: oxaloacetate carboxy-lyase [transphosphorylating]) which catalyzed the reversible formation of phosphoenolpyruvate from oxaloacetate with transfer of phosphate from GTP:

$$\text{Oxaloacetate} + \text{GTP} \rightarrow \text{phosphoenolpyruvate} + \text{carbon dioxide} + \text{GDP}$$

The free energy change in this reaction is small and overcomes a thermodynamic problem in the biosynthesis of hexose from two and three carbon precursers since the formation of phosphoenolpyruvate by the pyruvate kinase reaction is not practical because the equilibrium of the reaction greatly favors ATP formation. Phosphoenolpyruvate carboxykinase is known to participate in gluconeogenesis in animal tissues (Krebs, 1964) and is assumed to do so in the bacterial process although relatively little is known of its distribution in bacteria (Wood and Stjernholm, 1962). Once phosphoenolpyruvate is formed, hexose can be synthesized by reversal of glycolysis as outlined in Chapter 14 in the discussion of the reductive pentose cycle.

II. Formation of Monosaccharides

A. Hexoses and Pentoses

Glucose-6-phosphate is converted to fructose-6-phosphate and mannose-6-phosphate by phosphoglucose isomerase and phosphomannose isomerase respectively and to UDP-galactose by UDP-glucose-4-epimerase in bacteria (see Fig. 6.2).

Ribose formation in bacteria has been studied in a number of instances by isolation of RNA-ribose after growth on ^{14}C-carbon sources (see review of Sable, 1966). The labeling patterns of ribose obtained in these studies are consistant with the interpretation that two pathways for the formation of ribose from glucose-6-phosphate operate in bacteria; (a) oxidative decarboxylation of glucose-6-phosphate to pentose phosphate and (b) conversion of hexose phosphate to pentose phosphate by way of the pentose phosphate pathway, that is, by reversal of the reactions shown in Fig. 6.11. In *E. coli* both the oxidative pathway and the pentose phosphate pathway participate in ribose formation although the relative participation depends on the conditions of growth (Bernstein, 1956; Bagatell *et al.* 1959). In *P. saccharophila* virtually all the pentose is synthesized by the pentose phosphate pathway (Fossitt and Bernstein, 1963). In *S. faecalis* ribose is synthesized by both the oxidative pathway and the version of the pentose phosphate pathway found in this organism, where aldolase functions in place of transaldolase (Sokatch, 1960).

B. Formation of Hexosamines

Glucosamine-6-phosphate is the central intermediate in the formation of other hexosamines and its formation is catalyzed by L-glutamine: D-fructose-6-phosphate aminotransferase:

$$\text{D-Fructose-6-phosphate} + \text{L-glutamine} \rightarrow \text{D-glucosamine-6-phosphate} + \text{L-glutamate}$$

Ghosh *et al.* (1960) purified the enzyme from *E. coli*, rat liver and *Neurospora crassa* and found that all preparations were specific in their requirement for fructose-6-phosphate.

Kornfield and Glaser (1962) found two routes for the conversion of glucosamine-6-phosphate to nucleotide diphosphate *N*-acetylhexosamines in *P. aeruginosa*, one leading to the formation of deoxythymidine diphosphate derivatives and the other leading to the formation of uridine diphosphate derivatives (Fig. 15.3). Extracts of their organisms contained a mutase which catalyzed the formation of glucosamine-1-phosphate from glucosamine-6-phosphate. dTDPglucosamine pyrophosphorylase catalyzed the formation of dTDPglucosamine from dTTP and glucosamine-1-phosphate. This enzyme is an example of a nucleoside diphosphate-sugar pyrophosphorylase, enzymes which catalyze the following general reaction:

$$\text{XTP} + \text{sugar-1-phosphate} \rightarrow \text{XDPsugar} + \text{pyrophosphate}$$

These enzymes are important in biosynthetic pathways because they are responsible for activation of sugar phosphates prior to formation of glycoside bonds. Acetylation of hexosamine by acetyl-CoA in the dTDP pathway occurred at the stage of dTDP-glucosamine and was effected by TDPglucosamine acetylase. One enzyme fraction from *P. aeruginosa* catalyzed the epimerization of dTDP-*N*-acetylglucosamine to dTDP-*N*-acetylgalactosamine. In the UDP pathway, acetylation occurred at the stage of glucosamine-1-phosphate and the resulting *N*-acetylglucosamine-1-phosphate was converted to UDP-*N*-acetylglucosamine with UTP by an unfractionated extract of *P. aeruginosa*. Enzyme preparations of *P. aeruginosa* also catalyzed the epimerization of UDP-*N*-acetylglucosamine to UDP-*N*-acetylgalactosamine.

Ghosh and Roseman (1965a) isolated a strain of *Aerobacter cloacae* from an enrichment culture with *N*-acetylmannosamine as the carbon source and found that the organism contained an inducible enzyme

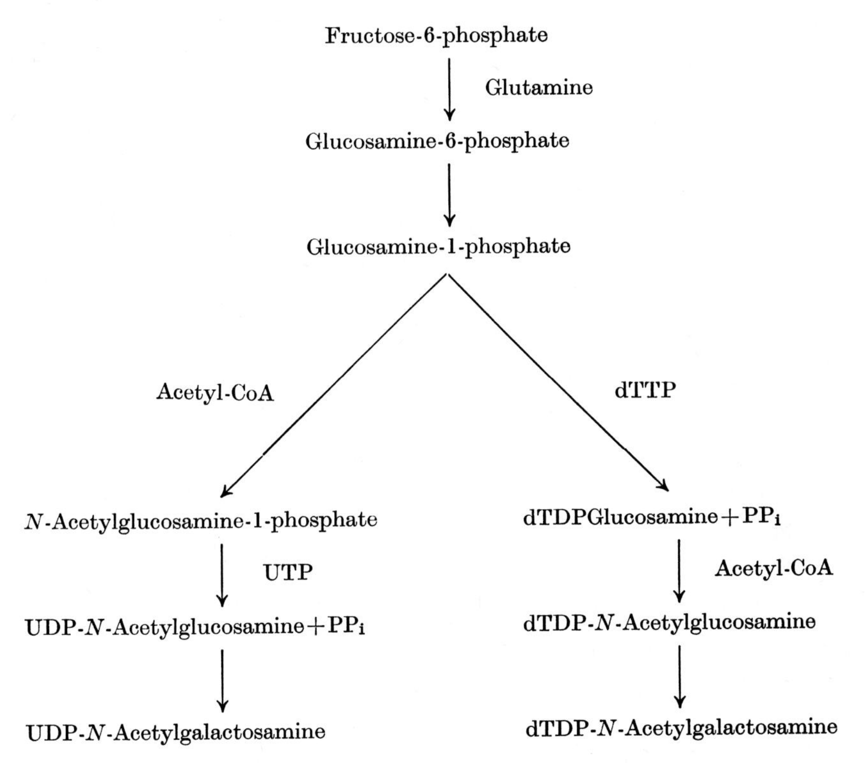

FIG. 15.3. Formation of hexosamines in *E. coli* and *P. aeruginosa*. dTTP = deoxythymidine triphosphate; UTP = uridine triphosphate.

which catalyzed an epimerization at carbon 2 of *N*-acetylmannosamine-6-phosphate resulting in the formation of *N*-acetylglucosamine-6-phosphate. Considering the conditions of growth and the information in Fig. 15.3, it seems likely that the 2-epimerase of Ghosh and Roseman is a catabolic enzyme although it may have a biosynthetic function in the formation of *N*-acetylneuraminic acid as well. It is interesting to compare the action of the bacterial enzyme with the corresponding mammalian enzyme which acts on the unphosphorylated sugars and requires catalytic amounts of ATP (Ghosh and Roseman, 1965b).

Blacklow and Warren (1962) extracted an enzyme from *Neisseria meningitidis* which catalyzed the condensation of *N*-acetylmannosamine and phosphoenolpyruvate to *N*-acetylneuraminic acid (Fig. 15.4). The bacterial route to *N*-acetylneuraminic acid differs from that in mammalian tissues where *N*-acetylmannosamine-6-phosphate is a substrate and *N*-acetylneuraminic acid-9-phosphate is the product.

$$
\begin{array}{c}
\text{COOH} \\
| \\
\text{C—O—PO}_3\text{H}_2 \\
\| \\
\text{CH}_2 \\
+ \\
\text{CHO} \\
| \\
\text{AcNCH} \\
| \\
\text{HOCH} \\
| \\
\text{HCOH} \\
| \\
\text{HCOH} \\
| \\
\text{CH}_2\text{OH}
\end{array}
\longrightarrow
\begin{array}{c}
\text{COOH} \\
| \\
\text{C=O} \\
| \\
\text{CH}_2 \\
| \\
\text{HCOH} \\
| \\
\text{AcNCH} \\
| \\
\text{HOCH} \\
| \\
\text{HCOH} \\
| \\
\text{HCOH} \\
| \\
\text{CH}_2\text{OH}
\end{array}
+ \; \text{H}_3\text{PO}_4
$$

N-Acetyl-D-mannosamine *N*-Acetyl-D-neuraminic acid

FIG. 15.4. Formation of *N*-acetylneuraminic acid in *Neisseria meningitidis.*

Warren and Blacklow (1962) also described cytidine monophosphate (CMP) *N*-acetylneuraminic acid pyrophosphorylase, the enzyme responsible for the activation of *N*-acetylneuraminic acid, from the same organism. CMP-*N*-acetylneuraminic acid pyrophosphorylase catalyzes a rather unusual activating reaction since CMP becomes attached through its phosphate to the hemiacetal hydroxyl group on carbon 2 of *N*-acetylneuraminic acid.

C. FORMATION OF DEOXYSUGARS

1. *Deoxyribose*

Studies with *E. coli* (Bernstein and Sweet, 1958; Bagatell *et al.*, 1959; see also review of Sable, 1966) and with *Pseudomonas saccharophila* (Fossitt and Bernstein, 1963) grown with isotopic substrates yielded the information that deoxypentose of nucleic acids reflected the labeling pattern of RNA-ribose and led to the belief that deoxyribose was produced by the direct reduction of ribose. Reichard (1962) demonstrated such a reaction with cell-free preparations of *E. coli* which catalyzed the reduction of CDP to dCDP. Reduced lipoic acid was the best electron donor tested and ATP stimulated the reaction. Laurent *et al.* (1964) subsequently purified a sulfur-containing protein from *E coli* which they named thioredoxin and which was the natural electron donor for reduction of CDP. Thioredoxin itself was reduced by NADPH in a reaction catalyzed by another protein from *E. coli*, thioredoxin reductase (Moore *et al.*, 1964). In addition to thioredoxin and thioredoxin reductase, two other protein fractions from *E. coli*, B_1 and B_2 were required for reduction of CDP. The substrate specificity of the system was markedly altered by the addition of ATP or deoxyribonucleotides (Larsson and Reichard, 1966); addition of ATP stimulated the reduction of pyrimidine nucleotides while addition of dATP inhibited reduction of all four ribonucleotides. Although the effects of the added nucleotides were complex, it seems likely that these effects were related to the regulation of activity of the ribonucleotide reductase system of *E. coli*.

The reduction of ribonucleotides by *Lactobacillus leichmanii* is similar to the process in *E. coli* in that reduced lipoic acid and ATP were required by a crude enzyme preparation for activity, but strikingly different in that 5′deoxyadenosylcobalamin was essential for reduction (Blakley, 1965). Goulian and Beck (1966) showed that nucleoside triphosphates were the actual substrates for a purified enzyme from *L. leichmanii* and that all four ribonucleotides were reduced. The presence of deoxyribonucleotides in the reaction mixture altered the activity of the enzyme towards the individual ribonucleotides, and again it is believed that deoxyribonucleotides exert a regulatory effect on ribonucleotide reductase of *L. leichmanii* (Beck *et al.*, 1966).

2. 6-*Deoxyhexoses*

All deoxyhexoses synthesized in bacteria are formed from nucleoside diphosphate-4-oxo-6-deoxyhexoses. The first such pathway elucidated was that for the synthesis of L-fucose which occurs in the lipopolysaccharide of several Gram negative bacteria (Chapter 3). Several investigators showed that bacteria of the Klebsiella-Aerobacter group converted glucose-^{14}C to L-fucose-^{14}C without rearrangement of the hexose molecule (Wilkinson, 1957; Segal and Topper, 1957; Heath and Roseman, 1958). Ginsburg (1960) uncovered the enzymes responsible for this conversion using cell-free preparations of *Aerobacter aerogenes* which converted GDP-D-mannose to GDP-L-fucose. Ginsburg (1961) isolated an intermediate which he characterized as GDP-4-oxo-6-deoxy-D-mannose (Fig. 15.5) and found that NADPH was required for the reduction of the intermediate to GDP-L-fucose. Elbein and Heath (1965) later purified GDP-D-mannose-4-oxido-6-reductase from *E. coli* and established that NAD was required for this intramolecular oxidation-reduction reaction.

The biosynthetic pathway for L-rhamnose in group A streptococci

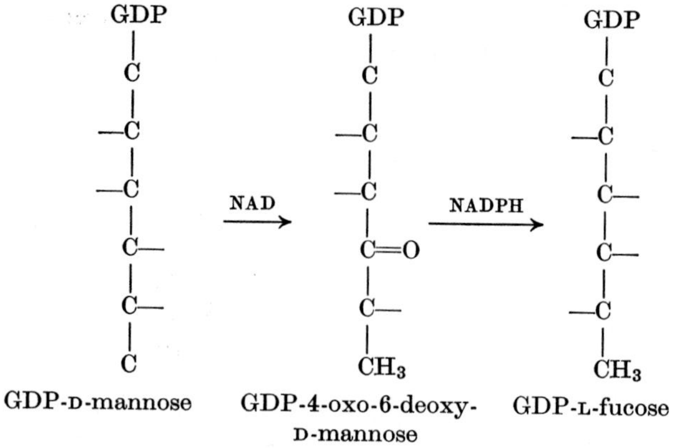

```
GDP                    GDP                   GDP
 |                      |                     |
 C                      C                     C
 |                      |                     |
—C                     —C                    —C
 |                      |                     |
—C        NAD          —C       NADPH         C—
 |       ----->         |      ------->       |
 C—                     C=O                   C—
 |                      |                     |
 C—                     C—                   —C
 |                      |                     |
 C                      CH3                   CH3
```

GDP-D-mannose GDP-4-oxo-6-deoxy-D-mannose GDP-L-fucose

FIG. 15.5. Conversion of GDP-D-mannose to GDP-L-fucose.

(Southard *et al.*, 1959) and *S. faecalis* (Pazur and Shuey, 1961) likewise involves a direct conversion of the glucose molecule. Pazur and Shuey (1961) showed that extracts of *S. faecalis* formed dTDP-D-glucose from dTTP and D-glucose-1-phosphate and that the extracts converted dTDP-D-glucose to dTDP-L-rhamnose. Kornfield and Glaser (1961) purified dTDP-D-glucose pyrophosphorylase from *P. aeruginosa* as well as another protein which converted dTDP-D-glucose to dTDP-L-rhamnose (Glaser and Kornfield, 1961). The enzyme fraction responsible for the conversion of dTDP-D-glucose to dTDP-L-rhamnose required NAD and NADPH and produced an intermediate compound which had the properties of dTDP-4-oxo-6-deoxy-D-glucose (Fig. 15.6). Okazaki *et al.* (1962) identified dTDP-4-oxo-6-deoxy-D-glucose as the inter-

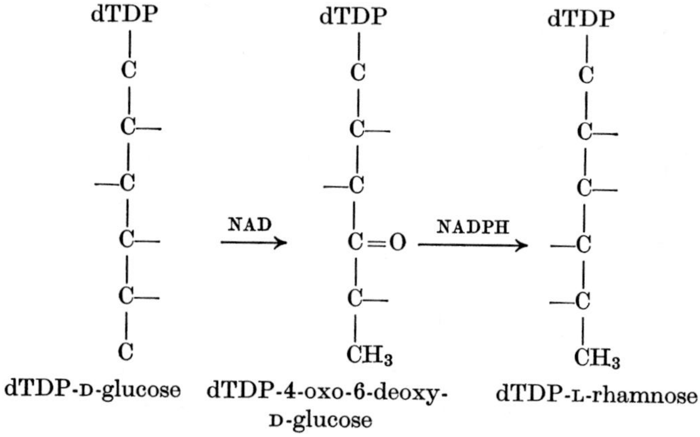

```
dTDP                   dTDP                  dTDP
 |                      |                     |
 C                      C                     C
 |                      |                     |
 C—                     C—                    C—
 |                      |                     |
—C                     —C                     C—
 |        NAD           |       NADPH         |
 C—      ----->         C=O    ------->      —C
 |                      |                     |
 C—                     C—                   —C
 |                      |                     |
 C                      CH3                   CH3
```

dTDP-D-glucose dTDP-4-oxo-6-deoxy-D-glucose dTDP-L-rhamnose

FIG. 15.6. Conversion of dTDP-D-glucose to dTDP-L-rhamnose.

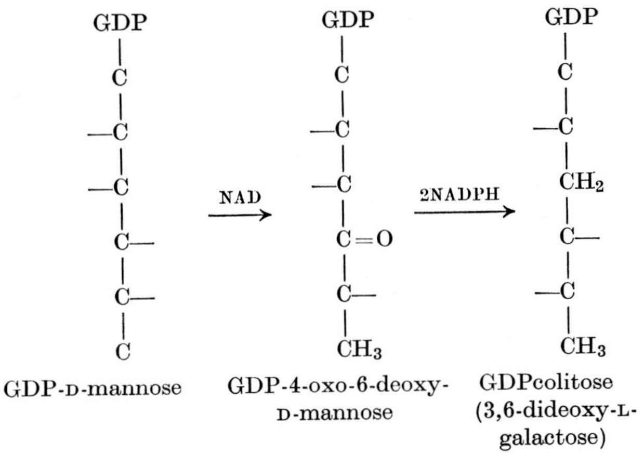

```
 GDP                     GDP                      GDP
  |                       |                        |
  C                       C                        C
  |                       |                        |
 —C                      —C                       —C
  |                       |                        |
 —C          NAD         —C         2NADPH        CH2
  |        ------>        |        -------->       |
  C—                      C=O                      C—
  |                       |                        |
  C—                      C—                      —C
  |                       |                        |
  C                       CH3                      CH3
```

GDP-D-mannose — GDP-4-oxo-6-deoxy-D-mannose — GDPcolitose (3,6-dideoxy-L-galactose)

FIG. 15.7. Conversion of GDP-D-mannose to GDPcolitose.

mediate in L-rhamnose formation in *E. coli* as well. NAD functioned in the formation of the intermediate 4-oxo-6-deoxy nucleotide and NADPH served as the electron donor for reduction of the intermediate to dTDP-L-rhamnose. The overall transformation of dTDP-D-glucose to dTDP-L-rhamnose thus involves epimerizations at carbons 3, 4 and 5 and reduction at carbon 6.

3. *Dideoxyhexoses*

Colitose (3,6-dideoxy-L-galactose) is the only one of the five dideoxyhexoses found in bacterial lipopolysaccahrides whose biosynthesis begins with GDP-D-mannose (Fig. 15.7). Elbein and Heath (1965) demonstrated the conversion of GDP-D-mannose to GDPcolitose by crude extracts of *E. coli* which was dependent on NADPH. They purified GDP-D-mannose-4-oxido-6-reductase several fold and found that NAD was very tightly bound to the enzyme. Reduction of GDP-4-oxo-6-deoxy-D-mannose with NADPH to GDPcolitose was accomplished by another enzyme fraction from *E. coli*. This complex transformation required two equivalents of NADPH for the reductions at carbons 3 and 6 and resulted in epimerization at carbon 5.

The remaining four natural dideoxy hexoses are synthesized from CDP-D-glucose (Fig. 15.8). Mayer and Ginsburg (1965) purified CDP-glucose pyrophosphorylase from *Salmonella paratyphi A*, which contains paratose in its lipopolysaccharide. Matsuhashi *et al.* (1966b) and Hey and Elbein (1966) purified CDP-D-glucose-4-oxido-6-reductase from *Pasteurella pseudotuberculosis* and *Salmonella typhosa* respectively, and in both cases NAD was required for activity of the enzyme.

Nikaido and Nikaido (1966) found that extracts of *Salmonella typhimurium* and *Salmonella enteritidis* both possessed the ability to convert glucose-1-phosphate and CTP to CDP-D-glucose and on to CDP-4-oxo-6-deoxy-D-glucose. With enzyme preparations from *S. typhimurium* CDP-4-oxo-6-deoxy-D-glucose was converted to CDPabequose, while with preparations from *S. enteritidis* it was converted to CDPparatose and CDPtyvelose. Hey and Elbein (1966) also found that their enzyme preparations of *S. typhosa* converted CDP-4-oxo-6-deoxy-D-glucose to CDPparatose and CDPtyvelose. Matsuhashi *et al.* (1966a) found that cell-free extracts of *P. pseudotuberculosis* serotypes converted CDP-D-glucose to the CDP derivatives of the dideoxyhexose present in their respective lipopolysaccharides; types I and III produced CDPparatose, type II produced CDPabequose, type IV produced CDPtyvelose and type V produced CDPascarylose. The link between the biosynthesis of paratose and tyvelose was discovered by Matsuhashi (1966) when he isolated an enzyme from both *S. enteritidis* and *P. pseudotuberculosis* which catalyzed the epimerization of CDPparatose at carbon 2 to form CDPtyvelose (Fig. 15.8).

D. Hexuronic Acids

Smith *et al.* (1958a) found that extracts of a non-encapsulated mutant derived from a strain of *D. pneumoniae*, type II, had the ability to oxidize UDPglucose with NAD to UDPglucuronate. Type I *D. pneumoniae* contains galacturonate in the capsule and enzyme extracts catalyzed the epimerization of UDPglucuronate to UDPgalacturonate (Smith *et al.*, 1958b). UDPglucuronate in plant tissues is converted to UDPxylose by a specific decarboxylase (Feingold *et al.*, 1960), and this may be the biosynthetic route for xylose in bacteria as well. Preiss (1964) purified an NAD-dependent GDPmannose dehydrogenase which forms GDPmannuronate from an *Arthrobacter* species which forms a capsule containing mannuronate.

III. Formation of Polysaccharides by Bacteria

A. Homopolysaccharides

One of the earliest examples of polysaccharide formation by a cell-free system from a bacterial source was reported by Hehre (1946) using extracts of *Leuconostoc mesenteroides* to synthesize dextran from sucrose. The polymer produced in his experiments resembled the dextran produced by living cells, and Hehre showed that the reaction could be described as follows:

$$\text{n Sucrose} \rightarrow (\text{glucose})_n + \text{n fructose.}$$

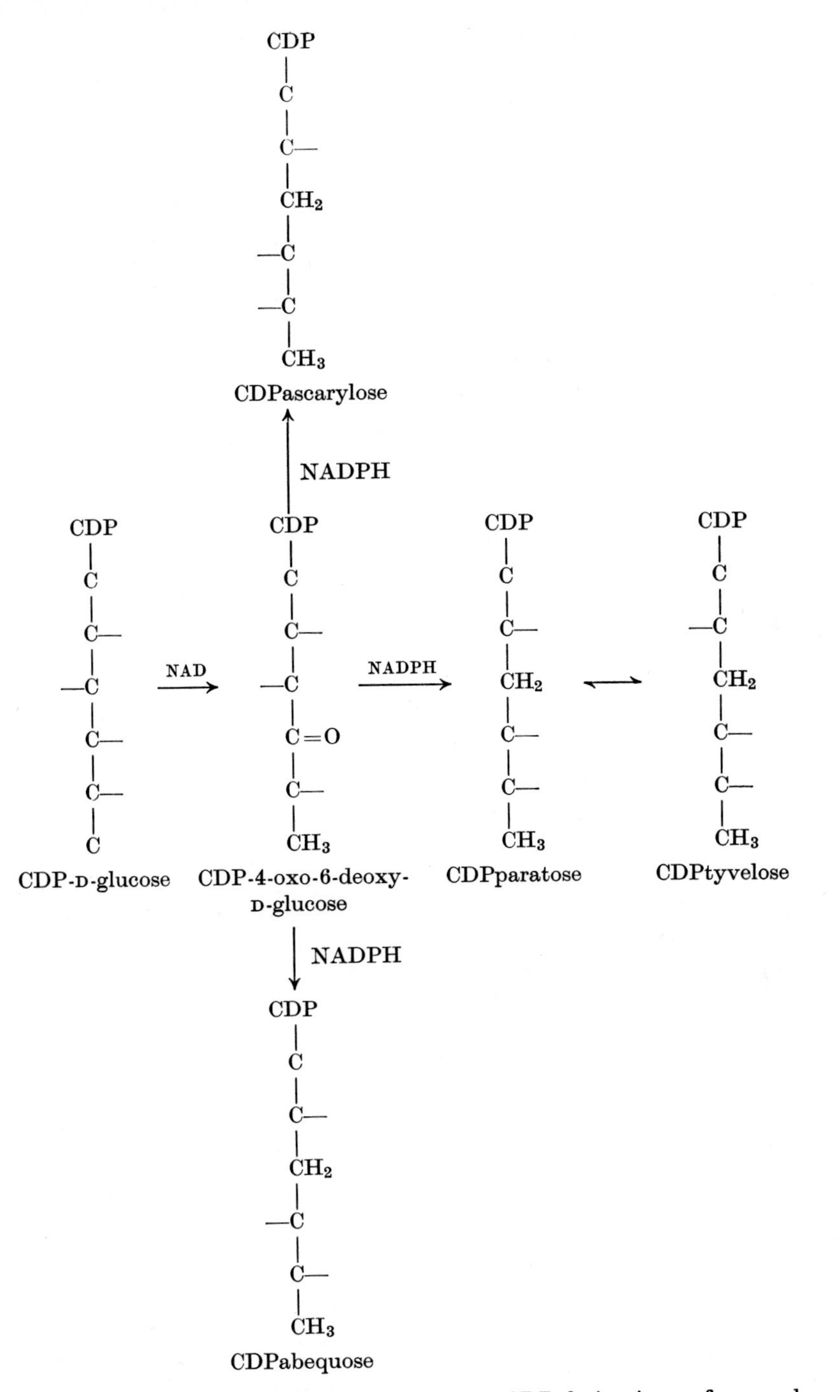

FIG. 15.8. Conversion of CDP-D-glucose to CDP-derivatives of ascarylose, paratose, tyvelose and abequose.

Koepsell *et al.* (1953) purified the enzyme, dextransucrase, from *L. mesenteroides* and found that the polymer size could be varied considerably by manipulation of the conditions. Oligosaccharides acted as primers and accepted glucosyl residues from sucrose. Dextransucrase (α-1,6-glucan:D-fructose 2-glucosyltransferase) is an enzyme that utilizes the energy of an already existing glycoside bond to form polymers, other examples being amylomaltase and amylosucrase mentioned in Chapter 4.

The biosynthesis of the great majority of oligosaccharides begins with activation of the hexose monomer by attachment of a nucleoside phosphate to the acetal hydroxyl as outlined in previous sections of this chapter. It is also true of bacterial preparations that, in general, the polysaccharide synthetases are found in the particulate fraction of the cell possibly because the product of their action is found near the cell surface or extracellularly.

The biosynthesis of cellulose by *Acetobacter xylinum* was studied by Glaser (1958) who found that the particulate fraction of the enzyme extract catalyzed the incorporation of UDPglucose into cellulose. Soluble cellodextrins stimulated the rate of incorporation.

Aminoff *et al.* (1963) achieved the net synthesis of colominic acid from CMP-*N*-acetylneuraminic acid by a particulate enzyme preparation from *E. coli*.

The biosynthesis of glycogen in bacteria is similar to the biosynthesis in animals except that ADPglucose is the starting material rather than UDPglucose. Shen and Preiss (1965) purified ADPglucose pyrophosphorylase from an Arthrobacter species and detected the enzyme in several other species of bacteria. Greenberg and Preiss (1965) purified ADPglucose:glycogen α-4-glucosyltransferase from Arthrobacter and showed that a number of maltodextrins from maltose to glycogen itself acted as acceptors or primers in the reaction. Isomaltose, sucrose and cellobiose were among several compounds not of the maltodextrin series that were tried and found to be inactive as acceptors. The final step in the formation of glycogen in animal tissues is catalyzed by the branching enzyme (α-1,4-glucan:α-1,4-oligoglucan 6-glycosyltransferase) which causes the formation of the 1,6-sidechains in glycogen. Preiss and his associates did not look for this enzyme in their species of Arthrobacter, but Zevenhuizen (1964) did find it in *Arthrobacter globigiformis*.

B. Heteropolysaccharides

Markovitz *et al.* (1959) prepared sonic extracts of a group A streptococcus that were capable of the net synthesis of hyaluronic acid from

UDPglucuronate and UDP-*N*-acetylglucosamine. These workers were unable to demonstrate requirement for a primer, even after treatment of their enzyme preparations with hyaluronidase. UDPglucose dehydrogenase and UDP-*N*-acetylglucosamine pyrophosphorylase were present in their enzyme preparations.

The studies of Mills and Smith (1962) pointed out the routes for formation of capsular polysaccharides by serotypes of *Diplococcus pneumoniae*. An enzyme preparation made from type III pneumococcus catalyzed incorporation of UDPglucose and UDPglucuronic acid into a polysaccharide which was precipitated by antiserum prepared against type III capsule (Smith *et al.*, 1960). The enzyme was partially characterized and found to be particulate (Smith *et al.*, 1961). Distler and Roseman (1964) found that the particle fraction of an enzyme extract of type XIV pneumococcus was able to synthesize a polysaccharide from UDPglucose, UDPgalactose and UDP-*N*-acetylglucosamine which reacted with antiserum prepared against the polysaccharide formed by this serotype. While the polysaccharide formed by enzyme action contained all the components of the native polysaccharide, it was not identical since it was possible to differentiate between the two by immunodiffusion techniques.

The biosynthesis of both polyglycerolphosphate and polyribitol phosphate teichoic acids begins with the formation of the CDP-derivatives. Shaw (1962) purified CDPglycerol pyrophosphorylase and CDPribitol pyrophosphorylase from *L. arabinosus* and *S. aureus* respectively and established that the reactions catalyzed were:

$$\text{CTP} + \alpha\text{-glycerolphosphate} \rightarrow \text{CDPglycerol} + \text{pyrophosphate}$$

$$\text{CTP} + \text{ribitol-5-phosphate} \rightarrow \text{CDPribitol} + \text{pyrophosphate}$$

The efforts of Burger and Glaser (1964) to demonstrate polyglycerolphosphate and polyribitolphosphate formation with cell-free preparations were the first to be successful. Particulate enzymes from *B. subtilis* and *B. licheniformis* both catalyzed the formation of polyglycerolphosphate from CDPglycerol with the release of CMP. The product was a linear polymer of about 30 units in length with 1,3-phosphodiester linkages and appeared to be identical with natural polyglycerolphosphate isolated from the organism.

Glaser (1964) reported on the formation of polyribitol phosphate with a particulate enzyme from *L. plantarum*. In this case the product was a chain of about 7 to 9 monomer units. Some unidentified acceptor, possibly a fragment of cell wall, was present in the enzyme preparations. Similar results were reported by Ishimoto and Strominger (1966) who studied polyribitolphosphate synthesis by a particulate enzyme from

S. aureus. Ishimoto and Strominger determined that one mole of phosphate from CDPribitol was transfered to the product for every mole of ribitol incorporated.

Addition of glycosyl units to both polyglycerolphosphate and polyribitolphosphate has been demonstrated. Glaser and Burger (1964) showed that a particulate enzyme preparation from *B. subtilis* catalyzed the addition of glycosyl units from UDPglucose to polyglycerolphosphate. Nathanson and Strominger (1963) treated teichoic acid from *S. aureus* with β-glucosaminidase leaving only the α-*N*-acetylglucosamine residues attached to polyribitolphosphate. The treated teichoic acid now acted as an acceptor for *N*-acetylglucosamine from UDP-*N*-acetylglucosamine in a reaction catalyzed by a particulate enzyme from this organism.

C. Complex Polysaccharides

1. *Mucopeptide*

Richmond and Perkins (1960) studied the biosynthesis of muramic acid from labeled precursers in *S. aureus* and found that glucose was incorporated into both the hexoseamine and lactyl side chain of muramic acid but that alanine was a somewhat better precurser of the lactyl side chain than was glucose. These results suggested that pyruvate or phosphoenolpyruvate was a precurser of the lactyl residue.

Formation of the mucopeptide begins with the formation of UDP-muramic pentapeptide (see review of Strominger, 1962) which is then incorporated into the growing cell wall. Ito and Strominger (1962a) demonstrated the sequential addition of L-alanine, D-glutamic acid and L-lysine to UDPmuramic acid by enzyme preparations of *S. aureus* (Fig. 15.9). ATP was required for the formation of each peptide linkage. The enzymes which added lysine (Ito and Strominger, 1964) and D-glutamic acid (Nathanson *et al.*, 1964) were purified and found to be highly specific for their respective substrates. Extracts of *S. aureus* also contained a specific synthetase for the formation of D-alanyl-D-alanine from D-alanine and ATP (Ito and Strominger, 1962b). A specific enzyme catalyzed the addition of D-alanyl-D-alanine to the lysine residue of the growing peptide. ATP was required for this last step as in the other cases, and D-alanine itself would not substitute for the dipeptide.

Further work uncovered a fascinating story in the biosynthesis of cell wall mucopeptide. Chatterjee and Park (1964) accomplished the polymerization of UDP-*N*-acetylmuramic pentapeptide with a particulate enzyme preparation from *S. aureus* and similar results were obtained by Meadow *et al.* (1964) with a particulate enzyme preparation from

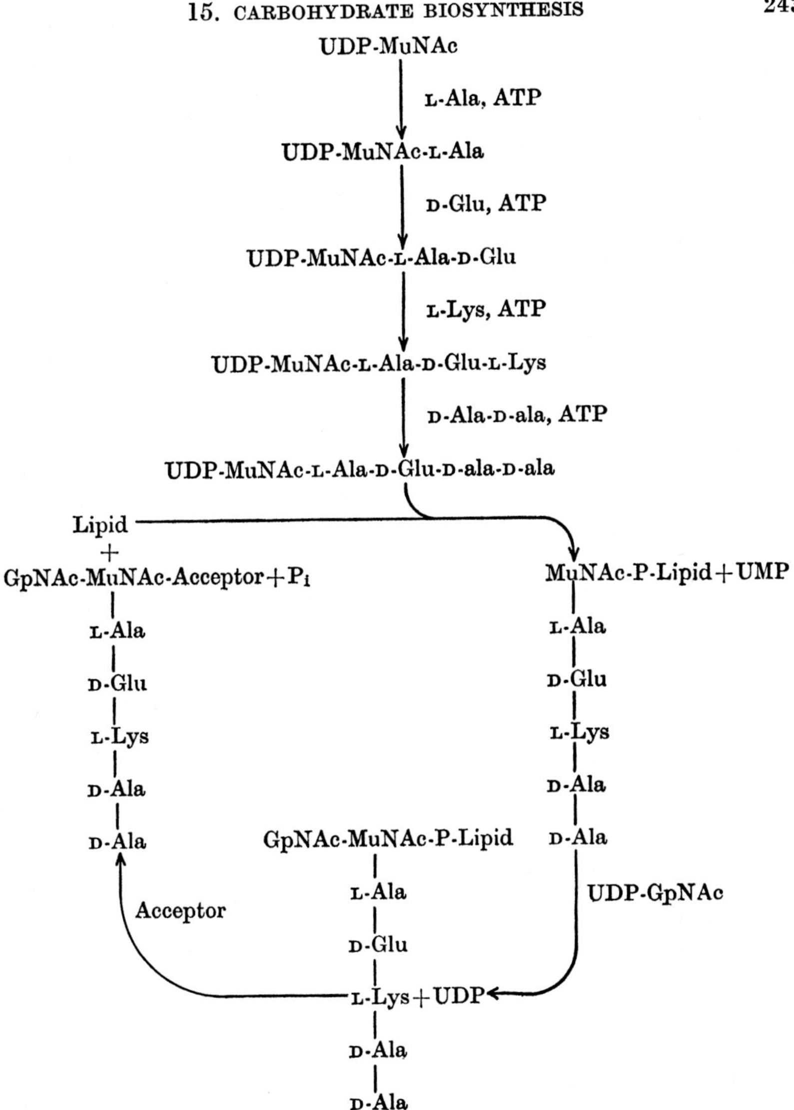

FIG. 15.9. Formation of glycopeptide in *S. aureus*.

another strain of *S. aureus*. Anderson *et al.* (1965) found that particulate enzyme preparations of *S. aureus* catalyzed the reversible attachment of UDP-*N*-acetylmuramic acid pentapeptide to a lipid acceptor with the release of UMP. The lipid was partially characterized as a phospholipid (Anderson and Strominger, 1965). In the next step *N*-acetylglucosamine was transferred from UDP-*N*-acetylglucosamine to the muramic acid pentapeptide-lipid complex to form the disaccharide pentapeptide lipid complex and UDP was released. The addition of *N*-acetylglucosamine was virtually nonreversible. The particulate enzyme preparations made by Anderson *et al.* (1965) also caused addition of the disaccharide pentapeptide to an acceptor present in their preparations, presumably the growing cell wall under physiological conditions, with the release of the lipid and inorganic phosphate.

The final process in cell wall formation is the cross-linking reaction which varies in detail with the organism. Wise and Park (1965) and Tipper and Strominger (1965) both presented evidence obtained with growing cultures of *S. aureus* that penicillin inhibited the cross-linking reaction and that the terminal D-alanine residue was eliminated in the process. Izaki *et al.* (1966) demonstrated both the cross-linking reaction and sensitivity of cross-linking to penicillin in enzyme preparations made from *E. coli*. The glycopeptide produced without penicillin was digested with lysozyme and analysis of the products revealed that cross-linking existed between D-alanine and *meso*-diaminopimelic acid and that only one mole of D-alanine was present in each peptide chain. When penicillin was added, glycopeptide was formed in the enzyme preparations, but analysis of the saccharides obtained from lysozyme digestion indicated that cross-linking was greatly reduced and that two moles of D-alanine were present in each peptide chain. These studies indicated that cross-linking was accomplished by a transpeptidation in which the penultimate D-alanine residue was attached to the epsilon amino group of *meso*-diaminopimelic acid with the release of the terminal D-alanine residue. The enzyme preparations made from *E. coli* also contained a carboxypeptidase which cleaved the terminal D-alanine residue from the pentapeptide chain and, interestingly enough, was also sensitive to penicillin. The alanine carboxypeptidase apparently functions in the living cell by removing D-alanine from uncross-linked peptide chains.

2. *Lipopolysaccharide*

a. *The Core Polysaccharide*

Considerable success has been achieved in studies of the formation of lipopolysaccharide by Gram negative bacteria by the use of mutants which lack enzymes involved in the formation of the constituents of lipopolysaccharide. Nikaido (1962) and Osborn *et al.* (1962) isolated mutants of *S. enteritidis* and *S. typhimurium* respectively which lacked UDPglucose 4-epimerase and were, therefore, unable to form galactose from glucose. When grown in galactose-free media these organisms formed lipolysaccharides which contained heptose and glucose but were deficient in galactose and the sugars of the O-antigens. Similarly, Rothfield *et al.* (1964) isolated a glucose-deficient mutant of *S. typhimurium* which lacked phosphoglucose isomerase and which formed a lipopolysaccharide containing heptose, but devoid of glucose, galactose and the sugars of the O-antigen when grown in glucose-free media. Nikaido (1962) discovered that the incomplete lipopolysaccharide formed by his mutant acted as an acceptor for galactose and, thus, discovered a valuable approach to the problem of lipopolysaccharide biosynthesis.

Subsequent studies led to discovery of the sequence of reactions in the formation of the core polysaccharide and provided information about the fine structure of the core as well (Fig. 15.10). Rothfield *et al.*

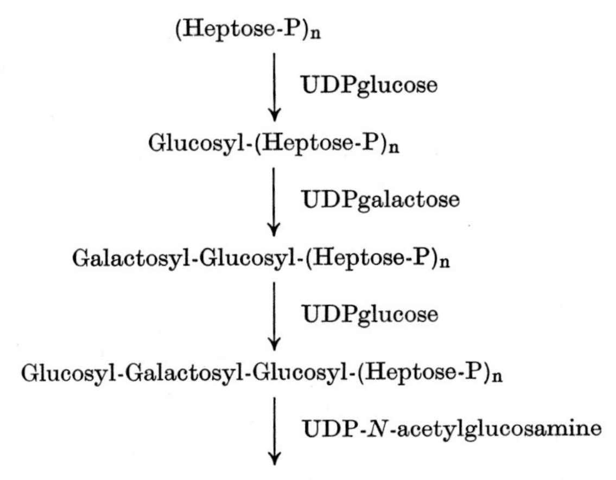

FIG. 15.10. Order of addition of hexoses in formation of core polysaccharide by *Salmonella typhimurium*.

(1964) also isolated a double mutant which lacked UDPglucose 4-epimerase as well as phosphoglucose isomerase and which proved valuable for enzyme studies since cell-free preparations were unable to interconvert glucose and galactose. Enzyme preparations of the double mutant catalyzed the specific incorporation of glucose from UDP-glucose into lipopolysaccharide obtained from the glucose-deficient mutant but not into lipopolysaccharide obtained from the galactose-deficient mutant. Conversely, enzyme preparations of the double mutant catalyzed transfer of galactose into lipopolysaccharide prepared from the galactose-deficient mutant but not into lipopolysaccharide prepared from the glucose-deficient mutant. Since glucose is present in the lipopolysaccharide isolated from galactose-deficient *S. typhimurium*, these studies indicated that glucose was added to the backbone before galactose. Further studies by Osborn and D'Ari (1964) established that the order of addition to the heptose phosphate backbone was glucose, galactose, glucose and *N*-acetylglucosamine (Fig. 15.10). The enzymes which catalyze these reactions have been named in order, UDPglucose-lipopolysaccharide transferase I, UDPgalactose-lipopolysaccharide transferase, UDPglucose-lipopolysaccharide transferase II and UDP-*N*-acetylglucosamine lipopolysaccharide transferase. A second galactosyl residue is known to be attached to the glucose molecule immediately adjacent to the heptose phosphate backbone (see Chapter 3), but the transferase responsible for its addition has not been studied. Edstrom and Heath (1964) reported that a mutant of *E. coli* which lacked UDP-glucose 4-epimerase transferred galactose, glucose and *N*-acetyl-glucosamine in the same order established for *S. typhimurium*.

Rothfield and Horecker (1964) uncovered a role for lipid in the biosynthesis of the core polysaccharide, possibly similar to that reported for lipid in the biosynthesis of mucopeptide. Lipopolysaccharide of either glucose or galactose-deficient strains of *S. typhimurium* treated with lipid solvents no longer functioned as acceptors in the transferase reactions. However, when lipid was added back to the solvent-treated lipopolysaccharide and the mixture was "annealed" by heating to 60°C followed by slow cooling, the lipopolysaccharide again acted as an acceptor for both glucose and galactose.

b. *The O-Antigen*

The O-antigen of *S. typhimurium* contains a repeating unit composed of galactose, mannose, rhamnose and abequose (see Fig. 15.11). The biosynthesis of the O-antigen in *S. typhimurium* was studied by Weiner *et al.* (1965) with cell-free preparations of their mutant lacking UDP-glucose 4-epimerase. Galactose, rhamnose and mannose were transferred to an insoluble product when all three sugars were present as their respective nucleoside diphosphates. The product in this case was not complete lipopolysaccharide, but it was a polymer of the sugars added and terminated in galactose-1-phosphate. The incomplete lipopolysaccharide formed by this mutant apparently could not act as an acceptor because it lacked the site of attachment of the O-antigen to the core. Instead, the sugars were found in a lipid fraction present in the extracts, and this lipid may act as a carrier after the fashion of the lipid involved in biosynthesis of mucopeptide. The order of addition of sugars appears to be galactose, rhamnose and mannose since rhamnosyl-galactosyl-1-phosphate and mannosyl-rhamnosyl-galactosyl-1-phos-

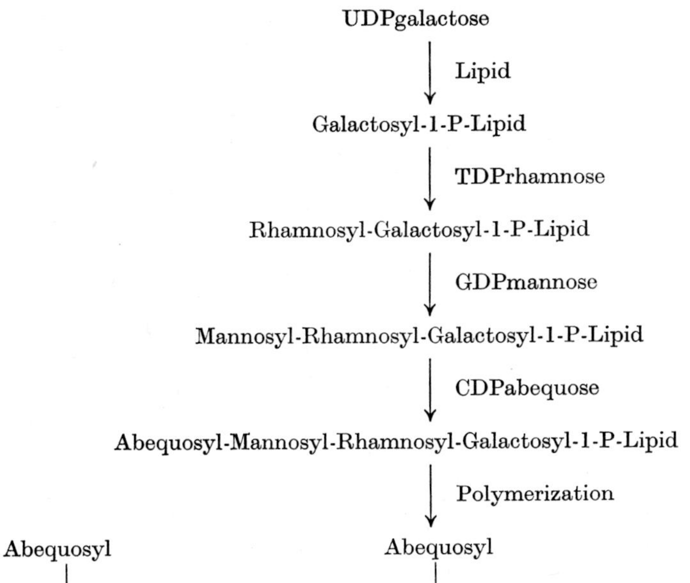

FIG. 15.11. Biosynthesis of O-antigen side chains in *S. typhimurium*. Following polymerization, the O-antigen is attached to the core polysaccharide.

phate were isolated from the lipid fraction of the extract. Later, Weiner *et al.* (1966) established that abequose was added after mannose, but before polymerization of the repeating unit although polymerization could occur without abequose being present. Although transfer of the preformed polysaccharide to the core was not observed in the experiments of Weiner *et al.* (1966), Nikaido *et al.* (1966) and Wright *et al.* (1965) reported incorporation of O-antigen carbohydrates into lipopoly-

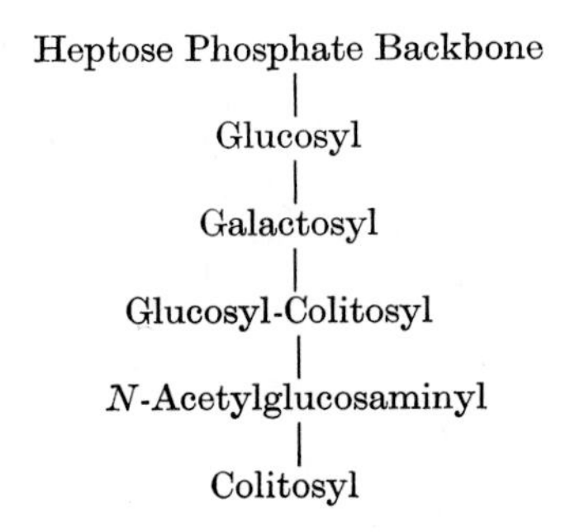

FIG. 15.12. Structure of lipopolysaccharide of *E. coli* 0111. (Heath *et al.* (1966).

saccharide of a rhamnose deficient strain of *S. typhimurium* and wild type *S. anatum* respectively.

Heath *et al.* (1966) also studied the biosynthesis of 2-keto-3-deoxyoctulosonic acid and the incorporation of this compound into lipopolysaccharide. Several years earlier Levin and Racker (1959) reported that 2-keto-3-deoxyoctulosonate was synthesized by an enzyme from *P. aeruginosa* which catalyzed the condensation of phosphoenolpyruvate and D-arabinose-5-phosphate:

D-Arabinose-5-phosphate + phosphoenolpyruvate →
2-keto-3-deoxyoctulosonate-8-phosphate + phosphate

2-Keto-3-deoxyoctulosonate is presumed to be dephosphorylated by cellular phosphatases since activation by CTP occurs with the free acid (Heath *et al.*, 1966):

2-Keto-3-deoxyoctulosonate + CTP → CMP-2-keto-3-deoxyoctulosonate + pyrophosphate

This reaction is analogous to the activation of neuraminic acid (Section II. B, this chapter). Heath *et al.* (1966) were able to demonstrate the transfer of 2-keto-3-deoxyoctulosonate to an acceptor which they prepared by alkaline hydrolysis of lipid A. The acceptor contained hexosamine and β-hydroxymyristic acid in a molar ratio of 1:1 as well as traces of phosphate.

References

Aminoff, D., Dodyk, F. and Roseman, S. (1963). *J. Biol. Chem.* **238**, PC1177.
Anderson, J. S., Matsuhashi, M., Haskin, M. A. and Strominger, J. L. (1965). *Proc. Natl. Acad. Sci. U.S.* **53**, 881.
Anderson, J. S. and Strominger, J. L. (1965) *Biochem. Biophys. Res. Commun.* **21**, 516.
Bagatell, F. K., Wright, E. M. and Sable, H. Z. (1959). *J. Biol. Chem.* **234**, 1369.
Beck, W. S., Goulian, M., Larsson, D. and Reichard, P. (1966). *J. Biol. Chem.* **241**, 2177.
Bernstein, I. A. (1956). *J. Biol. Chem.* **221**, 873.
Bernstein, I. A. and Sweet, D. (1958). *J. Biol. Chem.* **233**, 1194.
Blacklow, R. S. and Warren, L. (1962). *J. Biol. Chem.* **237**, 3520.
Blakley, R. L. (1965). *J. Biol. Chem.* **240**, 2173.
Burger, M. M. and Glaser, L. (1964). *J. Biol. Chem.* **239**, 3168.
Campbell, J. J. R., Smith, R. A. and Eagles, B. A. (1953). *Biochim. Biophys. Acta* **11**, 594.
Chatterjee, A. N. and Park, J. T. (1964). *Proc. Natl. Acad. Sci. U.S.* **51**, 9.
Distler, J. and Roseman, S. (1964). *Proc. Natl. Acad. Sci. U.S.* **51**, 897.
Edstrom, R. D. and Heath, E. C. (1964). *Biochem. Biophys. Res. Commun.* **16**, 576.
Elbein, A. D. and Heath, E. C. (1965). *J. Biol. Chem.* **240**, 1919.

Feingold, D. S., Neufeld, E. F. and Hassid, W. Z. (1960). *J. Biol. Chem.* **235,** 910.
Fossitt, D. D. and Bernstein, I. A. (1963). *J. Bacteriol.* **86,** 1326.
Ghosh, S., Blumenthal, H. J., Davidson, E. and Roseman, S. (1960). *J. Biol. Chem.* **235**, 1265.
Ghosh, S. and Roseman, S. (1965a). *J. Biol. Chem.* **240**, 1525.
Ghosh, S. and Roseman, S. (1965b). *J. Biol. Chem.* **240**, 1531.
Ginsburg, V. (1960). *J. Biol. Chem.* **235**, 2196.
Ginsburg, V. (1961). *J. Biol. Chem.* **236**, 2389.
Glaser, L. (1958). *J. Biol. Chem.* **232**, 627.
Glaser, L. (1964). *J. Biol. Chem.* **239**, 3178.
Glaser, L. and Burger, M. M. (1964). *J. Biol. Chem.* **239**, 3187.
Glaser, L. and Kornfield, S. (1961). *J. Biol. Chem.* **236**, 1795.
Goulian, M. and Beck, W. S. (1966). *J. Biol. Chem.* **241**, 4233.
Greenberg, E. and Preiss, J. (1965). *J. Biol. Chem.* **240**, 2341.
Heath, E. C., Mayer, R. M., Edstrom, R. D. and Beaudreau, C. A. (1966). *Ann. N.Y. Acad. Sci.* **133**, 315.
Heath, E. C. and Roseman, S. (1958). *J. Biol. Chem.* **230**, 511.
Hehre, E. J. (1946). *J. Biol. Chem.* **163**, 221.
Hey, A. D. and Elbein, A. D. (1966). *J. Biol. Chem.* **241**, 5473.
Horecker, B. L. (1966). *Ann. Rev. Microbiol.* **20**, 253.
Ishimoto, N. and Strominger, J. L. (1966). *J. Biol. Chem.* **241**, 639.
Ito, E. and Strominger, J. L. (1962a). *J. Biol. Chem.* **237**, 2689.
Ito, E. and Strominger, J. L. (1962b). *J. Biol. Chem.* **237**, 2696.
Ito, E. and Strominger, J. L. (1964). *J. Biol. Chem.* **239**, 210.
Izaki, K., Matsuhashi, M. and Strominger, J. L. (1966). *Proc. Natl. Acad. Sci. U.S.* **55**, 656.
Koepsell, H. J., Tsuchiya, H. M., Hellman, N. N., Kazenko, A., Hoffman, C. A., Sharpe, E. S. and Jackson, R. W. (1953). *J. Biol. Chem.* **200**, 793.
Kornberg, H. L. and Elsden, S. R. (1961). *Advan. Enzymol.*, **23**, 401.
Kornberg, H. L. and Krebs, H. A. (1957). *Nature*, **179**, 988.
Kornfield, S. and Glaser, L. (1961). *J. Biol. Chem.* **236**, 1791.
Kornfield, S. and Glaser, L. (1962). *J. Biol. Chem.* **237**, 3052.
Krebs, H. (1964). *Proc. Roy. Soc. (London) Ser. B.* **159**, 545.
Larsson, A. and Reichard, P. (1966). *Biochim. Biophys. Acta.* **113**, 407.
Laurent, T. C., Moore, E. C. and Reichard, P. (1964). *J. Biol. Chem.* **239**, 3436.
Levin, D. H. and Racker, E. (1959). *J. Biol. Chem.* **234**, 2532.
Markovitz, A., Cifonelli, J. A. and Dorfman, A. (1959). *J. Biol. Chem.* **234**, 2343.
Matsuhashi, S. (1966). *J. Biol. Chem.* **241**, 4275.
Matsuhashi, S., Matsuhashi, M. and Strominger, J. L. (1966a). *J. Biol. Chem.* **241**, 4267.
Matsuhashi, S., Matsuhashi, M., Brown, J. G. and Strominger, J. L. (1966b). *J. Biol. Chem.* **241**, 4283.
Mayer, R. M. and Ginsburg, V. (1965). *J. Biol. Chem.* **240**, 1900.
Meadow, P. M., Anderson, J. S. and Strominger, J. L. (1964). *Biochem. Biophys. Res. Commun.* **14**, 382.
Mills, G. T. and Smith, E. E. B. (1962). *Fed. Proc.* **21**, 1089.
Moore, E. C., Reichard, P. and Thelander, L. (1964). *J. Biol. Chem.* **239**, 3445.
Nathanson, S. G. and Strominger, J. L. (1963). *J. Biol. Chem.* **238**, 3161.
Nathanson, S. G., Strominger, J. L. and Ito, E. (1964). *J. Biol. Chem.* **239**, 1773.
Nikaido, H. (1962). *Proc. Natl. Acad. Sci. U.S.* **48**, 1337.

Nikaido, H. and Nikaido, K. (1966). *J. Biol. Chem.* **241**, 1376.
Nikaido, H., Naide, Y. and Makela, P. H. (1966). *Ann. N.Y. Acad. Sci.* **133**, 299.
Okazaki, R., Okazaki, J., Strominger, J. L. and Michelson, A. M. (1962). *J. Biol. Chem.* **237**, 3014.
Olson, J. A. (1961). *In* "The Enzymes" (P. D. Boyer, H. Lardy and K. Myrbäck, eds.) Academic Press, New York, Vol. **5**, p. 387.
Osborn, M. J. and D'Ari, L. (1964). *Biochem. Biophys. Res. Commun.* **16**, 568.
Osborn, M. J., Rosen, S. M., Rothfield, L. and Horecker, B. L. (1962). *Proc. Natl. Acad. Sci. U.S.* **48**, 1831.
Pazur, J. H. and Shuey, E. W. (1961). *J. Biol. Chem.* **236**, 1780.
Preiss, J. (1964). *J. Biol. Chem.* **239**, 3127.
Reichard, P. *J. Biol. Chem.* **237**, 3513 (1962).
Richmond, M. H. and Perkins, H. R. (1960). *Biochem. J.* **76**, 1P.
Rothfield, L. and Horecker, B. L. (1964). *Proc. Natl. Acad. Sci. U.S.* **52**, 939.
Rothfield, L., Osborn, M. J. and Horecker, B. K. (1964). *J. Biol. Chem.* **239**, 2788.
Sable, H. Z. (1966). *Advan. Enzymol.* **28**, 391.
Segal, S. and Topper, Y. J. (1957). *Biochim. Biophys. Acta.* **25**, 419.
Shaw, D. R. D. (1962). *Biochem. J.* **82**, 297.
Shen, L. and Preiss, J. (1965). *J. Biol. Chem.* **240**, 2334.
Smith, E. E. B., Mills, G. T., Bernheimer, H. P. and Austrian, R. (1958a). *Biochim. Biophys. Acta.* **28**, 211.
Smith, E. E. B., Mills, G. T., Bernheimer, H. P. and Austrian, R. (1958b). *Biochim. Biophys. Acta.* **29**, 640.
Smith, E. E. B., Mills, G. T., Bernheimer, H. P. and Austrian, R. (1960). *J. Biol. Chem.* **235**, 1876.
Smith, E. E. B., Mills, G. T. and Bernheimer, H. P. (1961). *J. Biol. Chem.* **236**, 2179.
Smith, R. A. and Gunsalus, I. C. (1957). *J. Biol. Chem.* **229**, 305.
Sokatch, J. R. (1960). *Arch. Biochem. Biophys.* **91**, 240.
Southard, W. H., Hayashi, J. A. and Barkulis, S. S. (1959). *J. Bacteriol.* **78**, 79.
Strominger, J. L. (1962). *In* "The Bacteria" (I. C. Gunsalus and R. Y. Stanier, eds.) Academic Press, New York, Vol. III, p. 413.
Tipper, D. J. and Strominger, J. L. (1965). *Proc. Natl. Acad. Sci. U.S.* **54**, 1133.
Utter, M. F. and Kurahashi, K. (1954). *J. Biol. Chem.* **207**, 821.
Warren, W. and Blacklow, R. S. (1962). *J. Biol. Chem.* **237**, 3527.
Weiner, I. M., Higuchi, T., Rothfield, L., Saltmarsh-Andrew, M., Osborn, M. J. and Horecker, B. L. (1965).*Proc. Natl. Acad. Sci. U.S.* **54**, 228.
Weiner, I. M., Higuchi, T., Osborn, M. J. and Horecker, B. L. (1966). *Ann. N.Y. Acad. Sci.* **133**, 391 (1966).
Wilkinson, J. F. (1957). *Nature* **180**, 995.
Wise, E. M. and Park, J. T. (1965). *Proc. Natl. Acad. Sci. U.S.* **54**, 75.
Wong, D. T. O. and Ajl, S. J. (1956). *J. Am. Chem. Soc.* **78**, 3220.
Wood, H. G. and Stjernholm, R. L. (1962). *In* "The Bacteria" (I. C. Gunsalus and R. Y. Stanier, eds.) Academic Press, New York, Vol. III, p. 41.
Wright, A., Dankert, M. and Robbins, P. W. (1965). *Proc. Natl. Acad. Sci. U.S.* **54**, 235.
Zevenhuizen, L. P. T. M. (1964). *Biochim. Biophys. Acta.* **81**, 608.

16. BIOSYNTHESIS OF AMINO ACIDS

I. Assimilation of Inorganic Nitrogen

A. Nitrate Assimilation

Many bacteria have the ability to grow with nitrate as the sole nitrogen source which means that nitrate must be reduced to ammonia, an eight-electron change in oxidation state. The enzymes of nitrate assimilation are distinct from those involved in nitrate respiration (Chapter 12), and are differentiated operationally by the fact that enzymes of nitrate assimilation are pyridine nucleotide-linked while the enzymes of nitrate respiration are connected to the cytochrome system.

The first step in nitrate assimilation is the reduction of nitrate to nitrite. Nicholas and Nason (1955) purified the first assimilatory nitrate reductase of a bacterial species in their study of the enzyme from *E. coli*. The *E. coli* enzyme resembled *Neurospora* nitrate reductase (Nason and Evans, 1953) in that the bacterial enzyme was a molybdoflavoprotein although it required NADH as the electron donor rather than NADPH. Taniguchi and Ohmachi (1960) studied an NAD-linked nitrate reductase from *Azotobacter vinelandii* which was associated with the particulate fraction but which was not reduced by cytochromes and therefore appears to be an assimilatory enzyme.

Kemp and Atkinson (1966) studied nitrite reductase of *E. coli* which appears to be an assimilatory enzyme. *E. coli* nitrite reductase utilized NADH as the reducing agent and three moles of NADH were consumed per mole of nitrite reduced to ammonia as required. Spencer *et al.* (1957) studied the soluble nitrite reductase of *Azotobacter agile* and concluded that this enzyme was a metallo-flavoprotein which utilized reduced pyridine nucleotides as electron donors.

B. Nitrogen Assimilation; Nitrogen Fixation

The reduction of molecular nitrogen to ammonia is a process principally of microorganisms, and occurs in bacteria of such widely diverse genera as *Clostridium*, *Azotobacter*, the symbiotic *Rhizobium*, and photosynthetic *Rhodospirillum* and *Chromatium* (see review of Takahashi *et al.*, 1963). Nitrogen fixation in a cell-free preparation was first demonstrated by Carnahan *et al.* (1960) with extracts of dried cells of *C. pasteurianum*. The product was identified as ammonia and the process required pyruvate and strictly anaerobic conditions. Subsequently nitrogen fixation was detected in extracts of other species of nitrogen-fixing organisms using the procedure of Carnahan *et al.* (see review of Mortenson, 1963). Further studies of the role of pyruvate in nitrogen fixation led Mortenson *et al.* (1963) to the discovery of ferredoxin and to the elucidation of its role as an electron carrier in the phosphoroclastic decomposition of pyruvate by *C. pasteurianum* (see Chapter 6). Mortenson (1964) found that acetylphosphate, or ATP, and hydrogen would substitute for pyruvate in nitrogen-fixing preparations of *C. pasteurianum*. Ferredoxin was still required for nitrogen fixation which means that it must play a role in nitrogen reduction by hydrogen as well as in pyruvate oxidation.

C. Ammonia Assimilation

Ammonia is the form of inorganic nitrogen most readily assimilated by bacteria able to grow with an inorganic nitrogen compound as the sole nitrogen source. The process has been studied in *E. coli*, and it seems likely that the route of ammonia fixation involves the reductive amination of glutamate by NADP-linked glutamate dehydrogenase and then dispersal of the amino group to other amino acids by transamination:

$$NH_3 + NADPH + 2\text{-oxoglutarate} \rightarrow \text{L-glutamate} + NADP$$

$$\text{L-Glutamate} + 2\text{-oxoacid} \rightarrow \text{L-amino acid} + 2\text{-oxoglutarate.}$$

Halpern and Umbarger (1960) studied the level of NADP-linked glutamate dehydrogenase in *E. coli* as a function of the composition of the growth medium and found that activity was high when *E. coli* was grown in a glucose-salts medium, but low when grown in a medium with casein hydrolysate. The interpretation was that glutamate dehydrogenase activity was elevated under conditions where ammonia fixation was important. Burchall *et al.* (1964a) studied the level of glutamate dehydrogenase in *Streptococcus bovis* and also observed that glutamate dehydrogenase activity was high when the organism was grown in a medium with ammonia as the sole nitrogen source.

Rudman and Meister (1953) characterized two transaminases from *E. coli* which catalyzed the reversible transfer of amino groups from several amino acids to 2-oxoglutarate. Transaminase A was active mainly with the aromatic amino acids and aspartate, and transaminase B was active mainly with the branched-chain amino acids and aliphatic amino acids (Table 16.1). Both enzymes also catalyzed amino transfer among amino acids which were substrates for the transaminases A and B as well as with glutamate. Except for alanine and serine all of the amino acids that are formed by transamination as the final biosynthetic step are represented in Table 16.1.

TABLE 16.1. Specificity of transaminases A and B of *E. coli* (Rudman and Meister, 1953). Enzyme activity was measured with the indicated amino acids as amino donors and 2-oxoglutarate as the amino acceptor. Specific activity is expressed as μmoles of L-glutamate formed per hour per mg protein.

	Specific Activity	
Amino Donor	Transaminase A	Transaminase B
L-Isoleucine	0	210
L-Valine	0	145
L-Leucine	33	250
L-Norvaline	33	286
L-Norleucine	72	196
L-Methionine	99	78·1
L-Phenylalanine	446	64·6
L-Tyrosine	257	15
L-Tryptophane	598	0
L-Aspartic Acid	1010	0

Glutamate dehydrogenase may not be the only enzyme involved in ammonia fixation. Halpern and Umbarger (1960) found that aspartase activity was high when glutamate was added to a glucose–salts medium and felt that a biosynthetic role for this enzyme could not be ruled out (see also section III.B.4 in Chapter 11). Vender and Rickenberg (1964) obtained mutants of *E. coli* which lacked glutamate dehydrogenase and found that these organisms were able to grow slowly in a glycerol–salts medium, and concluded that the activity of aspartase under these conditions was enough to account for ammonia assimilation. Hong *et al.* (1959) found that several species of *Bacillus* which lacked glutamate dehydrogenase were able to grow with ammonia as the sole nitrogen source. These organisms possessed an NAD-linked L-alanine dehydrogenase, and Shen *et al.* (1959) proposed a pathway of ammonia assimilation in *Bacillus* analogous to that for *E. coli* with the substitution of L-alanine dehydrogenase for L-glutamic dehydrogenase.

II. Biosynthesis of Amino Acids

A. The Glutamate Family

1. *Glutamate*

Cutinelli *et al.* (1951a) showed that *E. coli* synthesized the carbon skeleton of glutamate from acetate by way of the tricarboxylic acid cycle. Cutinelli and his associates used acetate labelled with ^{13}C in the methyl carbon and ^{14}C in the carboxyl carbon and determined the isotope distribution of glutamate and other amino acids of cellular protein. In the case of glutamate carbons 2, 3 and 4 were derived from the methyl carbon of acetate and both carboxyl carbons contained isotope from the carboxyl carbon of acetate, but the α-carboxyl carbon had a lower specific activity than the γ-carboxyl (Table 16.2). The unequal

Table 16.2. Derivation of glutamate carbons from bicarbonate (b), carboxyl carbon of acetate (c) and methyl carbon of acetate (m) during growth of *E. coli*, *C. kluyveri* and *R. rubrum*. [1]Acetate-2-^{14}C was not used in this experiment but these positions are assumed to be derived from the methyl carbon of acetate.

Glutamate carbon:	Origin of carbon of glutamate formed by: *E. coli*	*C. kluyveri*	*R. rubrum*
COOH	c, m	c	b
$CHNH_2$	m	m[1]	c
CH_2	m	c	m
CH_2	m	m[1]	m
COOH	c	b	c

specific activities of the carboxyl carbons of glutamate can be explained on the basis of the reactions of the tricarboxylic acid cycle. The portion of citrate derived from oxaloacetate is the first portion oxidized, hence the ^{14}C derived from the carboxyl carbon of acetate is retained at the stage of succinate and equilibrates between the two carboxyl carbons of succinate. Therefore, in the condensation of acetyl-CoA and oxaloacetate, the carboxyl carbon of citrate derived from acetate has approximately twice the specific activity of the other two carboxyls. This explanation implies that the α-carboxyl of glutamate should have a higher concentration of ^{13}C than the γ-carboxyl, and this was actually observed by Cutinelli *et al.* (1951 a). This pathway of glutamate formation is typical of aerobic organisms (Ehrensvärd, 1955). The final step in glutamate formation would be reductive amination or transamination as outlined in section I.C of this chapter.

Tomlinson (1954 a) discovered an interesting situation in the synthesis of L-glutamate by *C. kluyveri*. This organism incorporated labelled bicarbonate into the terminal carboxyl carbon of glutamate, while the α-carboxyl was derived mainly from the carboxyl carbon of acetate (Table 16.2). These findings could be explained if *C. kluyveri* synthesized 2-oxoglutarate by way of the tricarboxylic acid cycle but possessed citrate synthetase or aconitase of opposite stereospecificity to the enzymes in aerobic organisms. The fixation of CO_2 could be accounted for by the carboxylation of acetyl-CoA and pyruvate to oxaloacetate, along the lines suggested for those photosynthetic bacteria which use the reductive carboxylic acid cycle (see Fig. 14.3):

$$\text{Acetyl-CoA} + CO_2 + 2(H) \rightarrow \text{pyruvate} + \text{CoA}$$
$$\text{Pyruvate} + CO_2 \rightarrow \text{oxaloacetate}$$
$$\text{Oxaloacetate} + \text{acetyl-CoA} \rightarrow \text{citrate} + \text{CoA}$$
$$\text{Citrate} \rightarrow \text{2-oxoglutarate} + CO_2 + 2(H).$$

Citrate synthetase and aconitase of *C. kluyveri* were studied by Gottschalk and Barker (1966), who came to the conclusion that citrate synthetase did indeed possess an unorthodox stereospecificity, but in the same journal Stern *et al.* (1966) reported exactly the opposite result.

Hoare (1963) uncovered still a third pathway of glutamate synthesis in *R. rubrum* and *C. thiosulfatophilum* (Hoare and Gibson, 1964). The isotope distribution of glutamate from cells grown in the presence of radioactive acetate or bicarbonate was different from that reported for *E. coli* and *C. kluyveri* (Table 16.2). The pattern became understandable, however, with the discovery of 2-oxoglutarate synthetase in *C. thiosulfatophilum* by Buchanan and Evans (1965). 2-Oxoglutarate synthesized by the reactions of the reductive carboxylic acid cycle (Fig. 14.3) would account for the labelling pattern of the carbons of glutamate formed by *R. rubrum* and *C. thiosulfatophilum.*

2. *Glutamine*

Fry (1955) studied glutamine synthesis in *M. pyogenes* and found that the bacterial glutamine synthetase (L-glutamate: ammonia ligase [ADP]) catalyzed the same reaction as the well-known animal enzyme:

$$NH_3 + \text{L-Glutamate} + \text{ATP} \rightarrow \text{L-glutamine} + \text{ADP} + P_i$$

Woolfolk and Stadtman (1964) reported similar results with enzyme preparations from *E. coli*.

3. *Proline*

A metabolic relationship between glutamate and proline was established when Vogel and Davis (1952) found that one of their mutants of *E. coli* which was unable to synthesize proline, would grow if glutamate was added to the medium. Vogel and Davis identified an excretory product of the mutant as Δ^1-pyrroline-5-carboxylate, and postulated that this compound arose from the spontaneous cyclization of glutamate-γ-semialdehyde and was a precursor of proline by the route shown in Fig. 16.1. The only step of the pathway which has been studied with

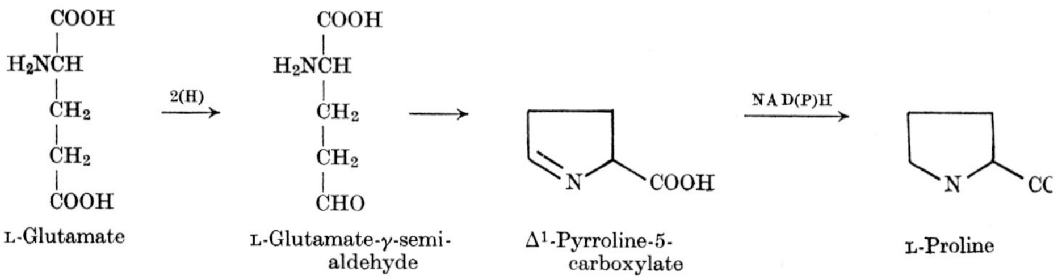

COOH–H$_2$NCH–CH$_2$–CH$_2$–COOH (L-Glutamate) $\xrightarrow{2(H)}$ COOH–H$_2$NCH–CH$_2$–CH$_2$–CHO (L-Glutamate-γ-semialdehyde) $\longrightarrow$ N, COOH (Δ^1-Pyrroline-5-carboxylate) $\xrightarrow{NAD(P)H}$ N, CC (L-Proline)

FIG. 16.1. Conversion of L-glutamate to L-proline.

cell-free preparations is the reduction of Δ^1-pyrroline-5-carboxylate to L-proline, demonstrated by Meister *et al.* (1957) with enzyme preparations of *E. coli* and *A. aerogenes* and by Yura and Vogel (1959) with similar preparations of *Neurospora crassa*.

4. *Arginine*

Vogel (1953) uncovered the key to the biosynthesis of arginine when he found that a mutant of *E. coli* which was unable to synthesize ornithine accumulated a compound which he identified as *N*-acetylglutamate-γ-semialdehyde. Vogel proposed that the initial step in the formation of ornithine was the formation of *N*-acetyl-L-glutamate (Fig. 16.2), which immediately distinguishes the arginine pathway from the proline pathway. The advantage to the organism of having *N*-acetyl-L-glutamate as an intermediate in the formation of ornithine is clear; blocking the amino group prevents cyclization of the γ-semialdehyde as as it occurs in the proline pathway.

Maas *et al.* (1953) demonstrated the acetylation of L-glutamate by enzyme preparations of *E. coli* using acetylphosphate as the acyl donor.

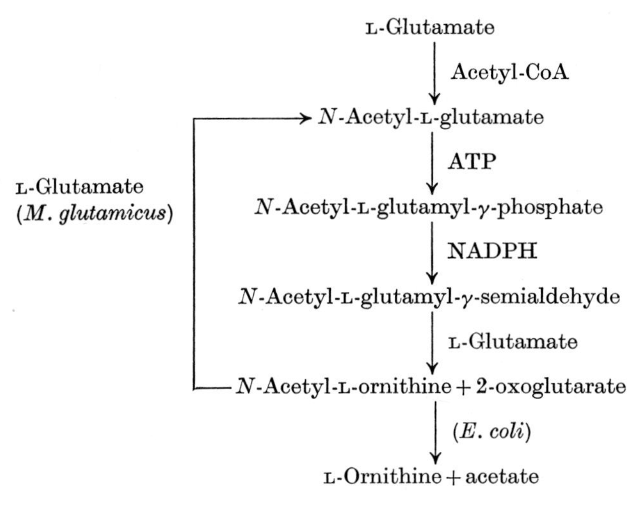

FIG. 16.2. Biosynthesis of arginine; conversion of L-glutamate to ornithine in *E. coli* and *M. glutamicus*.

Since CoA was present and *E. coli* is known to possess phosphotransacetylase, presumably acetyl-CoA was the acyl donor. However, the reaction was rather unspecific, several other amino acids being acetylated in addition to L-glutamate. A somewhat different system for the acetylation of glutamate was studied by Udaka and Kinoshita (1959), who discovered an enzyme in *M. glutamicus* which catalyzed the transfer of an acetyl residue from *N*-acetylornithine to L-glutamate (Fig. 16.2). The transacetylase apparently functions in the cyclic transfer of acetyl residues between these two compounds, and would reduce the ATP requirement for ornithine biosynthesis since only catalytic amounts of acetylated derivatives would be needed. Udaka and Kinoshita also demonstrated the conversion of *N*-acetyl-L-glutamate to *N*-acetyl-L-glutamate-γ-semialdehyde in extracts of *M. glutamicus* supplemented with ATP and NADPH. Baich and Vogel (1962) demonstrated that two enzymes were involved in these reactions, *N*-acetyl-L-glutamate kinase and *N*-acetyl-L-glutamyl-γ-semialdehyde dehydrogenase. Transamination between *N*-acetyl-L-glutamyl-γ-semialdehyde and L-glutamate was demonstrated by Udaka and Kinoshita in extracts prepared from *M. glutamicus*, and the enzyme *N*-acetyl-L-ornithine transaminase (α-*N*-acetyl-L-ornithine:2-oxoglutarate aminotransferase) was purified by Albrecht and Vogel (1964) from cell-free preparations of *E. coli*. The purified transaminase was specific for *N*-acetyl-L-ornithine and 2-oxoglutarate and required pyridoxal phosphate as a cofactor. Vogel and

Bonner (1956) purified acetylornithinase (α-*N*-acetyl-L-ornithine amidohydrolase) from *E. coli* and the enzyme which catalyzes the hydrolysis of *N*-acetyl-L-ornithine. In *M. glutamicus* the same result is achieved by means of the transacetylase reported by Udaka and Kinoshita.

The conversion of ornithine to citrulline (Fig. 16.3) is catalyzed by ornithine transcarbamylase, which is also involved in the degradation of arginine. Ornithine transcarbamylase of *E. coli* was highly purified by Rogers and Novelli (1962). The last two enzymes of the arginine pathway, argininosuccinate synthetase (L-citrulline:L-aspartate ligase [AMP]) and argininosuccinase (L-argininosuccinate arginine-lyase) have been highly purified from liver (see review of Ratner, 1962) but not from bacterial systems. The condensation of citrulline and aspartate requires energy which is supplied by ATP. The cleavage of argininosuccinate to arginine and fumarate is similar to the reaction catalyzed by aspartase.

Carbamoylphosphate synthesis in *E. coli* is catalyzed by a considerably different enzyme from the one involved in arginine degradation in *S. faecalis*. Kalman *et al.* (1966) purified the enzyme from *E. coli* and discovered that the ammonia donor is actually glutamine and that 2 moles of ATP were utilized per mole of carbamoyl phosphate formed:

$$HCO_3^- + \text{L-glutamine} + 2ATP \rightarrow NH_2\text{-}CO\text{-}OPO_3H_2 + 2ADP + H_3PO_4 + \text{L-glutamate.}$$

It is interesting that both the *E. coli* and animal carbamoylphosphate synthetases utilize two moles of ATP in the formation of carbamoylphosphate, since both are biosynthetic enzymes.

B. The Aspartate Family

1. *Aspartate*

Aspartate is an important amino acid which is involved in the biosynthesis of lysine, threonine, methionine and isoleucine and provides four of the carbon atoms of each of these amino acids. Cutinelli *et al.* (1951a) found, in contrast to their findings with glutamate, that aspartate isolated from protein of *E. coli* grown on acetate-1-^{14}C,-2-^{13}C was symmetrically labelled (Table 16.3). They explained this result by proposing that acetate was converted to oxaloacetate by the reactions of the tricarboxylic acid cycle, and that oxaloacetate would be symmetrically labelled because succinate is an intermediate in its formation. Wang *et al.* (1952) also found that aspartate synthesized by yeast from acetate-1-^{14}C was symmetrically labelled, but when synthesized from pyruvate-2-^{14}C the label was mainly in the α-amino carbon, which sug-

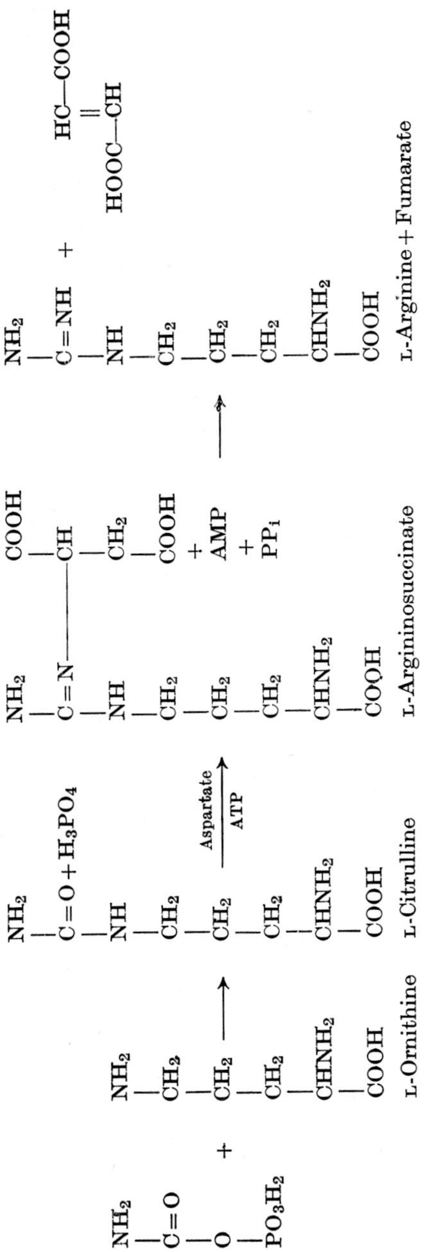

$$NH_2{-}C({=}O){-}O{-}PO_3H_2 + NH_2{-}CH_2{-}CH_2{-}CH_2{-}CHNH_2{-}COOH \longrightarrow NH_2{-}C({=}O){-}NH{-}CH_2{-}CH_2{-}CH_2{-}CHNH_2{-}COOH + H_3PO_4$$

L-Ornithine → L-Citrulline

$$\xrightarrow[\text{ATP}]{\text{Aspartate}} NH_2{-}C({=}N{-}CH(COOH){-}CH_2{-}COOH){-}NH{-}CH_2{-}CH_2{-}CH_2{-}CHNH_2{-}COOH + AMP + PP_i$$

L-Argininosuccinate

$$\longrightarrow NH_2{-}C({=}NH){-}NH{-}CH_2{-}CH_2{-}CH_2{-}CHNH_2{-}COOH + HOOC{-}CH{=}CH{-}COOH$$

L-Arginine + Fumarate

FIG. 16.3 Biosynthesis of arginine; conversion of L-ornithine to L-arginine.

TABLE 16.3. Derivation of aspartate carbons from biocarbonate (b), carboxyl carbon of acetate (c) and methyl carbon of acetate (m) during growth of *E. coli*, *R. rubrum*, *C. kluyveri* and *C. thiosulfatophilum*. [1]Acetate-2-^{14}C was not used in the experiment with *C. kluyveri* but this position is assumed to be derived from the methyl carbon of acetate.

	Origin of carbon of aspartate formed by:	
Aspartate carbon	*E. coli*	*R. rubrum, C. kluyveri, C. thiosulfatophilum*
COOH	c	b
$CHNH_2$	m	c
CH_2	m	m[1]
COOH	c	b

gested that carboxylation of pyruvate was the principal route of aspartate formation under these conditions. Aspartate biosynthesis is completed by transamination between oxaloacetate and L-glutamate.

In anaerobic bacteria, the isotope distribution of aspartate formed from acetate is different from that formed by aerobes (Table 16.3). In *R. rubrum* (Cutinelli *et al.*, 1951b), *C. kluyveri* (Tomlinson, 1954a) and *C. thiosulfatophilum* (Hoare and Gibson, 1964) the labelling pattern is the same and can best be explained by a stepwise addition of carbon dioxide to acetate and to pyruvate (see also Fig. 14.3 and section II.A.1 of this chapter):

$$\text{Acetyl-CoA} + CO_2 + 2(H) \rightarrow \text{pyruvate} + \text{CoA}$$
$$\text{Pyruvate} + CO_2 \rightarrow \text{oxaloacetate.}$$

2. *Asparagine*

Ravel *et al.* (1962) purified the asparagine synthetase (L-aspartate: ammonia ligase [AMP]) ten-fold from extracts of *L. arabinosus* and determined that the enzyme catalyzed the following reaction:

$$\text{L-Aspartate} + \text{ATP} + NH_3 \rightarrow \text{L-asparagine} + \text{AMP} + PP_i$$

Burchall *et al.* (1964b) purified a similar enzyme from *S. faecalis*. In the formation of asparagine, pyrophosphate and AMP are the products of ATP cleavage, whereas in the formation of glutamine, ADP and phosphate are the products.

3. *Lysine*

Gilvarg (1958) provided an important clue to the nature of the route for lysine biosynthesis (Fig. 16.4) when he found that fortified extracts of a mutant of *E. coli*, unable to synthesize lysine, formed 2,6-diaminopimelate from aspartate and pyruvate. It was already known that diaminopimelate was a constituent of the cell wall of certain bacteria and that it could be converted to lysine by the action of a specific decarboxylase (Hoare and Work, 1955).

Cohen *et al.* (1954) first provided evidence for the initial reactions in the conversion of aspartate to amino acids of the aspartate family in *E. coli*. They prepared extracts which catalyzed the formation of a hydroxamic acid when aspartate, ATP and hydroxylamine were added, and these results were interpreted to mean that aspartyl phosphate had been formed. When NADP was added as well, homoserine, a precursor of threonine, was formed. Black and Wright (1955a, b) characterized the enzymes which are responsible for the formation of aspartate β-semialdehyde from yeast, aspartyl kinase (ATP:L-aspartate 4-phosphotransferase) and aspartate-β-semialdehyde dehydrogenase (L-aspartic-β-semialdehyde:NADP oxidoreductase [phosphorylating]). These enzymes are involved in threonine biosynthesis in yeast, but not in lysine biosynthesis since this amino acid is formed by a pathway where α-aminoadipate is a key intermediate (see review of Meister, 1965). Yugari and Gilvarg (1965) purified the enzyme which catalyzed the condensation of L-aspartate-β-semialdehyde and pyruvate from extracts of *E. coli*, and characterized the product as one of the isomers of dihydropicolinate, probably the one shown in Fig. 16.4. The intermediate shown in brackets in Fig. 16.4 has the structure expected of the product formed in the condensation reaction, but appears to cyclize rapidly since the formation of the six-membered ring is highly favored. Farkas and Gilvarg (1965) purified an enzyme from *E. coli* which catalyzed the reduction of the intermediate formed in the condensation reaction by NADPH, and characterized the product of this reaction as Δ^1-piperideine-2,6-dicarboxylate. Gilvarg (1962) reported preliminary evidence for *N*-succinyl-L-2-amino-6-oxopimelate with succinyl-CoA as the donor of the succinyl residue. Apparently the cyclic and open chain compounds are in equilibrium, and it is the latter that is the substrate for an enzyme which adds the succinyl residue. A specific transaminase from *E. coli* (*N*-succinyl-L-2,6-diaminopimelate:2-oxoglutarate aminotransferase) catalyzes the transfer of the amino group from L-glutamate to *N*-succinyl-L-2-amino-6-oxopimelate and yields *N*-succinyl-L,L-2,6-diaminopimelate (Peterkofsky and Gilvarg, 1961).

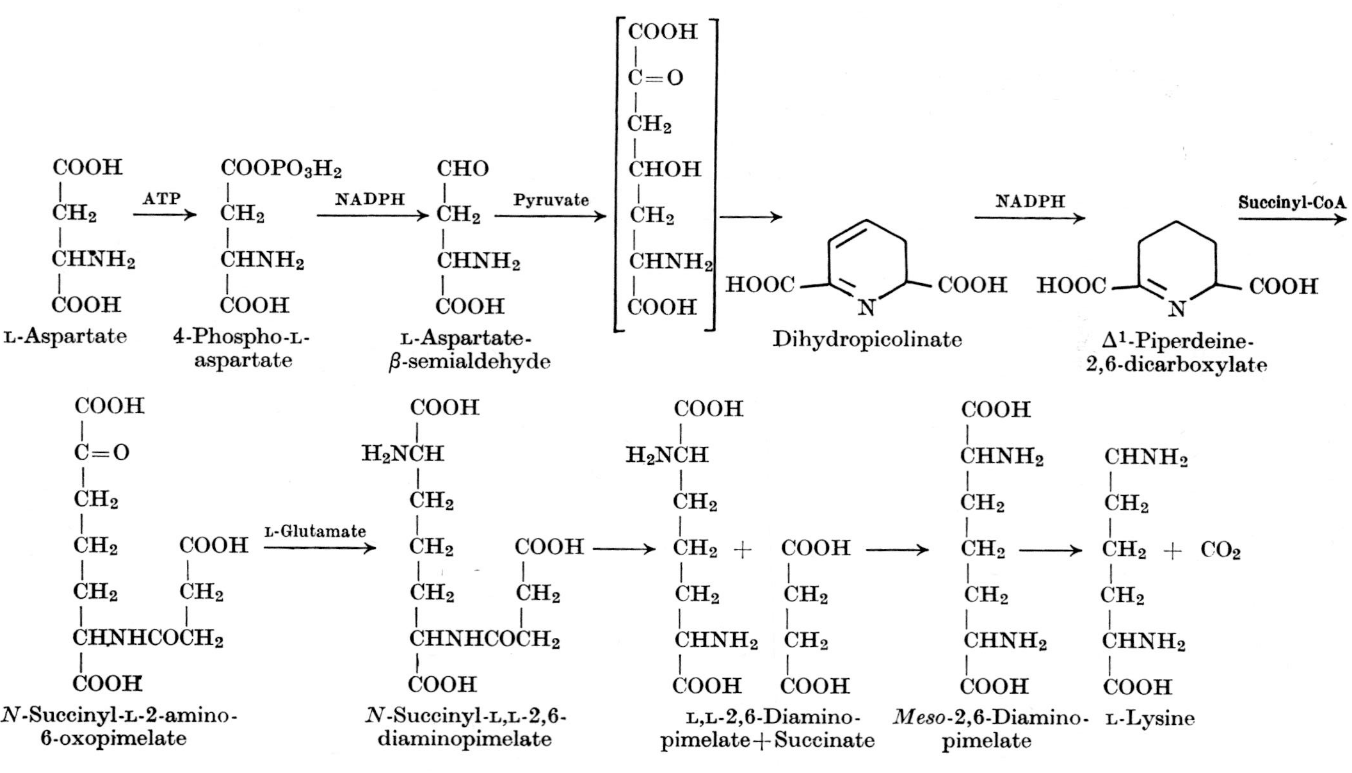

Fig. 16.4. Biosynthesis of L-lysine in *E. coli*.

Kindler and Gilvarg (1960) purified *N*-succinyl-2,6-diaminopimelate deacylase (*N*-succinyl-L,L-2,6-diaminopimelate amidohydrolase) from *E. coli* and showed that the enzyme was distinct from acetylornithinase although both enzymes catalyzed a similar reaction and were activated by cobalt. Antia *et al.* (1957) had already studied a racemase from *A. aerogenes* which catalyzed the interconversion of L,L-2,6-diaminopimelate and *meso*-2,6-diaminopimelate (2,6-L,L-diaminopimelate 2-epimerase), and Hoare and Work (1955) described a decarboxylase from the same organism which specifically catalyzed the decarboxylation of the *meso* isomer of diaminopimelate to L-lysine (*meso*-2,6-diaminopimelate carboxy-lyase).

4. *Threonine*

Cutinelli *et al.* (1951a) provided one of the early links between aspartate and threonine when they found that these two amino acids formed by *E. coli* during growth on doubly labelled acetate were similarly labelled. The finding of Cohen *et al.* (1954) that extracts of *E. coli* were able to convert aspartate to homoserine when supplemented with ATP and NADP was discussed in the previous section. These same extracts converted homoserine to threonine when ATP and pyridoxal phosphate were added.

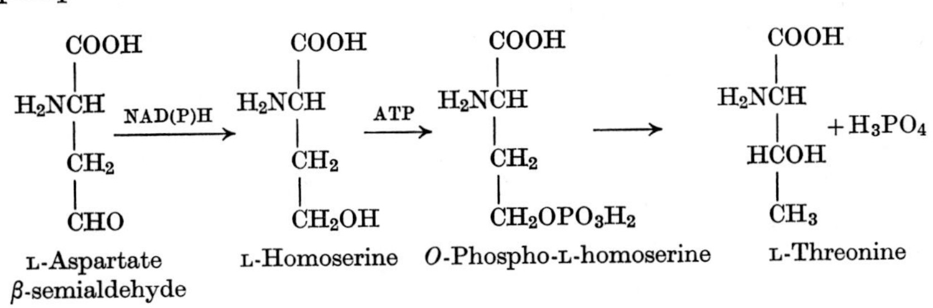

FIG. 16.5. Biosynthesis of threonine in yeast and *E. coli*.

All the reactions of the threonine pathway (Fig. 16.5) have been demonstrated in cell-free preparations of bacteria. Black and Wright (1955c) were the first to purify homoserine dehydrogenase (L-homoserine:NAD oxidoreductase) in their study of threonine biosynthesis in yeast. Datta and Gest (1965) later purified homoserine dehydrogenase from *R. rubrum* and found it to be similar to the yeast enzyme except that NADPH was a better hydrogen donor than NADH. The phosphorylation of homoserine to *O*-phosphohomoserine and its conversion to L-threonine were first demonstrated by Watanabe *et al.* (1955) with preparations of yeast enzymes. Wormser and Pardee (1958) demonstrated the existence of both enzymes, homoserine kinase (ATP:L-homoserine *O*-phosphotransferase) and threonine synthetase (*O*-phosphohomoserine phospholyase [adding water]), in cell-free preparations of *E. coli*. Flavin and Slaughter (1960) succeeded in purifying the threonine synthetase of *N. crassa* to a state of homogeneity and found that pyridoxal phosphate was a cofactor for the reaction.

5. *Methionine*

From the results of isotope competition studies with *E. coli*, Abelson *et al.* (1953) recognized that methionine was one of the amino acids of the aspartate family and that homoserine was a precursor of methionine. This is different from the situation in animal tissues where methionine is supplied by the diet and is a precursor of cystathionine and cysteine. Wijesundera and Woods (1962) found that enzyme preparations of *E. coli* cleaved cystathionine by a β-elimination reaction to homocysteine and pyruvate. Thus it became apparent that while cystathionine was an intermediate in formation of sulfur amino acids in both animal and bacterial cells, it led to the formation of cysteine in one case and methionine in the other.

Much work has been done on the biosynthesis of methionine by bacteria in the laboratory of D. D. Woods, where it was discovered that succinate, ATP and coenzyme A as well as homoserine and cysteine were required for cystathionine formation by extracts of *E. coli* (Rowbury and Woods, 1964b). A mutant, unable to form cystathionine, accumulated an intermediate produced from succinyl-CoA and homoserine which Rowbury and Woods, identified as *O*-succinylhomoserine (Fig. 16.6). A second mutant, also unable to synthesize cystathionine, converted *O*-succinylhomoserine to cystathionine when L-cysteine was present. Flavin *et al.* (1964) also reported that extracts of a *Salmonella* mutant converted *O*-succinylhomoserine into cystathionine. Kaplan and Flavin (1966) purified the cystathionine γ-synthetase from *S. typhimurium* to the stage of a homogenous protein and found that pyridoxal phosphate was a cofactor for the reaction and was bound to the enzyme. Cystathionase, the cystathionine cleavage enzyme of *E. coli*, was studied by Wijesundera and Woods (1962), who established that one mole each of homocysteine, pyruvate and ammonia were produced from cystathionine. Delavier-Klutchko and Flavin (1965) purified the enzyme approximately 100 fold from *E. coli*.

Woods and his associates found evidence for two pathways for the methylation of homocysteine in *E. coli*. In a mutant which required cobalamin for growth, the reaction was a complex one which required ATP, *S*-adenosylmethionine, a system for generating NADH, (which

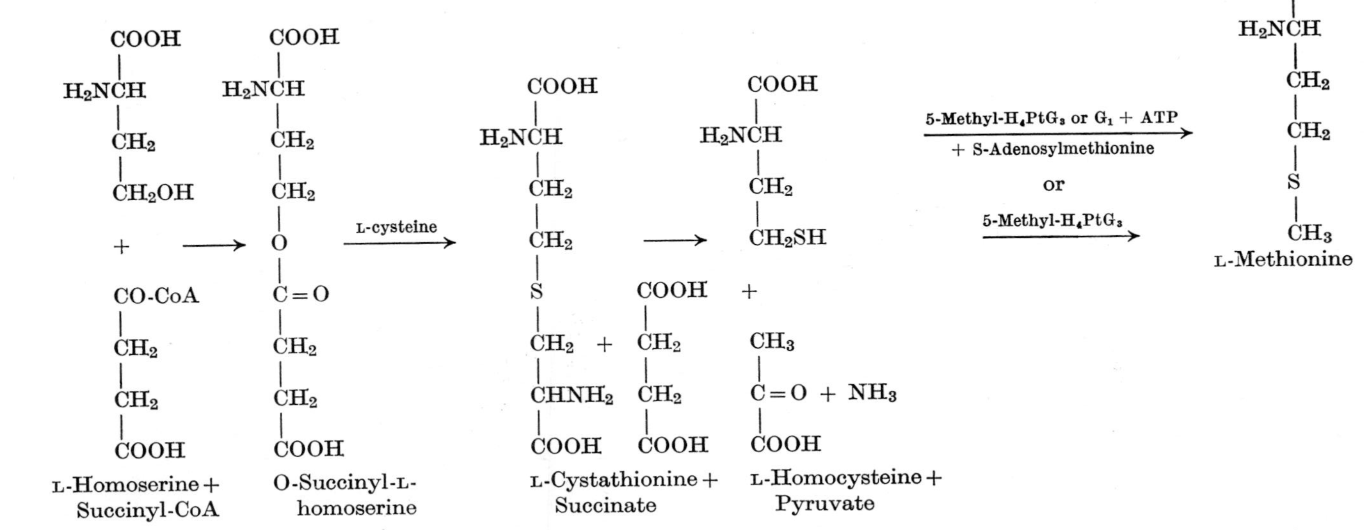

FIG. 16.6. Methionine formation in *E. coli* and *S. typhimurium*. 5-Methyl-H_4-PtG_3, or G_1 are the tri- and monoglutamate derivatives of N^5-Methyltetrahydrofolate respectively.

they supplied as ethanol dehydrogenase, ethanol and NAD), a methyl donor and the enzyme (Guest *et al.*, 1964b). In this enzyme system, N^5-methyltetrahydropteroyl mono- or triglutamate would function as the methyl donor. The enzyme which catalyzed the reaction contained cobalamin, a form of vitamin B_{12}, as the cofactor. NADH presumably acted as a reducing system, but its function and the function of *S*-adenosylmethionine were not apparent. Guest *et al.* (1964a) also identified an enzyme in *E. coli* which catalyzed the reduction of N^5, N^{10}-methylenetetrahydrofolate to N^5-methyltetrahydrofolate with $FADH_2$ as the reducing agent. N^5,N^{10}-Methylenetetrahydrofolate is a product of the cleavage of serine by serine aldolase (section G), and thus carbon 3 of serine can function as a methyl donor in methionine biosynthesis.

The second system for the methylation of homocysteine was present in methionine-sufficient strains of *E. coli*, and the purified enzyme was found to be free of cobalamin. Only the triglutamate derivative of N^5-methyltetrahydrofolate would serve as a methyl donor for this enzyme. The only other cofactor required was magnesium.

Studies with mutants led Foster *et al.* (1964) to the conclusion that the cobalamin-independent system was the primary one for the synthesis of methionine, and that the cobalamin-dependent system was a secondary system which was evoked when mutation resulted in loss of the primary system.

C. Cysteine

1. *Assimilatory Sulfate Reduction*

It was pointed out in the previous section that cysteine is the source of sulfur for methionine in bacteria while in animal tissues the converse is true. Many bacteria are able to grow with sulfate as the sole source of sulfur and, hence, are able to reduce it to sulfide (Peck, 1962). Studies with mutants unable to make cysteine have shown that several of them are deficient in enzymes involved in the reduction of sulfate to sulfide; thus it is apparent that cysteine is involved in the early reactions in the fixation of inorganic sulfur into the cell (Dreyfuss and Monty, 1963).

Assimilatory sulfate reduction begins with the activation of sulfate by the enzyme ATP-sulfurylase (ATP: sulfate adenyl transferase). This enzyme was first isolated from yeast extracts by Robbins and Lipmann (1958), who showed that the product of the reaction between ATP and sulfate was adenosine-5′-phosphosulfate (Table 16.4). Robbins and Lipmann also described the second reaction which is catalysed by APS-phosphokinase (ATP:adenylylsulfate 3′-phosphotransferase) and which results in the phosphorylation of adenosine-5′-

TABLE 16.4. Reactions in the assimilation of sulfate and the formation of L-cysteine in bacteria.

$$ATP + SO_4^{2-} \longrightarrow \text{adenosine-5'-phosphosulfate} + PP_i$$

$$ATP + \text{adenosine-5'-phosphosulfate} \longrightarrow ADP + \text{adenosine-3'-phosphate-5'-phosphosulfate}$$

$$2\ RSH + \text{adenosine-3'-phosphate-5'-phosphosulfate} \longrightarrow SO_3^{2-} + \text{adenosine-3',5'-diphosphate} + RSSR$$

$$SO_3^{2-} + 3\ NADPH + 2H^+ \longrightarrow H_2S + 3\ NADP$$

$$\text{L-Serine} + \text{acetyl-CoA} \longrightarrow \textit{O}\text{-acetyl-L-serine} + CoA$$

$$\textit{O}\text{-Acetyl-L-serine} + H_2S \longrightarrow \text{L-cysteine} + H_2O + \text{acetate}$$

Sum: $2\ ATP + SO_4^{2-} + 2\ RSH + 3\ NADPH + 2H^+ + \text{L-serine} + \text{acetyl-CoA} \longrightarrow ADP + \text{adenosine-3',5'-diphosphate} + RSSR + 3\ NADP + PP_i + \text{acetate} + CoA + H_2O + \text{L-cysteine}$

phosphosulfate to adenosine-3′-phosphate-5′-phosphosulfate. Hilz and Kittler (1960) showed that reduction of adenosine-3′-phosphate-5′-phosphosulfate to sulfite required a thiol such as lipoic acid, and that the thiol itself was reduced by their enzyme preparations when NADPH was added. Bandurski *et al.* (1960) isolated a sulfhydryl protein from yeast which had a molecular weight of 10,000 and which appeared to be the thiol responsible for the physiological reduction of adenosine-3′-phosphate-5′-phosphosulfate. An enzyme fraction from yeast catalyzed the reduction of the disulfide form of the protein RSSR of Table 16.4 to the sulfhydryl form. Mager (1960) isolated a sulfite reductase (hydrogen sulfide:NADP oxidoreductase) from *E. coli* which catalyzed the reduction of sulfite with NADPH to sulfide (Table 16.4). A similar enzyme was reported in *S. typhimurium* by Siegel *et al.* (1964).

2. *Formation of Cysteine*

Abelson (1954) showed that serine was a precursor of the carbon skeleton of cysteine and glycine in *E. coli*, by use of the isotope competition technique. Several years later, Kredich and Tomkins (1966) isolated both enzymes involved in serine biosynthesis from *S. typhimurium.* They purified serine transacetylase 1,000-fold and demonstrated that the acetylation of serine by acetyl-CoA resulted in the formation of *O*-acetyl-L-serine and free CoA as shown in Table 16.4. During purification of serine transacetylase, they became aware of a second enzymic reaction which resulted in the formation of L-cysteine when sulfide was added to their reaction mixtures containing L-serine and acetyl-CoA. Kredich and Tomkins found that the substrate for the second reaction was *O*-acetyl-L-serine and named the enzyme *O*-acetylserine sulfhydrylase.

The synthesis of cysteine in bacteria is similar to the pathway in yeast where cysteine formation is catalyzed by serine sulfhydrylase (L-serine hydro-lyase [adding hydrogen sulfide]) from free serine and hydrogen sulfide:

$$\text{L-Serine} + H_2S \rightarrow \text{L-cysteine} + H_2O$$

while in animal tissues, cysteine is formed from homocysteine and serine by way of the cystathionine pathway (see review of Meister, 1965). Thus, serine is a precursor of cysteine in all organisms studied.

D. Histidine

The earliest clues to the pathway of histidine formation in microorganisms were obtained by isolation of products from culture filtrates of mutants deficient in the ability to synthesize histidine. Vogel *et al.*

(1951) isolated L-histidinol from culture filtrates of a mutant of *E. coli*, and Ames (1955) isolated imidazole glycerol and imidazole acetol as well as the phosphorylated derivatives of the latter two compounds from culture filtrates and mycelia of *Neurospora crassa*. Moyed and Magasanik (1960) used cell-free preparations of three species of Gram negative bacteria to show that the nitrogen at position 3 of the imidazole ring and the carbon at position 2 were derived from adenine of ATP. The remainder of the ATP molecule was recovered as 5-amino-imidazole-4-carboxamide ribonucleotide. Neidle and Waelsch (1959) showed that the other nitrogen of the imidazole ring was derived from the amide nitrogen of glutamine.

These observations outlined the form of the pathway for histidine biosynthesis and all the reactions shown in Fig. 16.7 have been demonstrated. Ames *et al.* (1961) characterized the first reaction of the histidine pathway using a mutant of *S. typhimurium* unable to synthesize histidine. A partially purified enzyme preparation catalyzed the condensation of phosphoribosylpyrophosphate (PRPP) to phosphoribosyl-ATP and pyrophosphate. In this reaction, the nitrogen at position 1 of the purine ring became attached to carbon 1 of ribose. The enzyme which catalyzed this reaction was named phosphoribosyl-ATP pyrophosphorylase. Kornberg *et al.* (1955) had already shown the formation of phosphoribosylpyrophosphate by an enzyme (ATP:D-ribose-5-phosphate pyrophosphotransferase) from pigeon liver:

$$\text{ATP} + \text{ribose-5-phosphate} \rightarrow \text{AMP} + \text{5-phosphoribosylpyrophosphate}.$$

Phosphoribosylpyrophosphate is an intermediate in purine, pyrimidine, tryptophane and histidine biosynthesis. Smith and Ames (1965) demonstrated the conversion of phosphoribosyl-ATP to phosphoribosyl-AMP by extracts of a mutant of *Salmonella* and characterized the products as phosphoribosyl-AMP and pyrophosphate. The enzyme which catalyzed this reaction was designated phosphoribosyl-ATP phosphohydrolase. Smith and Ames (1964) isolated the third intermediate of the histidine pathway, phosphoribosylformimino-5-amino-imidazole-4-carboxamide ribonucleotide, from reaction mixtures containing ATP and phosphoribosylpyrophosphate and extracts of a mutant of *Salmonella* which accumulated this particular intermediate. The reaction resulted in the hydrolytic cleavage of the adenine ring, and the enzyme which catalyzed this cleavage was designated cyclohydrolase by the above authors. They also isolated and characterized the intermediate phosphoribulosylformimino-5-aminoimidazole-4-carboxamide ribonucleotide (Smith and Ames, 1964). The enzyme which catalyzed this isomerization was later highly purified by Margolies and

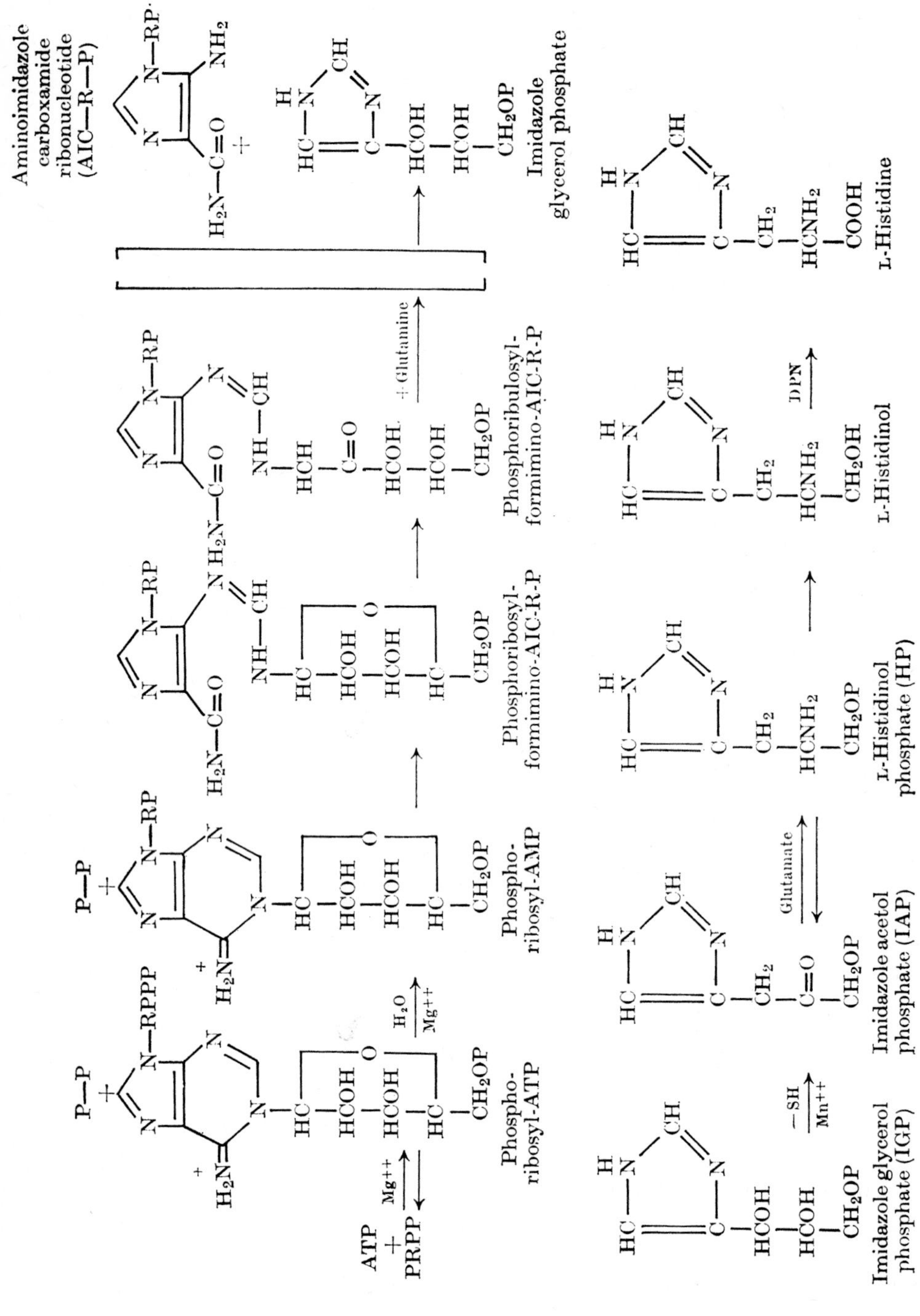

ATP + PRPP $\xrightarrow{Mg^{++}}$ Phosphoribosyl-ATP + P—P $\xrightarrow[Mg^{++}]{H_2O}$ Phosphoribosyl-AMP + P—P → Phosphoribosylformimino-AIC-R-P → Phosphoribulosylformimino-AIC-R-P $\xrightarrow{+\text{ Glutamine}}$ Imidazole glycerol phosphate + Aminoimidazole carboxamide ribonucleotide (AIC—R—P)

Imidazole glycerol phosphate (IGP) $\xrightarrow[Mn^{++}]{-SH}$ Imidazole acetol phosphate (IAP) $\xrightarrow{\text{Glutamate}}$ L-Histidinol phosphate (HP) → L-Histidinol $\xrightarrow{DPN}$ L-Histidine

FIG. 16.7. Biosynthesis of histidine. (From Broquist and Tropin, 1966.)

Goldberger (1966) from extracts of *S. typhimurium*, and goes by the name of phosphoribosylformimino-phosphoribosylaminoimidazole carboxamide ketol isomerase. Smith and Ames (1964) demonstrated the conversion of phosphoribulosylformimino-5-aminoimidazole-4-carboxamide ribonucleotide to imidazole glycerol phosphate and 5-aminoimidazole-4-carboxamide ribonucleotide. This reaction required glutamine and resulted in the addition of the amide nitrogen to the carbonyl carbon of the ribulosyl moiety, release of 5-aminoimidazole-4-carboxamide ribonucleotide and cyclization to form the imidazole ring. The sequence of events in this conversion has not been established nor has the structure of the intermediate in brackets in Fig. 16.7; however, genetic studies indicate that two enzymic steps are involved in this transformation (Smith and Ames, 1964). Ames (1957b) purified the dehydrase from *N. crassa* which catalyzed the conversion of imidazole glycerol phosphate to imidazole acetol phosphate (D-*erythro*-imidazole-glycerolphosphate hydro-lyase). He established the configuration of the substrate as D-*erythro*-imidazole glycerol phosphate by chemical synthesis. Ames and Horecker (1956) purified imidazole acetol phosphate transaminase from *N. crassa.* The transaminase utilized glutamate as the amino donor and required pyridoxal phosphate as a cofactor. Ames (1957a) purified a phosphatase about 13-fold from *N. crassa* which was specific for histidinol phosphate. Adams (1954) purified an enzyme from a soil organism isolated from histidinol enrichment culture. The enzyme, histidinol dehydrogenase (L-histidinol:NAD oxidoreductase) catalyzed the oxidation of histidinol to histidine with the utilization of two equivalents of NAD and without the intermediate formation of histidinal. In a later paper (Adams, 1955) the dehydrogenase was purified from yeast and, although the enzyme catalyzed the oxidation of histidinal, there was no evidence for separation into two separate enzyme activities. Loper and Adams (1965) purified histidinol dehydrogenase of *S. typhimurium* and obtained a homogenous preparation which still catalyzed the formation of histidine from histidinol.

E. The Branched Chain Amino Acids

1. *Valine and Isoleucine*

The pathway for the biosynthesis of these two amino acids is unique among biosynthetic pathways in that one set of enzymes catalyzes the formation of intermediates of both pathways. Strassman *et al.* (1953) studied the formation of valine by *Torulopsis utilis* from radioactive precursors, and found that lactate was an excellent source of valine carbon but that the three carbons of lactate did not appear in order in the valine molecule (Fig. 16.8). They proposed that a condensation between acetaldehyde and pyruvate occurred, with the formation of α-acetolactate followed by migration of a methyl carbon to yield the carbon skeleton of valine. On the basis of similar evidence, Strassman *et al.* (1954) later proposed that the same mechanism applied to the formation of the isoleucine carbon skeleton starting with 2-oxobutyrate instead of pyruvate. Adelberg (1955) further showed that threonine

```
      ᶜCOOH
        |
      ʰC—NH₂
        |
      ʰC
      / \
   ᵐC     ᵐC
```

FIG. 16.8. Origin of valine carbons from lactate; c, h, and m represent the carboxyl, hydroxyl and methyl carbons of lactate respectively (Strassman *et al.*, 1953).

was incorporated into carbons 1, 2, 4 and 5 of isoleucine by *N. crassa* apparently after conversion to 2-oxobutyrate. The fundamental correctness of Strassman's scheme has been borne out by subsequent enzymic studies (Fig. 16.9).

Strassman *et al.* (1960) established the role of α-acetolactate as an intermediate in the biosynthesis of valine by demonstrating the conversion of synthetic α-acetolactate to 2-oxoisovalerate by extracts of baker's yeast. Umbarger and Brown (1958) studied the formation of α-acetolactate from pyruvate by extracts of *E. coli* and found that the enzyme was similar to the catabolic enzyme of *A. aerogenes* studied earlier by Juni and Heym (Chapter 6); thiamine pyrophosphate was a cofactor for both enzymes, but the pH optimum of the enzyme from *E. coli* was 8·0 as opposed to 6·0 for the *Aerobacter* enzyme. Leavitt and Umbarger (1961) reported that extracts of *E. coli* catalyzed the formation of α-aceto-α-hydroxybutyrate when incubated with 2-oxobutyrate and pyruvate. The requirements for the formation of α-aceto-α-hydroxybutyrate were identical to those for the formation of α-acetolactate by the pH 8·0 enzyme, and since mutants deficient in α-acetolactate synthetase were unable to form either valine or isoleucine, they suggested that a single enzyme catalyzed both reactions.

The first step in isoleucine biosynthesis is actually the conversion of threonine to 2-oxobutyrate. Umbarger and Brown (1957) discovered that *E. coli* possessed two separate enzymes which brought about the deamination of threonine; one appeared to be identical to the one reported by Wood and Gunsalus (Chapter 11) which was active on both

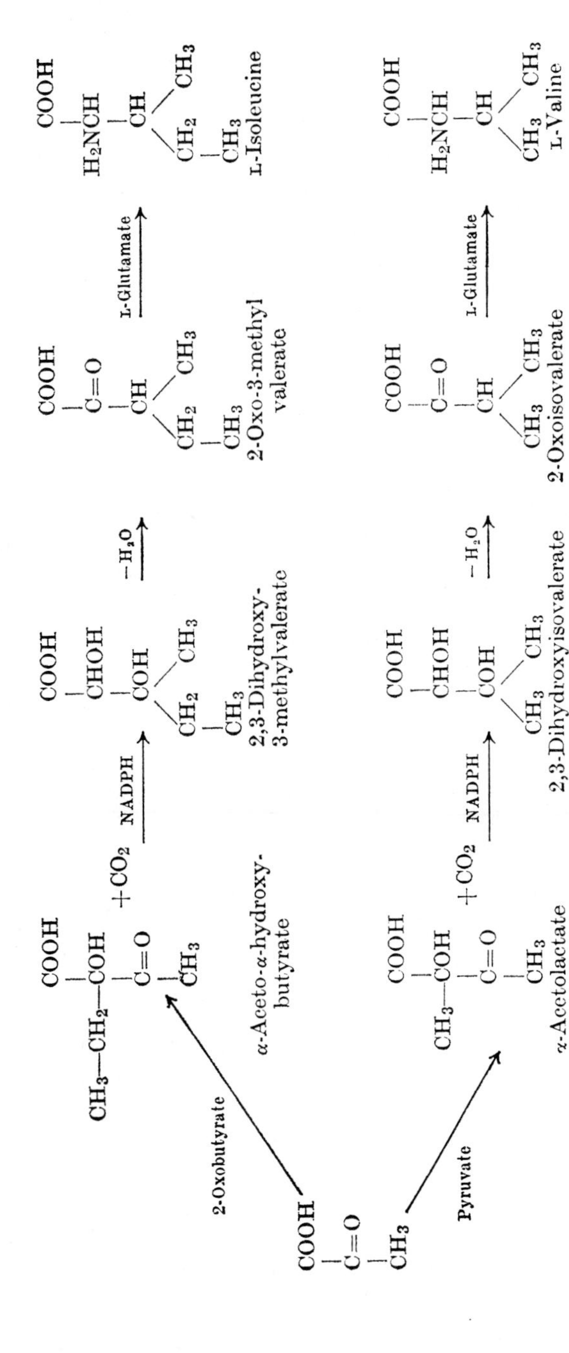

FIG. 16.9. Biosynthesis of isoleucine and valine.

serine and threonine, and the other was a biosynthetic enzyme (L-threonine hydro-lyase [deaminating]) which was lacking in isoleucine-less mutants of *E. coli* which responded to 2-oxobutyrate.

Umbarger *et al.* (1960) demonstrated that extracts of *E. coli* and *A. aerogenes* converted α-acetolactate and α-aceto-α-hydroxybutyrate to 2,3-dihydroxyisovalerate and 2,3-dihydroxy-3-methylvalerate respectively and that the transformation required NADPH. Radhakrishnan *et al.* (1960) purified an enzyme from *E. coli* which catalyzed the reduction and migration of the alkyl residue of both acetohydroxy acids, and they named the enzyme reductoisomerase. Radhakrishnan and his associates synthesized 2-oxo-3-hydroxyisovalerate and 2-oxo-3-hydroxy-3-methylvalerate and found that the purified enzyme had no activity against these compounds. Strassman *et al.* (1960) found that both of the stereoisomers obtained by the direct reduction of α-acetolactate were inactive with an enzyme preparation which converted α-acetolactate to 2-oxoisovalerate. However, the compounds tested by Radhakrishnan *et al.* (1960) and Strassman *et al.* (1960) might still be intermediates in the reactions catalyzed by the reductoisomerase but might be very tightly bound to the enzyme.

Myers (1961) partially purified a dihydroxyacid dehydrase from *E. coli* which catalyzed the removal of water from 2,3-dihydroxy-3-methylvalerate and 2,3-dihydroxyisovalerate to yield the oxoacid precursors of isoleucine and valine respectively.

The final step in isoleucine and valine biosynthesis is transamination with L-glutamate catalyzed by transaminase B of *E. coli* (Rudman and Meister, 1953). A mutant of *E. coli* studied by Rudman and Meister which lacked transaminase B was found to have an absolute requirement for isoleucine but not for the other two branched-chain amino acids.

2. *Leucine*

Abelson (1954) implicated valine and 2-oxoisovalerate as precursors of a portion of the carbon skeleton of leucine when he found that *E. coli* formed leucine with a lowered specific activity when these compounds were added to a medium with glucose-U-^{14}C as the main carbon source He proposed that acetate and 2-oxoisovalerate condensed with the loss of carbon dioxide to form the carbon skeleton of leucine. Strassman *et al.* (1956) extended these observations in their studies of leucine biosynthesis in yeast and proposed that acetyl-CoA and 2-oxoisovalerate condensed to form 3-hydroxy-3-carboxyisocaproate, which was then dehydrated and rehydrated in a fashion analogous to the formation of citrate in its conversion to isocitrate by enzymes of the tricarboxylic acid cycle. Later studies with cell-free extracts have provided ample evidence in support of this proposal (Fig. 16.10).

Jungwirth *et al.* (1963) isolated and identified 3-hydroxy-3-carboxyisocaproate (also known as 2-isopropylmalate and β-carboxy-β-hydroxyisocaproate) from culture filtrates of mutants of *N. crassa* and *S. typhimurium* which required leucine for growth. They also demonstrated the formation of 3-hydroxy-3-carboxyisocaproate from 2-oxoisovalerate and acetyl-CoA by extracts of both *N. crassa* and *S. typhimurium.* Shortly thereafter, Strassman and Ceci (1963) reported that extracts of baker's yeast also catalyzed the condensation of acetyl-CoA and 2-oxoisovalerate to give 3-hydroxy-3-carboxyisocaproate. Gross *et al.* (1963) demonstrated the conversion of 3-hydroxy-3-carboxyisocaproate to 2-hydroxy-3-carboxyisocaproate by extracts of *N. crassa* and *S. typhimurium.* These authors also purified the isomerase from *N. crassa* and characterized the reaction. They found that the isomerase catalyzed the conversion of 3-hydroxy-3-carboxyisocaproate to an equilibrium mixture of the starting material, isopropylmaleate and 2-hydroxy-3-carboxyisocaproate and, therefore, possessed a catalytic activity analogous to aconitase. They were unable to obtain evidence for the action of more than one enzyme in the isomerization reaction. To complete the analogy to aconitase, Calvo *et al.* (1964) reported that the absolute configuration of 2-hydroxy-3-carboxyisocaproate was *threo*-$\mathrm{D_S}$ which is the same as isocitrate produced by aconitase. The enzyme which catalyzed the oxidative decarboxylation of 2-hydroxy-3-carboxyisocaproate to 2-oxoisocaproate was partially purified by Burns *et al.* (1963), and was found to function with NAD as the electron acceptor. The counterpart of this enzyme in the tricarboxylic acid cycle would be isocitrate dehydrogenase. The final step in leucine biosynthesis is transamination with L-glutamate catalyzed by transaminase B in *E. coli.*

F. Alanine

In their experiments with *E. coli* grown on acetate-1-^{14}C-2-^{13}C, Cutinelli *et al.* (1951a) found that the carboxyl carbon of alanine was derived mainly from carboxyl of acetate and that carbons 2 and 3 were derived equally from the methyl carbons of acetate (Table 16.5). This type of labelling pattern would be expected if acetate was oxidized by way of the tricarboxylic acid cycle to oxaloacetate which was then decarboxylated to pyruvate. In contrast to this, several anaerobic bacteria such as *C. kluyveri* (Tomlinson, 1954a), *R. rubrum* (Cutinelli *et al.* 1951b) and *C. thiosulfatophilum* (Hoare and Gibson, 1964) synthe-

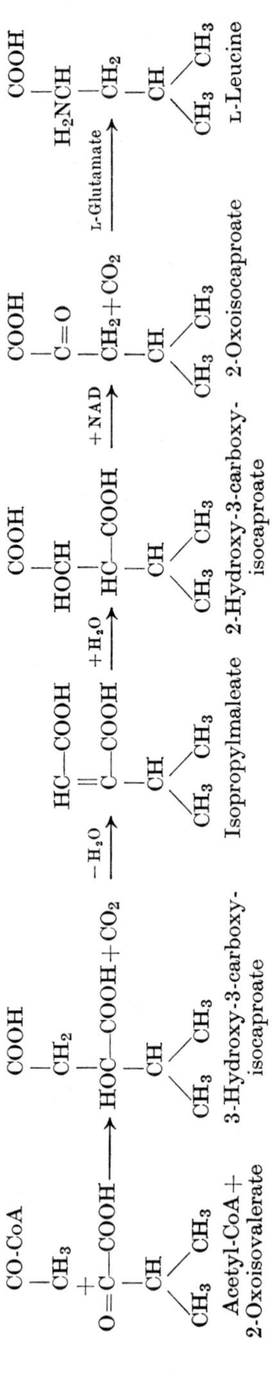

Fig. 16.10. Biosynthesis of leucine.

TABLE 16.5. Derivation of alanine carbons from bicarbonate (b), the carboxyl carbon of acetate (c) and the methyl carbon of acetate (m) during growth of *E. coli*, *C. kluyveri*, *R. rubrum* and *C. thiosulfatophilum*.

Alanine carbon:	*E. coli*	*C. kluyveri*, *R. rubrum*, *C. thiosulfatophilum*
COOH	c	b
| $CHNH_2$	m	c
| CH_3	m	m

sized alanine such that the acetate molecule was incorporated intact into carbons 2 and 3 of alanine while the carboxyl carbon was derived from bicarbonate (Table 16.5). This type of labelling pattern provided strong support for a pathway which involved the carboxylation of acetate to pyruvate. Bachofen *et al.* (1964) discovered the enzyme responsible for the reductive carboxylation of acetyl-CoA, namely pyruvate synthetase, in extracts of the nitrogen-fixing sporeformer *C. pasteurianum*:

$$\text{Acetyl-CoA} + CO_2 + 2\text{ferredoxin}_{red} \rightarrow \text{pyruvate} + 2\text{ferredoxin}_{ox} + \text{CoA}$$

Pyruvate synthetase was also reported in the photosynthetic organisms *Chromatium* strain D (Buchanan *et al.*, 1964) and *C. thiosulfatophilum* (Evans and Buchanan, 1965) where it plays a key role in the reductive carboxylic acid cycle in this organism (Chapter 14). Pyruvate synthetase has also been reported in *C. kluyveri* (Andrew and Morris, 1965).

Rudman and Meister (1953) described an alanine-valine transaminase distinct from transaminases A and B of *E. coli* which would account for the amination of pyruvate in this organism. The reductive amination of pyruvate by alanine dehydrogenase was discussed in section I.C. of this chapter. The epimerization of alanine is an important physiological reaction which provides D-alanine for synthesis of the mucopeptide (Chapter 15). Wood and Gunsalus (1951) discovered the enzyme responsible for this process, alanine racemase, in *S. faecalis*.

G. GLYCINE AND SERINE

Abelson (1954) found that *E. coli* grown in the presence of glucose-U-^{14}C and unlabelled serine, synthesized serine, glycine and cysteine at a specific activity lower than that formed by the organism with glucose-U-^{14}C alone. When glucose-U-^{14}C and unlabelled glycine were provided, only the specific activity of glycine was markedly lowered. Abelson, therefore, proposed that serine was a precursor of both glycine and cysteine, and this proposal has been substantiated by enzyme studies.

Ichihara and Greenberg (1957) identified enzymes in rat liver which catalyzed the conversion of phosphoglycerate to phosphohydroxypyruvate, phosphoserine and serine. Umbarger *et al.* (1963) reported evidence for the same pathway in extracts of *E. coli* and *S. typhimurium* (Table 16.6). The first two reactions were demonstrated in the reverse

TABLE 16.6. Reactions in the biosynthesis of serine in *E. coli* and *S. typhimurium*.

3-Phosphoglycerate + NAD → 3-phosphohydroxypyruvate + NADH
3-Phosphohydroxypyruvate + L-glutamate → 3-phosphoserine + 2-oxoglutarate
3-Phosphoserine + H_2O → serine + P_i

direction, namely by showing the oxidation of NADH with phosphohydroxypyruvate and the formation of glutamate from phosphoserine and 2-oxoglutarate. Umbarger *et al.* (1963) also demonstrated the presence of a phosphatase for phosphoserine in extracts of *E. coli* and *S. typhimurium*. Mutants were isolated which responded to either serine or glycine and were found to be deficient in either phosphoglycerate dehydrogenase or phosphoserine phosphatase. The fact that these mutants required serine or glycine was strong evidence in support of this pathway of serine biosynthesis being the major pathway as opposed to a similar pathway involving the corresponding unphosphorylated intermediates. Pizer (1963) reported similar findings with *E. coli* very shortly after the appearance of the publication by Umbarger *et al.*, and in addition partially purified the phosphatase and found it to be specific for phosphoserine.

Serine aldolase catalyzes the transfer of the hydroxymethyl group from serine to tetrahydrofolate with pyridoxal phosphate as a cofactor and is responsible for the formation of glycine:

Serine + tetrahydrofolate → N^5,N^{10}-methylenetetrahydrofolate + glycine.

Serine aldolase is widely distributed in nature (see review of Rabinowitz, 1960) and was first studied in bacteria by Wright and Stadtman (1956) with a species of clostridium. N^5,N^{10}-Methylenetetrahydrofolate is active in methyl transfer to several cellular constituents. Pizer (1965) isolated a mutant of *E. coli* which required only glycine for growth and was found to be deficient in serine aldolase. The existence of this mutant provides evidence that serine is the precursor to glycine and not vice versa.

Large and Quayle (1963) studied the synthesis of serine during growth of a pseudomonad on one-carbon compounds such as methanol and formate. Methanol was oxidized to formate and Large and Quayle proposed that formate was incorporated into the terminal carbon of serine as outlined in Table 16.7, since all the required enzymes were present in extracts of the organism.

H. The Aromatic Amino Acids

1. *The Common Pathway to Chorismic Acid*

(a) *Shikimic Acid as a Key Intermediate.* Davis (1951) isolated a mutant of *E. coli* which required phenylalanine, tyrosine, tryptophane, *p*-aminobenzoic acid and *p*-hydroxybenzoic acid for growth. Davis found that the mutant would grow when shikimic acid (Fig. 16.12) was added to the growth medium and thereby implicated shikimic acid as an intermediate in the biosynthesis of aromatic compounds by *E. coli*. Srinivasan *et al.* (1956) studied the isotope distribution in shikimic acid formed by a mutant of *E. coli* from labelled presursors (Fig. 16.11). The results were interpreted as evidence for the condensation of a four-carbon compound, derived mainly from the bottom four carbons of glucose, and a three-carbon compound, derived equally from the top and bottom halves of glucose.

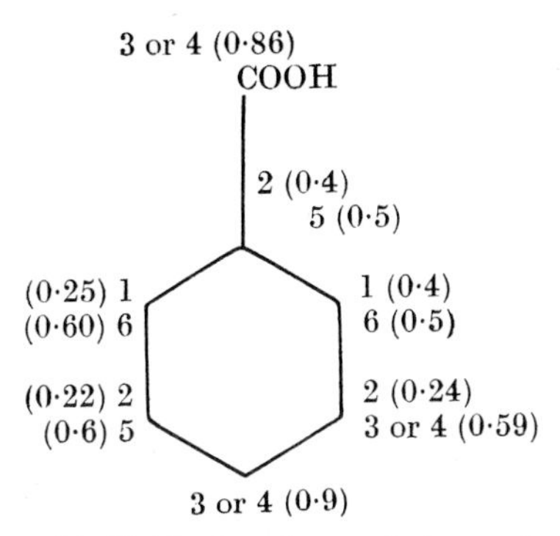

FIG. 16.11. Distribution of individual carbons of glucose in shikimic acid formed by a mutant of *E. coli* (Srinivasan *et al.*. 1956). The numbers in parentheses represent the proportion of the carbon of that position of shikimate supplied by glucose as measured by the incorporation of ^{14}C from glucose-1-^{14}C, glucose-2-^{14}C, glucose-3,4-^{14}C and glucose-6-^{14}C.

TABLE 16.7. Reactions in the incorporation of formate into carbon 3 of serine by a pseudomonad (Large and Quayle, 1963).

Reaction:	Enzyme:
H_4-Folate + formate + ATP → N^{10}-formyl-H_4-folate + ADP + P_i + H_2O	Tetrahydrofolate formylase
N^{10}-Formyl-H_4-folate → N^5,N^{10}-methenyl-H_4-folate + H_2O	Methenyltetrahydrofolate cyclohydrolase
N^5,N^{10}-Methenyl-H_4-folate + NADPH → N^5,N^{10}-methylene-H_4-folate + NADP	Methylenetetrahydrofolate dehydrogenase
N^5,N^{10}-Methylene-H_4-folate + glycine → serine + H_4-folate	Serine aldolase

(b) *Enzymic reactions in the biosynthesis of chorismic acid.* Srinivasan and Sprinson (1958) purified the first enzyme of the aromatic pathway, 3-deoxy-D-*arabino*heptulosonate-7-phosphate synthetase (7-phosphate-2-keto-3-deoxy-D-*arabino*-heptonate D-erythrose-4-phosphate-lyase [pyruvate phosphorylating]), from extracts of *E. coli*. This enzyme catalyzed the formation of its namesake from D-erythrose-4-phosphate and phosphoenolpyruvate (Fig. 16.12) and accounted for the results of Srinivasan *et al.* (1956) discussed in the previous paragraph. Srinivasan *et al.* (1963) purified 5-dehydroquinate synthetase, which catalyzed the cyclization of 3-deoxy-D-*arabino*heptulosonate-7-phosphate to 5-dehydroquinate. The reaction required NAD and cobalt ion although there was no overall change in oxidation state of the product. Although the 2,6-dioxo compound illustrated in Fig. 16.12 appeared to be a logical intermediate in this conversion, Srinivasan and his collaborators were unable to obtain any evidence for its formation. However, it is still possible for the 2,6-dioxo compound to be an intermediate if it is very tightly bound to the enzyme during catalysis. Mitsuhashi and Davis (1954) partially purified 5-dehydroquinase (5-dehydroquinate hydro-lyase), the enzyme which catalyzed the dehydration of 5-dehydroquinate to 5-dehydroshikimate in *E. coli*. No cofactors were required for the dehydration. Yaniv and Gilvarg (1955) purified 5-dehydroshikimate reductase from *E. coli*. The enzyme specifically required NADPH and 5-dehydroshikimate and was detected in *A. aerogenes* and *S. cerevisiae* as well. Fewster (1962) demonstrated the phosphorylation of shikimic acid to shikimate-5-phosphate by ATP with extracts of *E. coli*. Levin and Sprinson (1964) studied the condensation of phosphoenolpyruvate with shikimate-5-phosphate and gathered evidence to support the proposal that the product was 3-enolpyruvylshikimate-5-phosphate (Fig. 16.12). The dephosphorylated derivative was known to accumulate in culture fluids of certain mutants, and the studies of Levin and Sprinson established that the phosphorylated compound was the biologically active material and that dephosphorylation occurred in crude enzyme preparations. Gibson and Gibson (1964) isolated three mutants of *A. aerogenes* which were able to form anthranilate, the precursor to tryptophane, but not prephrenate, the precursor to phenylalanine and tyrosine, when grown in minimal media. Gibson and Gibson prepared cell-free extracts of one of the mutants and caused the accumulation of a new intermediate in the aromatic pathway in reaction mixtures containing shikimic acid and several supplements. The new intermediate was named chorismic acid and was converted to anthranilate when incubated with enzyme preparations supplemented with glutamine, and to prephrenate by extracts of wild-type *A. aero-*

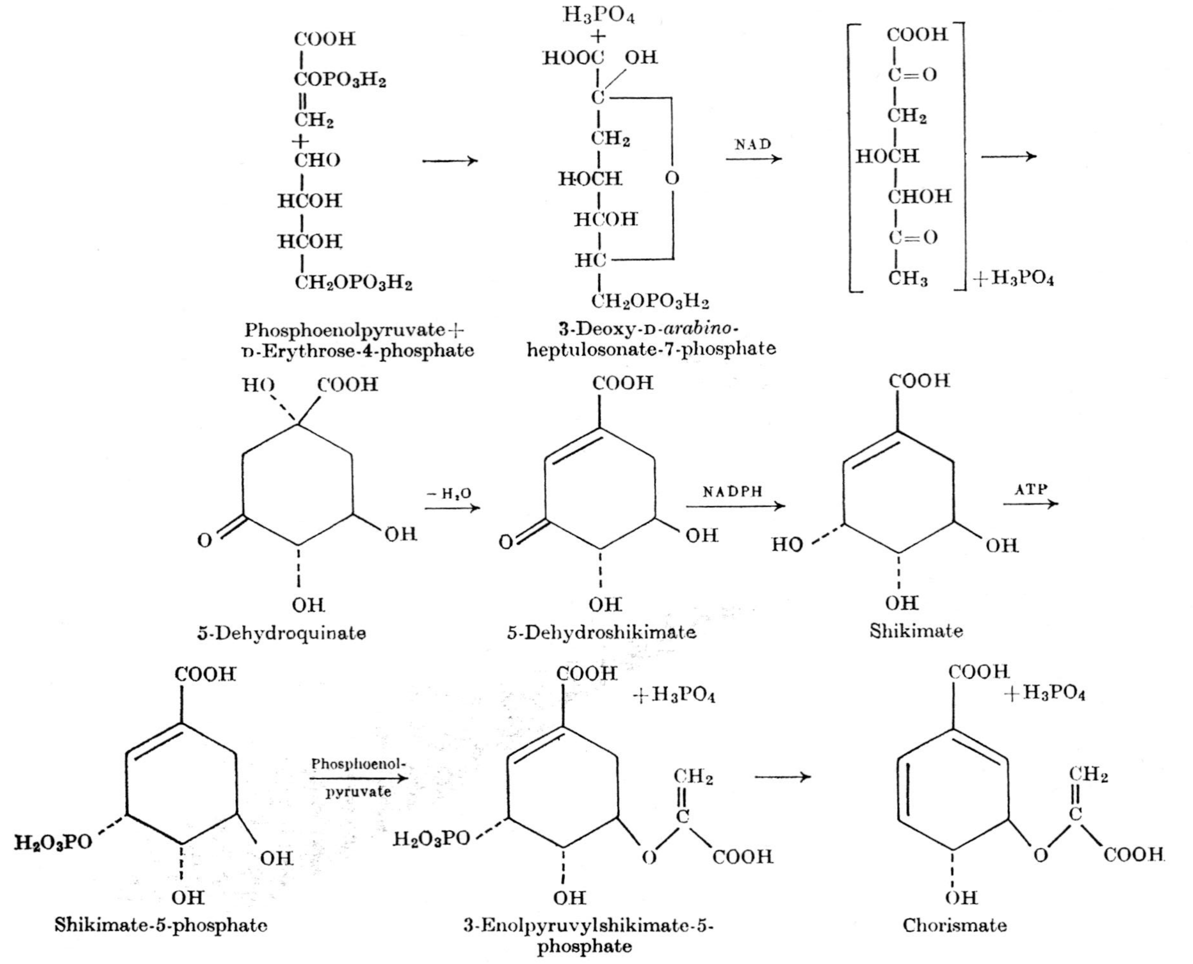

FIG. 16.12. The common pathway in the biosynthesis of aromatic compounds leading to chorismic acid.

genes. Gibson (1964) characterized chorismic acid chemically and proposed the structure shown in Fig. 16.12.

2. *Conversion of Chorismic Acid to Phenylalanine and Tyrosine*

Weiss *et al.* (1954) found that a mutant of *E. coli*, which required phenylalanine, excreted a compound into the medium which they identified as prephrenic acid (Fig. 16.13). They implicated prephrenic acid as an intermediate in phenylalanine biosynthesis, since extracts of wild-type *E. coli* converted prephrenate to phenylpyruvate, while extracts of the mutant were unable to do so. When the role of chorismic acid in the aromatic pathway became known, Cotton and Gibson (1965) studied the conversion of chorismic acid to prephrenate by an enzyme from *E. coli* and *A. aerogenes* which they designated chorismate mutase. The pyruvyl side-chain migrates from position 3 to position 1 of the ring during this transformation. These workers made the interesting discovery that two peaks of chorismate mutase activity were obtained when cell-free extracts were chromatographed on DEAE cellulose columns. They also looked for enzymes responsible for the conversion of prephrenate to phenylpyruvate, which they named prephrenate dehydratase, and for the conversion of prephrenate to *p*-hydroxyphenylpyruvate, which they designated prephrenate dehydrogenase (Fig. 16.13). They discovered that prephrenate dehydratase was eluted coincidentally with one of the peaks of chorismate mutase, while prephrenate dehydrogenase was eluted coincidentally with the other peak of chorismate mutase activity. Thus, in *E. coli* and *A. aerogenes*, two proteins appear to be associated with chorismate mutase activity, one involved in the formation of prephrenate leading to phenylpyruvate, and the other in the formation of prephrenate leading to *p*-hydroxyphenylpyruvate. In *E. coli*, transaminase A appears to be mainly responsible for amination of phenylpyruvate and *p*-hydroxyphenylpyruvate, although transaminase B may also play a role (Table 16.1).

3. *Conversion of Chorismic Acid to Tryptophane*

Rivera and Srinivasan (1963) established that anthranilate could be formed from shikimate-5-phosphate by extracts of *E. coli* supplemented with glutamine. Later Edwards *et al.* (1964) demonstrated the conversion of chorismate to anthranilate by extracts of *A. aerogenes* (Fig. 16.13). They demonstrated the incorporation of the amide nitrogen of glutamine into anthranilate, although ammonium ion also served as a source of nitrogen. Srinivasan (1965) proved that the nitrogen was incorporated into position 2 of the ring, that is, the carbon between the carboxyl and the pyruvyl side chain of chorismate (Fig. 16.13). Baker

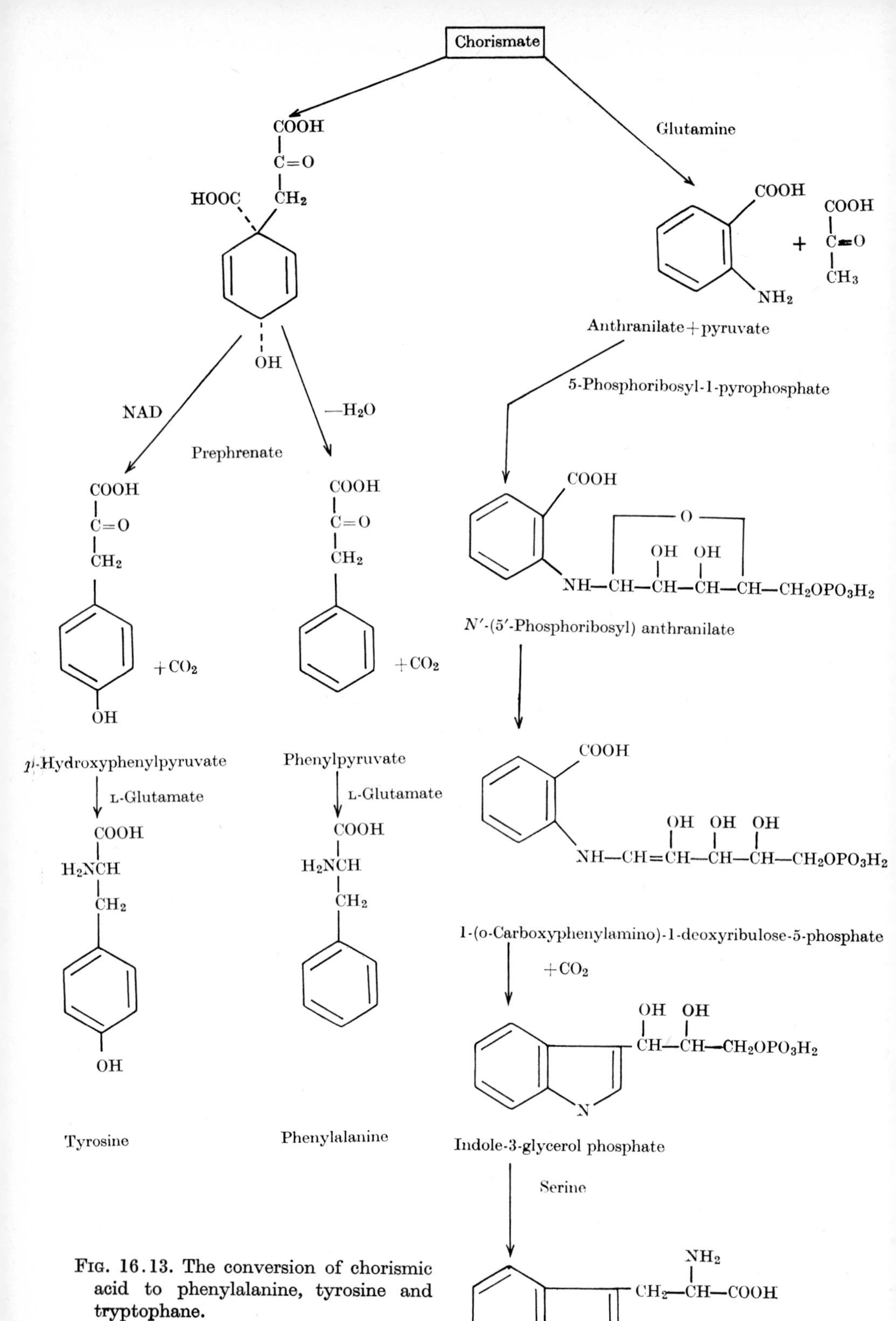

FIG. 16.13. The conversion of chorismic acid to phenylalanine, tyrosine and tryptophane.

and Crawford (1966) purified anthraniliate synthetase from *E. coli* and came to the conclusion that one enzyme was involved in this highly complex reaction.

Yanofsky (1956) prepared an enzyme from *E. coli* by ammonium sulfate fractionation which catalyzed the conversion of anthranilate and 5-phosphoribosyl-1-pyrophosphate to indole-3-glycerol phosphate. The carboxyl carbon of anthranilate was lost as carbon dioxide in this reaction. Subsequently, Smith and Yanofsky (1960) isolated and studied mutants of *E. coli* and *S. typhimurium* unable to convert anthranilate to indole-3-glycerol phosphate which they used to elucidate the reactions involved in this conversion. The mutants of both organisms fell into two groups: extracts of one class of mutants converted anthranilate and 5-phosphoribolsylpyrophosphate to an unknown compound, while the second class of mutants was inactive in this reaction but was able to convert the unknown intermediate to indole-3-glycerol phosphate. Smith and Yanofsky suspected that the intermediate which accumulated was 1-(*O*-carboxyphenylamino)-1-deoxyribulose-5-phosphate. They synthesized the proposed intermediate and found that the synthetic compound and natural intermediate were very similar in their chemical properties. Smith and Yanofsky proposed that the initial condensation between anthranilate and phosphoribosylpyrophosphate resulted in the formation of *N*-(5′-phosphoribosyl) anthranilate (Fig. 16.13). Doy *et al.* (1961) obtained some evidence in favor of this idea when they were able to isolate mutants of *E. coli* and *A. aerogenes* whose extracts catalyzed the conversion of anthranilate and phosphoribosylpyrophosphate to an unstable compound with the spectral properties of *N*-(5′-phosphoribosyl) anthranilate. The conversion of *N*-(5′-phosphoribosyl) anthranilate to 1-(*o*-carboxyphenylamino)-1-deoxyribulose-5-phosphate occurs by an Amadori rearrangement.

The conversion of indole-3-glycerol phosphate and serine to tryptophane catalyzed by the enzyme system tryptophane synthetase (Fig. 16.13) is the final step in trytophane biosynthesis, and the discovery of the mechanism of this reaction is an extremely interesting story. Much of the early work on the characterization and genetics of the tryptophane synthetase system has been reviewed by Yanofsky (1960). Crawford and Yanofsky (1958) chromatographed the enzyme from *E. coli* on DEAE cellulose and found that the enzyme separated into two components, proteins A and B. When proteins A and B were combined and chromatographed, a third peak representing a complex formed by A and B was obtained as well as the peaks for A and B. The complex formed by A and B catalyzed the conversion of indole-3-glycerol-phosphate and serine to tryptophane, but neither A nor B was able to catalyze this reaction when tested separately (Yanofsky, 1960). The complex also catalyzed the reversible cleavage of indole-3-glycerol phosphate to indole and glyceraldehyde-3-phosphate and the condensation of indole and serine to tryptophane:

Indole-3-glycerol phosphate→indole + glyceraldehyde-3-phosphate
Indole + serine→tryptophane.

Pyridoxal phosphate was required for the formation of tryptophane from either indole or indole-3-glycerol phosphate but not for the cleavage of indole-3-glycerol phosphate to indole. Fildes (1940) had shown several years earlier that many organisms which require tryptophane for growth could use indole instead and it seemed possible that indole might be an intermediate in the tryptophane synthetase reaction:

Indole-3-glycerol phosphate→Indole→tryptophane.

However, the conversion of indole-3-glycerol phosphate and serine to tryptophane by the complete tryptophane synthetase system was 10–20 times the rate of cleavage of indole-3-glycerol-phosphate to indole and glyceraldehyde-3-phosphate, which appeared to rule out this possibility. Other enzymic and isotopic experiments by Yanofsky (1960) failed to detect any free indole as an intermediate in the reaction, although there is always the possibility that indole is very tightly bound to the enzyme. Another interesting finding was that the A protein catalyzed a slow cleavage of indole-3-glycerol phosphate to indole and glyceraldehyde-3-phosphate and the B protein catalyzed a slow condensation of indole and serine to tryptophane, but neither protein catalyzed the other two reactions catalyzed by the complete system.

III. Regulation of Metabolic Processes

A. Feedback Control

The end product of a biosynthetic pathway frequently causes the enzymes of that pathway to be formed in quantities less than would be observed in the absence of the end product. This phenomenon is called "repression" and the molecular basis for repression is discussed in Chapter 19. Omission of the end product of a pathway from the growth medium results in the formation of the enzymes of that pathway, a phenomenon called "derepression". Another type of pathway control involves the regulation of activity of fully formed enzymes. In biosynthetic pathways, the regulator is again the end product and the effect is usually the inhibition of the enzyme. In catabolic pathways, however, the regulator may activate the enzyme (see review of Atkinson, 1966).

Umbarger (1956) pointed out in a brief note that regulatory mechanisms for metabolic processes must exist in living organisms in order to explain such occurrences as the orderly transformation of a group of cells to a definite organized structure, and the maintenance of body temperature and blood sugar levels within rather narrow limits. He also pointed to Abelson's study (1954) with *E. coli* which showed that threonine was a precursor of a portion of the carbon skeleton of isoleucine, but that when isoleucine was present in the medium, threonine was not incorporated into isoleucine. Umbarger discovered that L-isoleucine acted as a competitive inhibitor of the biosynthetic threonine deaminase of *E. coli*, the first enzyme involved in isoleucine biosynthesis, and thus described the first instance of regulation at the molecular level. Umbarger termed this "feedback control", in analogy with industrial processes which used automated controls to prevent overproduction of any component. Umbarger's work stimulated further study by other investigators and now several examples of regulated enzymes are known, particularly in pathways of amino acid biosynthesis (see review of Cohen, 1965).

Monod and Jacob (1961) pointed out that the inhibitor or regulator was frequently not a steric analogue of the substrate, and proposed the term "allosteric" inhibitor. Since enzyme activity is stimulated by the regulator, particularly in the case of many catabolic enzymes, the term allosteric effector or modulator is frequently used as a general term for the regulator (Atkinson, 1966).

B. Action of Allosteric Effectors

The mode of action of allosteric effectors has been the subject of a large number of studies, and several characteristics of their action are known. Umbarger (1956) noted that the biosynthetic threonine deaminase did not exhibit conventional first-order kinetics with respect to substrate and velocity. Changeux (1961) studied the action of threonine deaminase in detail and confirmed and extended Umbarger's observations. The effect was particularly noticeable when the double reciprocal plot of Michaelis and Menten was used (Fig. 16.14). In this case a curved line was obtained rather than the expected linear plot. In a plot of velocity versus substrate concentration, a sigmoidal curve resulted rather than a first-order curve (see Fig. 16.15). Such a finding is characteristic of allosteric enzymes (Atkinson, 1966). In the case of threonine deaminase, Changeux (1961) showed that inhibition by isoleucine was

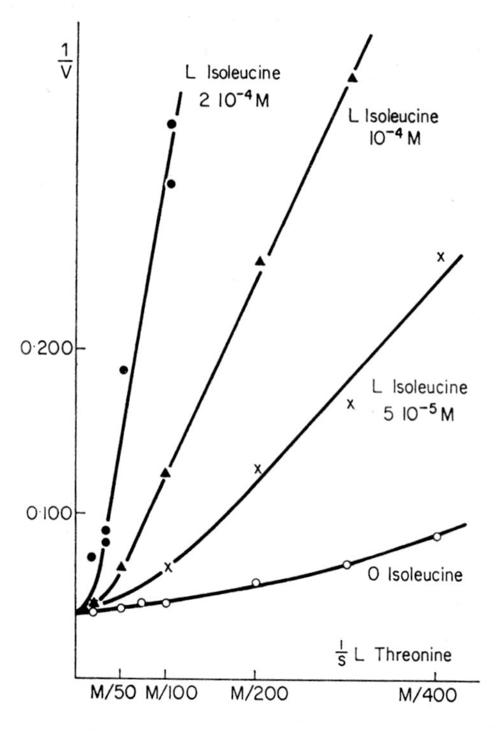

FIG. 16.14. Michaelis–Menton plot of threonine deaminase action on L-threonine alone and with various concentrations of isoleucine. (Changeux, 1961).

competitive, but that isoleucine and threonine were bound to different parts of the enzyme. He was able to demonstrate the latter point by treatment of the enzyme with heat or *p*-chloromercuribenzoate. Both treatments yielded a desensitized enzyme capable of deaminating threonine, but no longer inhibited by isoleucine.

Another characteristic of regulated enzymes is that they are composed of subunits. One very well documented case of the role of enzyme subunits in the regulation of enzyme activity was reported by Gerhart and Pardee (1962) and Gerhart and Schachman (1965). These investigations studied aspartate carbamoyltransferase, which is the first enzyme of the pyrimidine biosynthetic pathway and which catalyzes the following reaction:

$$\text{L-Aspartate} + \text{carbamoylphosphate} \rightarrow N\text{-carbamoyl-L-aspartate} + P_i.$$

The enzyme was inhibited by CTP, the end product of the pyrimidine pathway, and a sigmoidal plot resulted when velocity was plotted as a

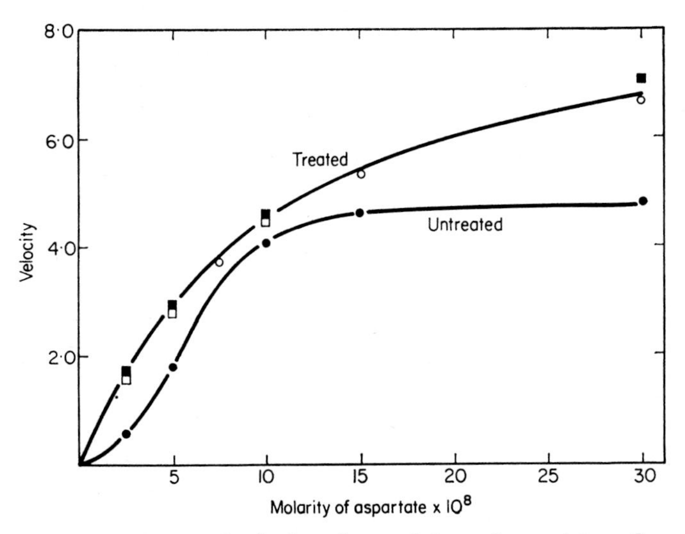

FIG. 16.15. Dependence of velocity of aspartate carbamoyl transferase on substrate concentration with native and desensitized enzyme. The untreated enzyme (●) exhibits the sigmoidal curve characteristic of allosteric enzymes when velocity is plotted as a function of substrate concentration. Enzyme treated with heat (○) or mercuric nitrate (■) exhibited ordinary first order kinetics and was no longer affected by CTP. (Gerhart and Pardee, 1962).

function of substrate concentration (Fig. 16.15). The enzyme was desensitized by heating at 60°C for 4 minutes or by incubation with mercuric ion; there was no longer any inhibition by CTP and the enzyme now exhibited normal first order kinetics. Aspartate carbamoyltransferase was highly purified by Gerhart and Schachman (1965), who found that the enzyme could be cleaved into a small and a large subunit by treatment with *p*-mercuribenzoate and that the two types of subunits could be separated by physical methods. The process was reversible and the native enzyme could be recovered by treatment of the enzyme with mercaptoethanol. The smaller subunit had no catalytic activity, but did bind 5-bromocytidine triphosphate (BrCTP), an analogue of CTP, and was considered to be the binding site for the effector. The larger subunit possessed all the catalytic activity and in addition was not inhibited by CTP. The small subunit was named the regulatory subunit and the large subunit was named the catalytic subunit. It is obvious from these results that the effector and substrate were bound to different sites. Several theories have been advanced in order to account for the action of the effector and have been reviewed by Atkinson (1966).

C. Regulation in Branched Pathways

In branched biosynthetic pathways where a portion of the pathway is common to the formation of several end products, the cell usually has several control points along the pathway to regulate the flow of intermediates and to direct them into branches where they are needed. One well-studied example is regulation of the formation of amino acids of the aspartate family (Fig. 16.16; see review of Stadtman, 1963). Stadtman *et al.* (1961) found that *E. coli* was able to form two and possibly three separate aspartokinases. One aspartokinase was inhibited noncompetitively by lysine (enzyme 1, Fig. 16.16) and a second was inhibited competitively by L-threonine (enzyme 3, Fig. 16.16). These two enzymes were separated physically by fractionation with ammonium

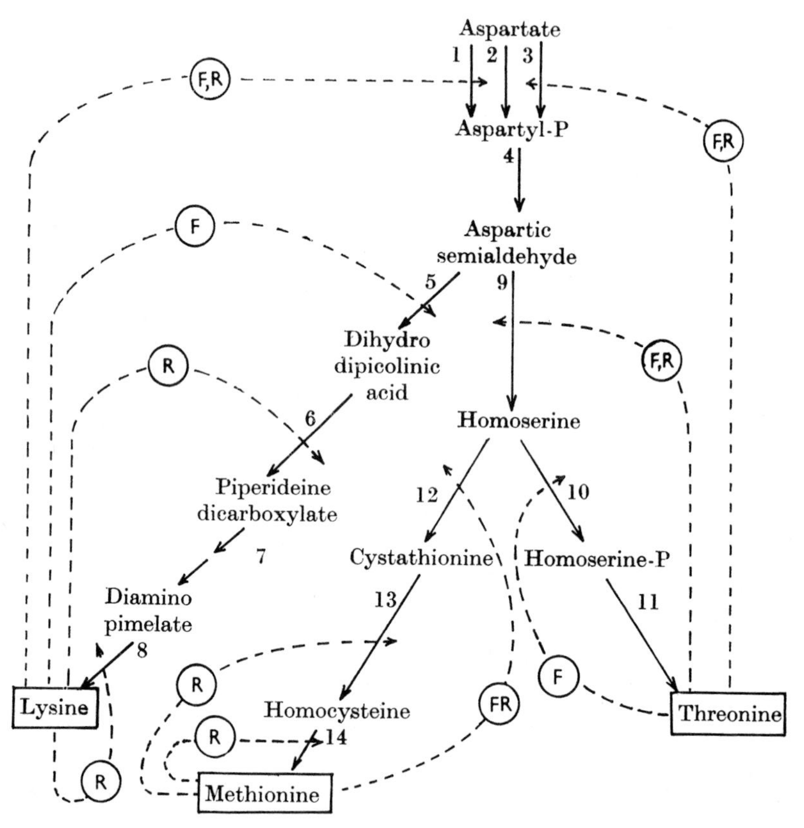

FIG. 16.16. End-product regulation of amino acids formed from aspartate. F = feedback inhibition, R = repression. The numbers refer to enzymes or enzyme systems which are discussed in the text. (From Stadtman, 1963.)

sulfate. Some evidence was obtained for a third aspartokinase which was sensitive to homoserine (enzyme 2, Fig. 16.16). Formation of the aspartokinase sensitive to lysine was repressed when lysine was present in the growth medium; however, the level of the threonine-sensitive enzyme was only slightly affected by growth on threonine. The result of this type of control is that the presence of either lysine or threonine in the medium results in a partial shut-down of the production of aspartyl phosphate but allows enough to be formed to provide for the synthesis of the other amino acids of the aspartate family. However, further regulatory mechanisms are required later in the pathway in order to shunt aspartyl phosphate into the proper branches of the pathway.

Other cases are known where multiple forms of an enzyme exist. The formation of two separate chorismate mutases was discussed in section II H.2 of this chapter. *E. coli* strain W forms three separate 3-deoxy-D-*arabino*heptulosonate-7-phosphate synthetases (Doy and Brown, 1965; Smith *et al.*, 1962); one is inhibited by L-phenylalanine, one by L-tyrosine and a third is repressed by growth on tryptophane.

The situation which exists in *E. coli* with respect to the multiplicity of aspartokinases is not universal, however. Datta and Gest (1964) found only one aspartokinase in *Rhodopseudomonas capsulatus*. The aspartokinase of this organism was affected only slightly by lysine and threonine when tested separately, but when both amino acids were present there was considerable inhibition of the enzyme. A similar situation seems to exist in *B. subtilis* (Paulus and Gray, 1964).

Freundlich *et al.* (1962) reported a third type of regulatory control of a branched pathway in *E. coli* and *S. typhimurium* which they named "multivalent repression". The presence of all three branched-chain amino acids in the medium caused the repression of the formation of threonine deaminase and dihydroxyacid dehydrase, the latter enzyme being involved in the formation of all three branched-chain amino acids in minimal medium. However, omission of any one of the amino acids from the medium resulted in derepression of these two enzymes. On the other hand, valine and isoleucine had no effect on the formation of 2-hydroxy-3-carboxyisocaproate decarboxylase, an enzyme specific for the leucine pathway, while inclusion of leucine in the growth medium resulted in the repression of the decarboxylase.

Other key control points in branched pathways are the first enzymes of the branches leading away from the common pathway. Yugari and Gilvarg (1962) purified the enzyme which catalyzed the condensation of aspartate-β-semialdehyde and pyruvate and showed that L-lysine inhibited the action of the enzyme. Thus when L-lysine was present in the medium, the production of aspartyl phosphate was curtailed and the flow of aspartate-β-semialdehyde into the lysine branch was restricted. Other enzymes of the lysine branch which are subject to regulation are the reductase which acts on dihydropicolinate (enzyme 6, Fig. 16.16; Stadtman, 1963) and *meso*-2,6-diaminopimelate decarboxylyase (enzyme 8, Fig. 16.16; Patte *et al.*, 1962), both of which are repressed by growth in the presence of lysine.

Rowbury and Woods (1966) reported that the initial enzyme of the methionine branch in *E. coli* which catalyzed the formation of *O*-succinylhomoserine (enzyme system 12, Fig. 16.16) was subject to both feedback inhibition and repression by methionine. None of the remaining enzymes was inhibited by methionine but all of those tested were repressed by growth in the presence of methionine. This included cystathionine-γ-synthetase (enzyme system 12, Fig. 16.16; Rowbury and Woods, 1966), cystathionase (enzyme 13, Fig. 16.16; Rowbury and Woods, 1964a) and two enzymes involved in the methylation of homocysteine (enzyme system 14, Fig. 16.16). The latter two enzymes were the reductase which catalyzed the reduction of N^5,N^{10}-methylenetetrahydrofolate to N^5-methyltetrahydrofolate and the cobamide-independent enzyme which catalyzed the methylation of homocysteine. Rowbury and Woods (1966) also state that the cobamide-dependent methylating system was repressed by growth on methionine.

Patte *et al.* (1963) found that homoserine dehydrogenase (enzyme 9, Fig. 16.16) of *E. coli* was both sensitive to L-threonine and repressed by the presence of L-threonine in the medium. An interesting sidelight was that the organism required methionine for growth when threonine was present apparently because threonine was such an effective regulator of homoserine dehydrogenase that not enough homoserine was formed to provide for methionine biosynthesis. Freundlich (1963) found that homoserine dehydrogenase of his mutants of *E. coli* and *S. typhimurium* and the two enzymes involved in the conversion of homoserine to threonine (enzymes 10 and 11, Fig. 16.16) were controlled by multivalent repression, a somewhat different finding than that of Patte *et al.* (1963). Wormser and Pardee (1958) found that homoserine kinase (enzyme 10, Fig. 16.16) was inhibited by L-threonine, but that threonine synthetase (enzyme 11, Fig. 16.16) was not.

References

Abelson, P. H. (1954). *J. Biol. Chem.* **206**, 335.

Abelson, P. H., Bolton, E., Britten, R., Cowie, D. B. and Roberts, R. B. (1953). *Proc. Natl. Acad. Sci. U.S.* **39**, 1020.

Adams, E. (1954). *J. Biol. Chem.* **209**, 829.

Adams, E. (1955). *J. Biol. Chem.* **217**, 325.
Adelberg, E. A. (1955). *J. Biol. Chem.* **216**, 431.
Albrecht, A. M. and Vogel H. J. (1964). *J. Biol. Chem.* **239**, 1872.
Ames, B. N. (1955). *In* "Amino Acid Metabolism" (W. D. McElroy and B. Glass, eds.). The Johns Hopkins Press, Baltimore, Maryland, p. 357.
Ames, B. N. (1957a). *J. Biol. Chem.* **226**, 583.
Ames, B. N. (1957b). *J. Biol. Chem.* **228**, 131.
Ames, B. N. and Horecker, B. L. (1956). *J. Biol. Chem.* **220**, 113.
Ames, B. N., Martin, R. G. and Garry, B. J. (1961). *J. Biol. Chem.* **236**, 2019.
Andrew, I. G. and Morris, J. G. (1965). *Biochim. Biophys. Acta* **97**, 176.
Antia, M., Hoare, D. S. and Work, E. (1957). *Biochem. J.* **65**, 448.
Atkinson, D. E. (1966). *Ann. Rev. Biochem.* **35**, 85.
Bachofen, R., Buchanan, B. B. and Arnon, D. I. (1964). *Proc. Natl Acad. Sci. U.S.* **51**, 690.
Baich, A. and Vogel, H. J. (1962). *Biochem. Biophys. Res. Commun.* **7**, 491.
Baker, T. I. and Crawford, I. P. (1966). *J. Biol. Chem.* **241**, 5577.
Bandurski, R. S., Wilson, L. G. and Asahi, T. (1960). *J. Am. Chem. Soc.* **82**, 3218.
Black, S. and Wright, N. G. (1955a). *J. Biol. Chem.* **213**, 27.
Black, S. and Wright, N. G. (1955b). *J. Biol. Chem.* **213**, 39.
Black, S. and Wright, N. G. (1955c). *J. Biol. Chem.* **213**, 51.
Broquist, H. P. and Trupin, J. S. (1966). *Ann. Rev. Biochem.* **35**, 231.
Buchanan, B. B. and Evans, M. C. W. (1965). *Proc. Natl Acad. Sci. U.S.* **54**, 1212.
Buchanan, B. B., Bachofen, R. and Arnon, D. I. (1964). *Proc. Natl Acad. Sci. U.S.* **52**, 839.
Burchall, J. J., Niederman, R. A. and Wolin, M. J. (1964a). *J. Bacteriol.* **88**, 1038.
Burchall, J. J., Reichelt, E. C. and Wolin, M. J. (1964b). *J. Biol. Chem.* **239**, 1794.
Burns, R. O., Umbarger, H. E. and Gross, S. R. (1963). *Biochemistry* **2**, 1053.
Calvo, J. M., Stevens, C. M., Kalyanpur, M. G. and Umbarger, H. E. (1964). *Biochemistry* **3**, 2024.
Carnahan, J. E., Mortenson, L. E., Mower, H. F. and Castle, J. E. (1960). *Biochim. Biophys. Acta* **44**, 520.
Changeux, J.-P. (1961). *Cold Spring Harbor Symp. Quant. Biol.* **26**, 313.
Cohen, G. N. (1965). *Ann. Rev. Microbiol.* **19**, 105.
Cohen, G. N., Hirsch, M.-L., Wiesendanger, S. G. and Nisman, M. B. (1954). *Compt. Rend.* **238**, 1746.
Cotton, R. G. H. and Gibson, F. (1965). *Biochem. Biophys. Acta* **100**, 76.
Crawford, I. P. and Yanofsky, C. (1958). *Proc. Natl Acad. Sci. U.S.* **44**, 1161.
Cutinelli, C., Ehrensvärd, G., Reio, L., Saluste, E. and Stjernholm, R. (1951a). *Acta Chem. Scand.* **5**, 353.
Cutinelli, C., Ehrensvärd, G., Reio, L., Saluste, E. and Stjernholm, R. (1951b). *Arkiv. Kemi* **3**, 315.
Datta, P. and Gest, H. (1964). *Proc. Natl Acad. Sci. U.S.* **52**, 1004.
Datta, P. and Gest, H. (1965). *J. Biol. Chem.* **240**, 3023.
Davis, B. D. (1951). *J. Biol. Chem.* **191**, 315.
Delavier-Klutchko, C. and Flavin, M. (1965). *Biochem. Biophys. Acta* **99**, 375.
Doy, C. H. and Brown, K. D. (1965). *Biochim. Biophys. Acta* **104**, 377.
Doy, C. H., Rivera, A. and Srinivasan, P. R. (1961). *Biochem. Biophys. Res. Comm.* **4**, 83.
Dreyfuss, J. and Monty, K. J. (1963). *J. Biol. Chem.* **238**, 1019.

Edwards, J. M., Gibson, F., Jackman, L. M. and Shannon, J. S. (1964). *Biochim. Biophys. Acta* **93**, 78.

Ehrensvärd, G. (1955). *Ann. Rev. Biochem.* **24**, 275.

Evans, M. C. W. and Buchanan, B. B. (1965). *Proc. Natl Acad. Sci. U.S.* **53**, 1420.

Farkas, W. and Gilvarg, C. (1965). *J. Biol. Chem.* **240**, 4717.

Fewster, J. A. (1962). *Biochem. J.* **85**, 388.

Fildes, P. (1940). *Brit. J. Exptl Pathol.* **21**, 67.

Flavin, M. and Slaughter, C. (1960). *J. Biol. Chem.* **235**, 1103.

Flavin, M., Delavier-Klutchko and C. Slaughter, C. (1964). *Science* **143**, 50.

Foster, M. A., Tejerina, G., Guest, J. R. and Woods, D. D. (1964). *Biochem. J.* **92**, 476.

Freundlich, M. (1963). *Biochem. Biophys. Res. Commun.* **10**, 277.

Freundlich, M., Burns, R. O. and Umbarger, H. E. (1962). *Proc. Natl Acad. Sci. U.S.* **48**, 1804.

Fry, B. A. (1955). *Biochem. J.* **59**, 579.

Gerhart, J. C. and Pardee, A. B. (1962). *J. Biol. Chem.* **237**, 891.

Gerhart, J. C. and Schachman, H. K. (1965). *Biochemistry* **4**, 1054.

Gibson, F. (1964). *Biochem. J.* **90**, 256.

Gibson, M. I. and Gibson, F. (1964). *Biochem. J.* **90**, 248.

Gilvarg, C. (1958). *J. Biol. Chem.* **233**, 1501.

Gilvarg, C. (1962). *Federation Proc.* **21**, 10.

Gottschalk, G. and Barker, H. A. (1966). *Biochemistry* **5**, 1125.

Gross, S. R., Burns, R. O. and Umbarger, H. E. (1963). *Biochemistry* **2**, 1046.

Guest, J. R., Foster, M. A. and Woods, D. D. (1964a). *Biochem. J.* **92**, 488.

Guest, J. R., Friedman, S., Foster, M. A., Tejerina, G. and Woods, D. D. (1964b). *Biochem. J.* **92**, 497.

Halpern, Y. S. and Umbarger, H. E. (1960). *J. Bacteriol.* **80**, 285.

Hilz, H. and Kittler, M. (1960). *Biochem. Biophys. Res. Commun.* **3**, 140.

Hoare, D. S. (1963). *Biochem. J.* **87**, 284.

Hoare, D. S. and Gibson, J. (1964). *Biochem. J.* **91**, 546.

Hoare, D. S. and Work, E. (1955). *Biochem. J.* **61**, 562.

Hong, M. M., Shen, S. C. and Braunstein, A. E. (1959). *Biochim. Biophys. Acta* **36**, 288.

Ichihara, A. and Greenberg, D. M. (1957). *J. Biol. Chem.* **224**, 331.

Jungwirth, C., Gross, S. R., Margolin, P. and Umbarger, H. E. (1963). *Biochemistry* **2**, 1.

Kalman, S. M., Duffield, P. H. and Brzozowski, T. (1966). *J. Biol. Chem.* **241**, 1871.

Kaplan, M. M. and Flavin, M. (1966). *J. Biol. Chem.* **241**, 4463.

Kemp, J. D. and Atkinson, D. E. (1966). *J. Bacteriol.* **92**, 628.

Kindler, S. H. and Gilvarg, C. (1960). *J. Biol. Chem.* **235**, 3532.

Kornberg, A., Lieberman, I. and Simms, E. S. (1955). *J. Biol. Chem.* **215**, 389.

Kredich, N. M. and Tomkins, G. M. (1966). *J. Biol. Chem.* **241**, 4955.

Large, P. J. and Quayle, J. R. (1963). *Biochem. J.* **87**, 386.

Leavitt, R. I. and Umbarger, H. E. (1961). *J. Biol. Chem.* **236**, 2486.

Levin, J. G. and Sprinson, D. B. (1964). *J. Biol. Chem.* **239**, 1142.

Loper, J. C. and Adams, E. (1965). *J. Biol. Chem.* **240**, 788.

Maas, W. K., Novelli, G. D. and Lipmann, F. (1953). *Proc. Natl Acad. Sci. U.S.* **39**, 1004.

Mager, J. (1960). *Biochim. Biophys. Acta* **41**, 553.

Margolies, M. N. and Goldberger, R. F. (1966). *J. Biol. Chem.* **241**, 3262.
Meister, A. (1965). *In* "Biochemistry of the Amino Acids". Academic Press, New York, Vol. II, p. 928.
Meister, A., Radhakrishnan, A. N. and Buckley, S. D. (1957) *J Biol Chem.* **229**, 789.
Mitsuhashi, S. and Davis, B. D. (1954). *Biochim. Biophys. Acta* **15**, 54.
Monod, J. and Jacob, F. (1961). *Cold Spring Harbor Symp. Quant. Biol.* **26**, 389.
Mortenson, L. E. (1963). *Ann. Rev. Microbiol.* **17**, 115.
Mortenson, L. E. (1964). *Proc. Natl Acad. Sci. U.S.* **52**, 272.
Mortenson, L. E., Valentine, R. C. and Carnahan, J. E. (1963). *J. Biol. Chem.* **238**, 794.
Moyed, H. S. and Magasanik, B. (1960). *J. Biol. Chem.* **235**, 149.
Myers, J. B. (1961). *J. Biol. Chem.* **236**, 1414.
Nason, A. and Evans, H. J. (1953). *J. Biol. Chem.* **202**, 655.
Neidle, A. and Waelsch, H. (1959). *J. Biol. Chem.* **234**, 586.
Nicholas, D. J. D. and Nason, A. (1955). *J. Bacteriol.* **69**, 580
Patte, J.-C., Loving, T. and Cohen, G. N. (1962). *Biochim. Biophys. Acta* **58**, 359.
Patte, J.-C., LeBras, G., Loving, T. and Cohen, G. N. (1963). *Biochim. Biophys. Acta* **67**, 16.
Paulus, H. and Gray, E. (1964). *J. Biol. Chem.* **239**, 4008.
Peck, H. D. (1962). *Bacteriol. Rev.* **26**, 67.
Peterkofsky, B. and Gilvarg, C. (1961). *J. Biol. Chem.* **236**, 1432.
Pizer, L. I. (1963). *J. Biol. Chem.* **238**, 3934.
Pizer, L. I. (1965). *J. Bacteriol.* **89**, 1145.
Rabinowitz, J. C. (1960). *In* "The Enzymes" (P. D. Boyer, H. Lardy and K. Myrbäck, eds.). Academic Press, New York, Vol. 2, p. 185.
Radhakrishnan, A. N., Wagner, R. P. and Snell, E. E. (1960). *J. Biol. Chem.* **235**, 2322.
Ratner, S. (1962). *In* "The Enzymes" (P. D. Boyer, H. Lardy and K. Myrbäck, eds.). Academic Press, New York, Vol. 6, p. 495.
Ravel, J. M., Norton, S. J., Humphreys, J. S. and Shive, M. (1962). *J. Biol. Chem.* **237**, 2845.
Rivera, A. and Srinivasan, P. R. (1963). *Biochemistry* **2**, 1063.
Robbins, P. W. snd Lipmann, F. (1958). *J. Biol. Chem.* **233**, 686.
Rogers, W. P. and Novelli, G. D. (1962). *Arch. Biochem. Biophys.* **96**, 398.
Rowbury, R. J. and Woods, D. D. (1964a). *J. Gen. Microbiol.* **35**, 145.
Rowbury, R. J. and Woods, D. D. (1964b). *J. Gen. Microbiol.* **36**, 341.
Rowbury, R. J. and Woods, D. D. (1966). *J. Gen. Microbiol.* **42**, 155.
Rudman, D. and Meister, A. (1953). *J. Biol. Chem.* **200**, 591.
Shen, S. C., Hong, M. M. and Braunstein, A. E. (1959). *Biochim. Biophys. Acta* **36**, 290.
Siegel, L. M., Click, E. M. and Monty, K. J. (1964). *Biochem. Biophys. Res. Commun.* **17**, 125.
Smith, D. W. E. and Ames, B. N. (1964). *J. Biol. Chem.* **239**, 1848.
Smith, D. W. E. and Ames, B. N. (1965). *J. Biol. Chem.* **240**, 3056.
Smith, L. C., Ravel, J. M., Lax, S. R. and Shire, W. (1962). *J. Biol. Chem.* **237**, 3566.
Smith, O. H. and Yanofsky, C. (1960). *J. Biol. Chem.* **235**, 2051.
Spencer, D., Takahashi, H. and Nason, A. (1957). *J. Bacteriol.* **73**, 553.
Srinivasan, P. R. (1965). *Biochemistry* **4**, 2860.

Srinivasan, P. R. and Sprinson, D. B. (1958). *J. Biol. Chem.* **234**, 716.

Srinivasan, P. R., Shigeura, H. T., Sprecher, M., Sprinson, D. B. and Davis, B. D. (1956). *J. Biol. Chem.* **220**, 477.

Srinivasan, P. R., Rothschild, J. and Sprinson, D. B. (1963). *J. Biol. Chem.* **238**, 3176.

Stadtman, E. R. (1963). *Bacteriol. Rev.* **27**, 170.

Stadtman, E. R., Cohen, G. N., LeBras, G. and DeRobichon-Szulmajster, H. (1961). *J. Biol. Chem.* **236**, 2033.

Stern, J. R., Hegre, C. S. and Bambers, G. (1966). *Biochemistry* **5**, 1119.

Strassman, M. and Ceci, L. N. (1963). *J. Biol. Chem.* **238**, 2445.

Strassman, M., Thomas, A. J. and Weinhouse, S. (1953). *J. Am. Chem. Soc.* **75**, 5135.

Strassman, M., Thomas, A. J., Locke, L. A. and Weinhouse, S. (1954). *J. Am. Chem. Soc.* **76**, 4241.

Strassman, M., Locke, L. A., Thomas, A. J. and Weinhouse, S. (1956). *J. Am. Chem. Soc.* **78**, 1599.

Strassman, M., Shatton, J. B. and Weinhouse, S. (1960). *J. Biol. Chem.* **235**, 700.

Takahashi, H., Taniguchi, S. and Egami, F. (1963). *In* "Comparative Biochemistry" (M. Florkin and H. S. Mason, eds.). Academic Press, New York, Vol. V, p. 91.

Taniguchi, S. and Ohmachi, K. (1960). *J. Biochem.* (*Tokyo*) **48**, 50.

Tomlinson, N. (1954a). *J. Biol. Chem.* **209**, 597.

Tomlinson, N. (1954b). *J. Biol. Chem.* **209**, 605.

Udaka, S. and Kinoshita, S. (1959). *J. Gen. Appl. Microbiol.* (*Tokyo*) **4**, 272.

Umbarger, H. E. (1956). *Science* **123**, 848.

Umbarger, H. E. and Brown, B. (1957). *J. Bacteriol.* **73**, 105.

Umbarger, H. E. and Brown, B. (1958). *J. Biol. Chem.* **233**, 1156.

Umbarger, H. E., Brown, B. and Eyring, E. J. (1960). *J. Biol. Chem.* **235**, 1425.

Umbarger, H. E., Umbarger, M. A. and Siu, P. M. L. (1963). *J. Bacteriol.* **85**, 1431.

Vender, J. and Rickenberg, H. V. (1964). *Biochim. Biophys. Acta* **90**, 218.

Vogel, H. J. (1953). *Proc. Natl Acad. Sci. U.S.* **39**, 578.

Vogel, H. J. and Bonner, D. M. (1956). *J. Biol. Chem.* **218**, 97.

Vogel, H. J. and Davis, B. D. (1952). *J. Am. Chem. Soc.* **74**, 109.

Vogel, H. J., Davis, B. D. and Mingioli, E. S. (1951). *J. Am. Chem. Soc.* **73**, 1897.

Wang, C. H., Thomas, R. C., Cheldelin, V. H. and Christensen, B. E. (1952). *J. Biol. Chem.* **197**, 663.

Watanabe, Y., Konishi, S. and Shimura, K. (1955). *J. Biochem.* (*Tokyo*) **42**, 837.

Weiss, U., Gilvarg, C., Mingioli, E. S. and Davis, B. D. (1954). *Science* **119**, 774.

Wijesundera, S. and Woods, D. D. (1962). *J. Gen. Microbiol.* **29**, 353.

Wood, W. A. and Gunsalus, I. C. (1951). *J. Biol. Chem.* **190**, 403.

Woolfolk, C. A. and Stadtman, E. R. (1964). *Biochem. Biophys. Res. Commun.* **17**, 313.

Wormser, E. H. and Pardee, A. B. (1958). *Arch. Biochem. Biophys.* **78**, 416.

Wright, B. E. and Stadtman, T. C. (1956). *J. Biol. Chem.* **219**, 863.

Yaniv, H. and Gilvarg, C. (1955). *J. Biol. Chem.* **213**, 787.

Yanofsky, C. (1956). *J. Biol. Chem.* **223**, 171.

Yanofsky, C. (1960). *Bacteriol. Rev.* **24**, 221.

Yugari, Y. and Gilvarg, C. (1962). *Biochim. Biophys. Acta* **62**, 612.

Yugari, Y. and Gilvarg, C. (1965). *J. Biol. Chem.* **240**, 4710.

Yura, T. and Vogel, H. J. (1959). *J. Biol. Chem.* **234**, 335.

17. BIOSYNTHESIS OF LIPIDS

I. Biosynthesis of Fatty Acids

A. Straight-chain Fatty Acids

One of the key discoveries in the study of lipid biosynthesis was made by Wakil (1958) when he found that an enzyme fraction prepared from pigeon liver and supplemented with ATP, manganese and bicarbonate converted acetyl-CoA to malonyl-CoA. Wakil also found that malonyl-CoA could be converted quantitatively to long-chain fatty acids when NADPH and a second enzyme fraction were added to the incubation mixture. Shortly thereafter, Kusunose *et al.* (1959) reported that fortified crude extracts of *Mycobacterium avium* catalyzed the conversion of acetyl phosphate to a derivative of malonate. Since then, the carboxylation of acetyl-CoA has been reported in several species of bacteria (see review of Kates, 1966). The reaction is catalyzed by acetyl-CoA carboxylase (acetyl-CoA: carbon dioxide ligase [ADP]), an enzyme which contains biotin as the prosthetic group:

$$\text{ATP} + \text{acetyl-CoA} + CO_2 + H_2O \rightarrow \text{ADP} + H_3PO_4 + \text{malonyl-CoA}.$$

Fatty acid synthetase preparations made from animal sources form palmitic acid as the main product and utilise 7 moles of malonyl-CoA and one mole of acetyl-CoA in the process. The bicarbonate which is used to form malonyl-CoA does not appear in the finished fatty acid (see review of Vagelos, 1964). The equation for the formation of palmitic acid is:

$$CH_3CO\text{-SCoA} + 7\ HOOC\text{-}CH_2\text{-CO-SCoA} + 14\ \text{NADPH} + 14\ H^+ \rightarrow$$
$$CH_3(CH_2)_{14}COOH + 7\ CO_2 + 8\ \text{CoASH} + 14\ \text{NADP} + 6\ H_2O.$$

The fatty acid synthetase which Goldman *et al.* (1963) obtained from *C. kluyveri* formed a mixture of palmitic, stearic and arachidic acids. However, preparations made from *E. coli* by Lennarz *et al.* (1962a) and Goldman *et al.* (1963) formed predominantly *cis*-vaccenic acid and only small amounts of C_{16} and C_{18} fatty acids.

In the studies of fatty acid synthesis by bacterial preparations, both Lennarz *et al.* (1962a) and Goldman *et al.* (1963) found that a heat-stable protein participated in the process. Majerus *et al.* (1964) obtained pure preparations of the heat-stable protein from *E. coli* and determined that the protein had a molecular weight of approximately 9500 and that there was a single free sulfhydryl group per mole of protein. They also showed that transacylases from *E. coli* fatty acid synthetase catalyzed the transfer of the acyl residues of acetyl-CoA and malonyl-CoA to the sulfhydryl group of the heat-stable protein. Majerus *et al.* (1964) proposed that the heat-stable protein acted as an acyl carrier during fatty acid synthesis and gave it the name acyl carrier protein (ACP). Majerus *et al.* (1965b) later made the important discovery that the binding site of ACP was 4′ phosphopantetheine (Fig. 17.1), which is the same as the binding site of CoA.

Vagelos and his associates isolated all but one of the enzymes involved in the utilization of the two carbon units of acetyl-CoA and malonyl-CoA during the biosynthesis of fatty acids (Fig. 17.1). They purified two enzymes which specifically catalyzed the transfer of the acetyl residue of acetyl-CoA and the malonyl residue of malonyl-CoA to ACP. Alberts *et al.* (1964) named these two enzymes acetyl transacylase and malonyl transacylase respectively. These were the two transacylases used by Majerus *et al.* (1964) to bind acyl residues to ACP in the study mentioned in the previous paragraph. Alberts *et al.* (1965) purified 3-ketoacyl acyl carrier protein synthetase of *E. coli* (Table 17.1) extensively and showed that the enzyme catalyzed the condensation of acetyl-ACP and malonyl-ACP to form acetoacetyl-ACP. They also noted that hexanoyl-ACP would serve in place of acetyl-ACP. Alberts *et al.* (1964) also purified 3-ketoacyl acyl carrier protein reductase of *E. coli* and established that acetoacetyl-ACP was the substrate for the enzyme. The reaction was reversible and since D-(-)-3-hydroxybutyryl-ACP was more rapidly oxidized than was L-(+)-3-hydroxybutyryl-ACP, they reasoned that the product of acetoacetyl-ACP reduction was the D isomer. It is now well established that D-3-hydroxyacyl thioesters are intermediates in fatty acid biosynthesis while the L-isomers are intermediates in fatty acid oxidation. Majerus *et al.* (1965a) purified enoyl-ACP hydrase from *E. coli* and showed that the enzyme specifically catalyzed that dehydration of D-(−)-3-hydroxybutyryl-ACP; the L isomer and the corresponding thioester of CoA were not attacked. The stereospecificity of this enzyme sets it apart from crotonase, which dehydrates the L isomers of 3-hydroxy fatty acyl CoA thioesters. The final enzyme of the cycle is enoyl-ACP reductase, which has not been studied with highly purified preparations, but whose presence has been

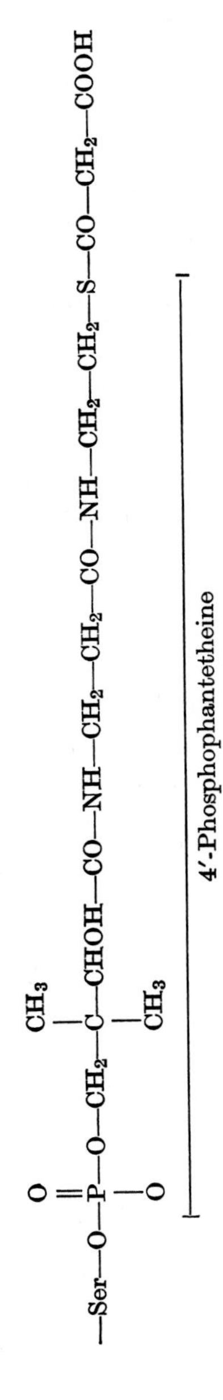

FIG. 17.1. Structure of the binding site for acyl residues on acyl carrier protein. Ser = a serine residue of the protein.

TABLE 17.1. Initial reactions of fatty acid synthesis catalyzed by enzymes of *E. coli*. Following the formation of butyryl-ACP another two-carbon unit derived from malonyl-ACP is added and this cycle is repeated until the long-chain fatty acid is complete.

Reaction:	Enzyme:
Acetyl-CoA + ACP → acetyl-ACP + CoA	Acetyl transacylase
Malonyl-CoA + ACP → malonyl-ACP + CoA	Malonyl transacylase
Acetyl-ACP + malonyl-ACP → acetoacetyl-ACP + ACP + CO_2	3-Ketoacyl acyl carrier protein synthetase
Acetoacetyl-ACP + NADPH + H^+ → D-(−)-3-hydroxybutyryl-ACP + NADP	3-Ketoacyl acyl carrier protein reductase
D-(−)-3-hydroxybutyryl-ACP → crotonyl-ACP + H_2O	Enoyl acyl carrier protein hydrase
Crotonyl-ACP + NADPH + H^+ → butyryl-ACP + NADP	Enoyl acyl carrier protein reductase

demonstrated in preparations of fatty acid synthetase of *E. coli* capable of reducing acetoacetyl-ACP to butyryl-ACP with NADPH (Goldman, 1964).

Formation of long-chain fatty acids involves successive addition of two-carbon units derived from malonyl CoA as described in the preceding paragraph. The two-carbon unit derived from acetyl-CoA is incorporated into the methyl end of the fatty acid (Lynen, 1961), which means that the two-carbon fragments derived from malonyl-CoA are inserted between the saturated acyl residue and ACP. Other pieces of evidence which substantiate this view are the observations that butyryl-ACP is incorporated into long-chain fatty acids by the *E. coli* fatty acid synthetase (Goldman, 1964), and that 3-ketoacyl acyl carrier protein synthetase can utilize hexanoyl-ACP as well as acetyl-ACP in the condensation reaction (Alberts *et al.*, 1965).

B. Branched-chain Fatty Acids

The branched-chain portion of the odd- and even-numbered branched-chain fatty acids (Chapter 3) is derived from the corresponding portion of the branched-chain amino acids. The carboxyl carbons of the branched-chain amino acids are not incorporated and this must mean that these carbons are removed by decarboxylation prior to incorporation of the remainder of the molecule. Allison *et al.* (1962) studied the incorporation of radioactive precursors by species of *Ruminococcus* which required volatile branched-chain fatty acids for growth. One species, *R. flavafaciens*, incorporated isotope from isovalerate-1-^{14}C mainly into C_{15} branched-chain fatty acids, but some ^{14}C was recovered in the fraction containing the C_{17} acids. Presumably the products in this case were the *iso* acids. This interesting organism also incorporated isovalerate-1-^{14}C into leucine. Another species which they studied, *R. albus*, incorporated isobutyrate-1-^{14}C into C_{14} and C_{16} branched-chain fatty acids. Lennarz (1961) showed that *M. lysodeikticus* incorporated isotope from isoleucine-U-^{14}C and 2-methyl-*n*-butyrate-U-^{14}C into C_{15} and C_{17} branched-chain fatty acids, in this case, apparently the *anteiso* acids. Isotope from isoleucine-1-^{14}C was not incorporated, signifying a loss of the carboxyl carbon. Wegner and Foster (1963) reported that *Bacteroides succinogenes* converted isobutyrate-1-^{14}C into radioactive C_{14} and C_{16} branched-chain fatty acids. In an interesting study, Kaneda (1963) showed that *B. subtilis* incorporated radioactive carbon from isobutyrate-1-^{14}C and valine-U-^{14}C into *iso* C_{14} and C_{16} branched-chain fatty acids. Isotope from valine-1-^{14}C was not incorporated. Isotope from isovalerate-1-^{14}C and 2-methyl-*n*-butyrate-1-^{14}C was recovered in a fraction which contained both *iso* and *anteiso* C_{15} branched-chain fatty acids. Kaneda also showed that the first 12 carbons of isopalmitic acid were derived from acetate. On the basis of these findings it seems reasonable to conclude that the branched-chain amino acids are oxidatively decarboxylated to the next lower branched-chain fatty acyl-CoA derivatives, which are then converted to the ACP derivatives and condense with malonyl-ACP to form branched long-chain fatty acids.

Gastambide-Odier *et al.* (1963) studied the biosynthesis of mycocerosic acid (see Fig. 3.2) by *M. tuberculosis* strain $H_{37}R_a$. They found that isotope from propionate-1-^{14}C was incorporated into carbons 1, 3, 5 and 7 of C_{32} mycocerosic acid. They proposed that the C_{32} mycocerosic acid arose by the condensation of a C_{20} long-chain fatty acid with four molecules of propionate (Fig. 17.2).

$$CH_3{-}(CH_2)_{18}{-}COOH + 4\ \underset{\displaystyle CH_3}{\underset{|}{CH_2}}{-}\overset{*}{C}OOH \rightarrow CH_3{-}(CH_2)_{19}{-}\left[\underset{\displaystyle CH_3}{\underset{|}{CH}}{-}\overset{*}{C}H_2\right]_3{-}\underset{\displaystyle CH_3}{\underset{|}{CH}}{-}\overset{*}{C}OOH$$

FIG. 17.2. The biosynthesis of mycocerosic acid in *M. tuberculosis* as proposed by Gastambide-Odier *et al.* (1963). The asterisk indicates the position of the labeled carbon.

Lennarz *et al.* (1962b) studied the biosynthesis of tuberculostearic acid (10-methyl stearate) by both growing and resting cells of *M. phlei*. These investigators found that stearate-1-^{14}C was converted to both oleate and tuberculostearate at a high specific activity. They also showed that isotope from methionine-methyl-^{14}C was incorporated into tuberculostearate and that the isotope was located in the 10-methyl position.

C. CYCLOPROPANE FATTY ACIDS

O'Leary (1959a) reported that *L. arabinosus* converted *cis*-vaccenic acid-1-^{14}C to radioactive lactobacillic acid, the C_{19} cyclopropane fatty acid. O'Leary (1959b) further demonstrated that the methyl group of methionine was incorporated into lactobacillic acid. Liu and Hofmann (1962) established that the methyl carbon of methionine was incorporated into the methylene bridge carbon of lactobacillic acid by *L. arabinosus*. Chalk and Kodicek (1961) showed that *E. coli* incorporated isotope from methionine-methyl-^{14}C into both C_{17} and C_{19} cyclopropane fatty acids.

Zalkin *et al.* (1963) were able to prepare cell-free preparations of *S. marcescens* and *C. kluyveri* which catalyzed the incorporation of isotope from *S*-adenosylmethionine-methyl-^{3}H into both C_{15} and C_{17} cyclopropane fatty acids. Extracts made from *S. marcescens* catalyzed the reaction using lipid acceptors present in the extract; however, enzyme extracts made from *C. kluyveri* did not contain enough endogenous lipid to do so. Dispersions of lipid extracted from early log phase cells of *S. marcescens* did act as an acceptor in the transfer of the methyl group catalyzed by enzymes of *C. kluyveri*. Zalkin *et al.* (1963) identified the principal lipid extracted from *S. marcescens* as phosphatidylethanolamine and showed that radioactivity from labeled *S*-adenosylmethionine was incorporated into this phospholipid but not into free fatty acids. The best lipid acceptor was phosphatidylethanolamine which contained unsaturated fatty acids. Neither palmitoleic acid nor palmitoleyl-CoA acted as substrates for the enzyme prepared from *C. kluyveri*. Chung and Law (1964) partially purified the cyclopropane fatty acid synthetase of *C. kluyveri* and showed that the substrate for the reaction was indeed phosphatidylethanolamine and that the purified enzyme formed both C_{17} and C_{19} cyclopropane fatty acids. These facts are summarized in the scheme shown in Fig. 17.3.

$$
\begin{array}{l}
CH_2—O—PO_3H—CH_2—CH_2—NH_2 \\
| \\
CH—O_2C—(CH_2)_x—CH{=}CH—(CH_2)_5—CH_3 \\
| \\
CH_2—O_2C—R
\end{array}
\quad + \quad
\begin{array}{l}
COOH \\
| \\
CHNH_2 \\
| \\
CH_2 \\
| \\
CH_2 \\
| \\
S\text{-adenosine} \\
| \\
CH_3
\end{array}
$$

Phosphatidylethanolamine ↓ *S*-adenosylmethionine

$$
\begin{array}{l}
CH_2—O—PO_3H—CH_2—CH_2—NH_2 \\
| \\
CH—O_2C—(CH_2)_x—CH—CH—(CH_2)_5—CH_3 \\
| \qquad\qquad\qquad\qquad\quad \diagdown\ \diagup \\
CH_2—O_2C—R \qquad\qquad\quad CH
\end{array}
\quad + \quad
\begin{array}{l}
COOH \\
| \\
CHNH_2 \\
| \\
CH_2 \\
| \\
CH_2 \\
| \\
S \\
| \\
\text{adenosine}
\end{array}
$$

FIG. 17.3. Biosynthesis of cyclopropane fatty acids in *C. kluyveri*. When palmitoleic acid is the substrate, x = 7 and the C_{17} cyclopropane fatty acid is the product. When *cis*-vaccenic acid is the substrate, x = 9 and the C_{19} cyclopropane fatty acid is the product.

D. Mycolic Acids

Lederer and his associates studied the conversion of labelled palmitic acid to the lower homologues of mycolic acid by *Corynebacterium* and *Nocardia* species. Gastambide–Odier and Lederer (1960) showed that *C. diphtheriae* utilized palmitic acid-$1^{14}C$ to produce corynemycolic acid-1,$3^{14}C$. On the basis of this evidence they suggested that corynemycolic acid was formed by the condensation of two moles of palmitic acid, as shown in Fig. 17.4. Etemadi and Lederer (1965) showed that *N. asteroides* incorporated two moles of palmitic acid-1-^{14}C into the nocardic acid synthesized by this organism. One atom of ^{14}C was located in the terminal carboxyl carbon, but the other was buried in the unsaturated chain. They proposed that one mole of palmitic acid underwent chain elongation by the addition of acetate, desaturation of the long-chain fatty acid and then condensation with another mole of palmitate (Fig. 17.4).

E. Unsaturated Fatty Acids

It was mentioned in Section I.A of this chapter that fatty acid synthetase preparations from *E. coli* formed *cis*-vaccenic acid as the major product (Lennarz *et al.*, 1962a; Goldman *et al.*, 1963). Fatty acid synthesis under these conditions was independent of oxygen, and this pathway is frequently referred to as the anaerobic pathway of unsaturated fatty acid synthesis. Bloch *et al.* (1961) showed that palmitic and stearic acids were not precursors of the unsaturated 16 and 18 carbon fatty acids present in *E. coli*, *L. plantarum* and *C. kluyveri*. Bloch *et al.* (1961) proposed a pathway (Fig. 17.5) for synthesis of unsaturated fatty acids that would explain the occurrence in bacteria of Δ^9 monounsaturated fatty acid, palmitoleic acid, and Δ^{11} monounsaturated fatty acid, *cis*-vaccenic acid. The important feature of this proposal is that a 3-hydroxydecanoic acid intermediate is dehydrated to form the Δ^3 decanoic acid, which then undergoes chain elongation to form either the C_{16} acid with the double bond at position 9 or the C_{18} acid with the double bond at position 11. A similar scheme was proposed to account for C_{18} fatty acids with the double bond at position 9, in which case a 12 carbon fatty acid was the species dehydrated. Norris *et al.* (1964) obtained strong experimental support for this proposal when she and her associates isolated an enzyme from fatty acid synthetase preparations of *E. coli* which catalyzed the dehydration of 3-hydroxydecanoyl-CoA. When the purified dehydrase was added back to fatty acid synthetase preparations, the proportion of unsaturated fatty acid in

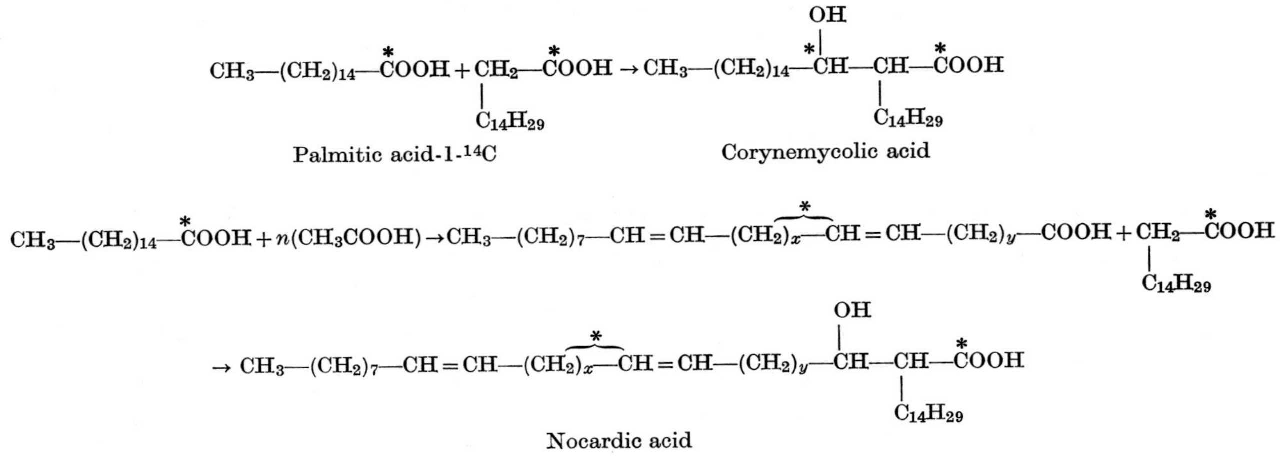

$$CH_3{-}(CH_2)_{14}{-}\overset{*}{C}OOH + \underset{\displaystyle C_{14}H_{29}}{\underset{|}{CH_2}}{-}\overset{*}{C}OOH \rightarrow CH_3{-}(CH_2)_{14}{-}\overset{\displaystyle OH}{\overset{|}{\overset{*}{C}H}}{-}\underset{\displaystyle C_{14}H_{29}}{\underset{|}{CH}}{-}\overset{*}{C}OOH$$

Palmitic acid-1-^{14}C　　　　Corynemycolic acid

$$CH_3{-}(CH_2)_{14}{-}\overset{*}{C}OOH + n(CH_3COOH) \rightarrow CH_3{-}(CH_2)_7{-}CH{=}CH{-}\overset{*}{\overbrace{(CH_2)_x{-}}}CH{=}CH{-}(CH_2)_y{-}COOH + \underset{\displaystyle C_{14}H_{29}}{\underset{|}{CH_2}}{-}\overset{*}{C}OOH$$

$$\rightarrow CH_3{-}(CH_2)_7{-}CH{=}CH{-}\overset{*}{\overbrace{(CH_2)_x{-}}}CH{=}CH{-}(CH_2)_y{-}\overset{\displaystyle OH}{\overset{|}{CH}}{-}\underset{\displaystyle C_{14}H_{29}}{\underset{|}{CH}}{-}\overset{*}{C}OOH$$

Nocardic acid

FIG. 17.4. Conversion of palmitic acid-1-^{14}C to corynemycolic acid and nocardic acid by *C. diphtheriae* and *N. asteroides* respectively. The asterisk denotes the position of the isotope.

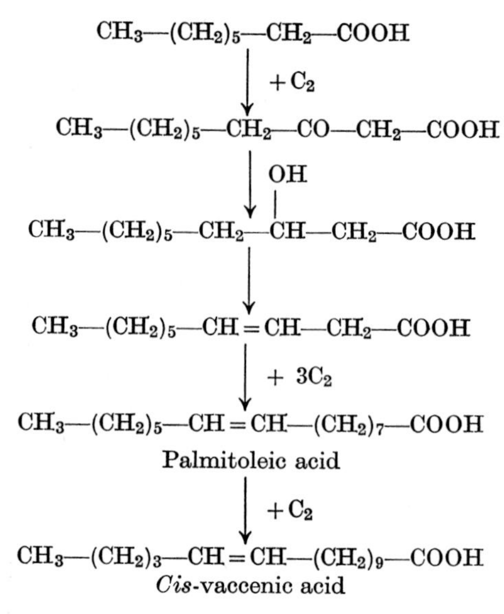

FIG. 17.5. The pathway for the anaerobic formation of unsaturated fatty acids in bacteria as proposed by Bloch *et al.* (1961).

the product was markedly increased. The dehydrase was specific for the D-(−)-3-hydroxydecanoyl-CoA and yielded a mixture of Δ^2 and Δ^3 unsaturated fatty acids. Norris *et al.* (1964) stated that the Δ^2 fatty acid was reduced and converted to long-chain fatty acid, while the Δ^3 acid was a precursor of palmitoleic and *cis*-vaccenic acid. The requirement of the dehydrase for the D isomer of 3-hydroxydecanoyl-CoA clearly places the enzyme on the pathway of fatty acid biosynthesis.

The proposal of Bloch *et al.* (1961) was made before the role of acyl carrier protein in fatty acid biosynthesis was known, and it now seems likely that acyl-ACP compounds are intermediates in the formation of monounsaturated fatty acids. Pugh *et al.* (1966) provided evidence in support of this idea when they separated the *E. coli* fatty acid synthetase into three fractions which they designated E_{II}, E_{III} and E_{IV}. Fraction E_{II} alone catalyzed the formation of 3-hydroxy fatty acids, the C_{10} acid being the principal product, but the C_{12} and C_{14} acids being formed as well. The requirements for this reaction were NADPH, malonyl-ACP and ACP. When fraction E_{III} was included in the reaction mixture, saturated long-chain fatty acids were the main product. Addition of fraction E_{IV} to E_{II} resulted in the formation of *cis*-vaccenic acid. Pugh *et al.* (1966) showed that fraction E_{IV} contained the dehydrase for 3-hydroxydecanoyl-CoA reported by Norris *et al.* (1964). The formation of the C_{10}, C_{12} and C_{14} 3-hydroxy acids is an interesting side-issue and might account for the formation of these acids which occur in the lipid fraction of several Gram negative bacteria (Chapter 3).

A pathway for the direct desaturation of long-chain fatty acids also exists in microorganisms and higher living forms. Bloomfield and Bloch (1960) were the first to demonstrate this reaction with cell-free preparations using extracts of *S. cerevisiae*. The reaction required molecular oxygen and NADPH and resulted in the formation of the Δ^9 derivatives of palmityl-CoA and stearyl-CoA. Fulco and Bloch (1964) studied the reaction with particulate preparations made from *M. phlei* and reported requirements for NADPH, oxygen and either FAD or FMN. Fulco *et al.* (1964) showed that whole cells of *C. diphtheriae* and *M. lysodeikticus* required oxygen for the conversion of palmitic acid-1-^{14}C to the corresponding Δ^9 monounsaturated fatty acid, and concluded that the aerobic desaturation pathway existed in these organisms. They also showed that oxygen was required for the desaturation of palmitic acid by *B. megaterium*, but that the product in this case was the Δ^5 monounsaturated acid.

II. Complex Lipids

A. Glycolipids

Burger *et al.* (1963) studied the biosynthesis of the rhamnolipid formed by *P. aeruginosa* (see Fig. 3.4). Crude extracts of their organism catalyzed the condensation of 2 moles of 3-hydroxydecanoyl-CoA to 3-hydroxydecanoyl-3-hydroxydecanoate (Fig. 17.6). The crude extract

2 (3-Hydroxydecanoyl-CoA) → 3-hydroxydecanoyl-3-hydroxydecanoate + 2CoA

TDP-L-rhamnose + 3-hydroxydecanoyl-3-hydroxydecanoate → L-rhamnosyl-3-hydroxydecanoyl-3-hydroxydecanoate

TDP-L-rhamnose + L-rhamnosyl-3-hydroxydecanoyl-3-hydroxydecanoate → L-rhamnosyl-L-rhamnosyl-3-hydroxydecanoyl-3-hydroxydecanoate

Fig. 17.6. Reactions in the biosynthesis of rhamnolipid of *P. aeruginosa*.

contained esterases which hydrolyzed the product and prevented the direct measurement of the reaction. In order to overcome this problem, Burger and his associates demonstrated the reaction by using radioactive 3-hydroxydecanoyl-CoA and including carrier 3-hydroxydecanoyl-3-hydroxydecanoate in the reaction mixture. The carrier was isolated at the termination of the experiment and shown to be radioactive. Burger obtained an enzyme fraction from *P. aeruginosa* which catalyzed the transfer of rhamnose from TDP-L-rhamnose-^{14}C to the ester yielding L-rhamnosyl-3-hydroxydecanoyl-3-hydroxydecanoate (Fig. 17.6). This enzyme preparation was specific in its substrate requirements and did not catalyze the transfer of the second rhamnosyl molecule. Burger *et al.* (1963) obtained a second enzyme fraction which catalyzed the transfer of rhamnose from TDP-L-rhamnose to the product of the preceding reaction, resulting in the formation of the complete rhamnolipid (Fig. 17.6).

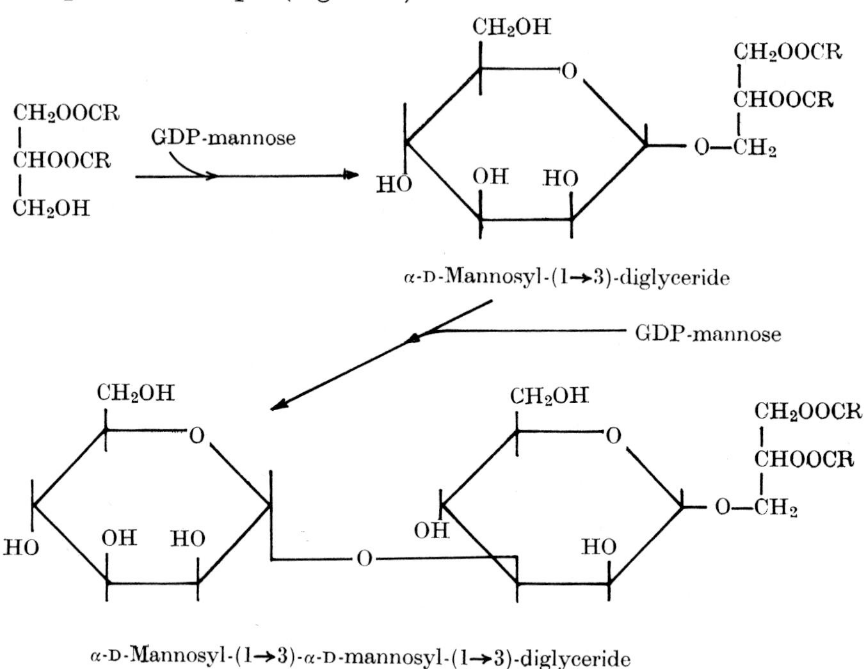

FIG. 17.7. Reactions in the biosynthesis of mannosyldiglycerides in *M. lysodeikticus* (Reproduced with permission from Lennarz and Talamo, 1966).

Lennarz and Talamo (1966) isolated a glycolipid from *M. lysodeikticus* which they characterized as α-D-mannosyl-(1-3)-α-D-mannosyl-(1-3)-diglyceride (Fig. 17.7). Crude extracts of *M. lysodeikticus* incubated with GDPmannose-^{14}C produced radioactive monomannosyldiglyceride, dimannosyldiglyceride and a third mannose derivative which they did not identify. An acetone powder was prepared from the crude cell-free extract and this preparation catalyzed the transfer of mannose from GDPmannose-^{14}C to a 1,2-diglyceride acceptor (Fig. 17.7). The reaction required magnesium ion, a high ionic strength and a fatty acid salt. Lennarz and Talamo found that the best acceptors were 1,2-diglycerides which contained branched-chain fatty acids and which they extracted from *M. lysodeikticus* and *B. megaterium*. Lennarz and Talamo separated a protein fraction from the acetone powder which catalyzed the transfer of mannose from GDPmannose-^{14}C to mono-mannosyldiglyceride to produce the dimannosyldiglyceride (Fig. 17.7).

B. Phospholipids

Ailhaud and Vagelos (1966) showed that a cell membrane fraction prepared from *E. coli* catalyzed the conversion of palmityl-ACP and glycerol-3-phosphate to monopalmitin. When palmityl-CoA and glycerol-3-phosphate were incubated with the cell membrane preparation, the products were mainly lysophosphatidic and phosphatidic acids. Many of the reactions in the biosynthesis of lipids are carried on in the cell membrane and it appears that this is the organelle responsible for the biosynthesis of structures near the cell surface.

Kanfer and Kennedy (1964) studied the biosynthesis of phosphatidylethanolamine and phosphatidylglycerol by extracts prepared from *E. coli* (Fig. 17.8). These investigators partially purified an enzyme which catalyzed the transfer of L-serine to CDP-dipalmitin to form phosphatidylserine and CMP. They studied the stoichiometry of the reaction and identified CMP as one of the products. Kanfer and Kennedy also identified a decarboxylase in extracts of *E. coli* which specifically catalyzed the decarboxylation of phosphatidylserine to phosphatidylethanolamine. Kennedy and his co-workers had earlier established the pathway for the biosynthesis of phosphatidylglycerol in animal tissues, and Kanfer and Kennedy (1964) demonstrated these reactions in extracts of *E. coli*. Specifically they showed that extracts catalyzed the conversion of CDP-diglyceride and glycerol-3-phosphate to phosphatidyglycerolphosphate and phosphatidylglycerol (Fig. 17.8).

Hill and Ballou (1966) studied the formation of mannophospholipids present in *M. phlei* (Fig. 17.9) with a particulate enzyme preparation. The enzyme catalyzed the incorporation of isotope from GDP-mannose-^{14}C to a phospholipid acceptor. Hill and Ballou characterized the product as 1-phosphatidyl-L-myoinositol 2-O-α-D-monomannoside and proposed that phosphatidymyoinositol was the mannose acceptor:

$$\text{GDPmannose} + \text{phosphatidylmyoinositol} \rightarrow \text{GDP} + \text{phosphatidylmyoinositol mannoside.}$$

Lennarz *et al.* (1966) made the extremely interesting discovery that lysyl-sRNA is the lysyl donor in the formation of *O*-lysylphosphatidylglycerol by *S. aureus* (Fig. 17.10). They found that incorporation of

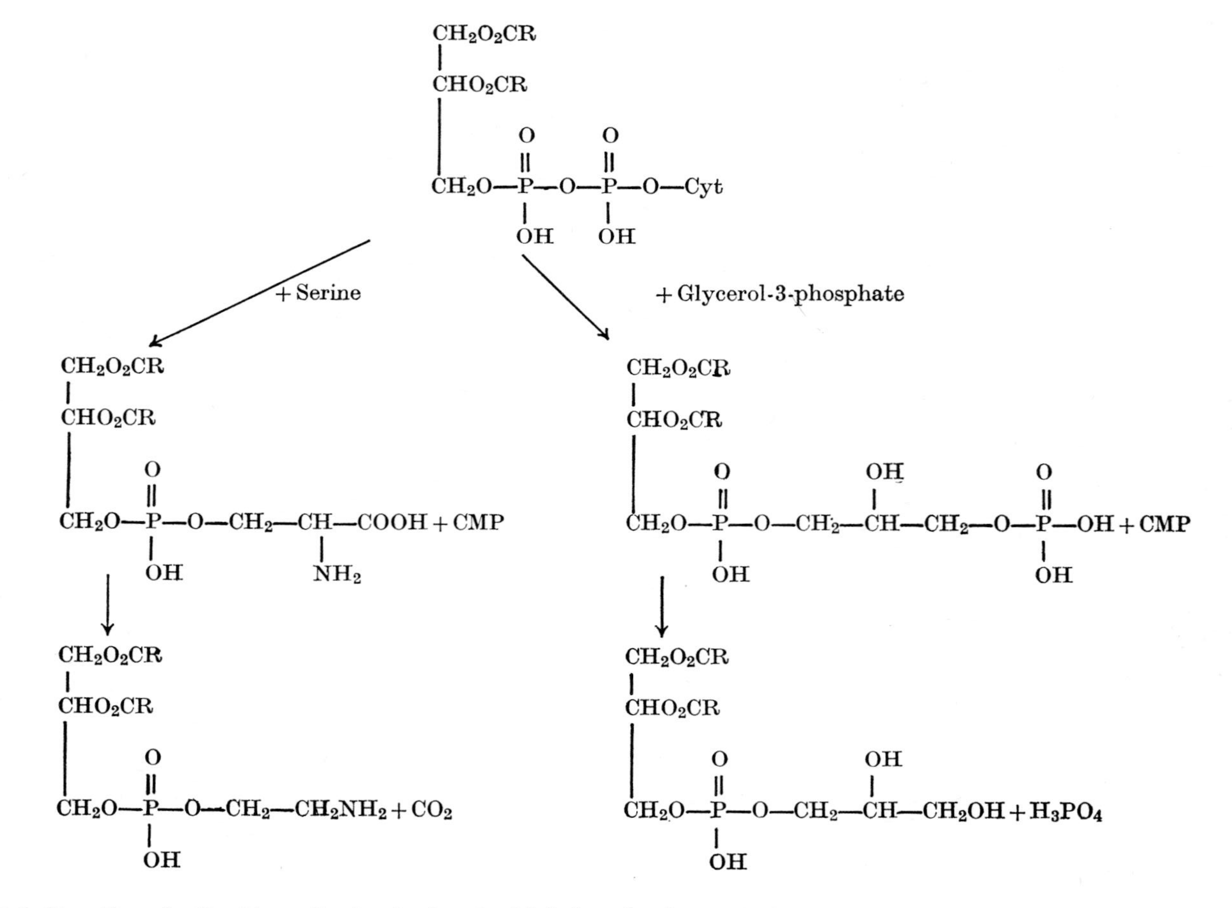

Fig. 17.8. Reactions in the biosynthesis of phosphatidylethanolamine and phosphatidylglycerol in *E. coli*. Cyt = cytidine

(Mannose)$_x$

O
‖
O—P—O—CH_2
|
OH
Mannose

$CH—O_2CR$
|
CH_2O_2CR

FIG. 17.9. Structure of the mannophospholipid formed by *M. phlei*. A single mannose residue is attached to position 2 of inositol and from zero to five mannose residues may be attached to position 6.

CH_2O_2CR
|
CHO_2CR
|
$CH_2O—P(=O)(OH)—O—CH_2—CH(OH)—CH_2—O_2C—CH(NH_2)—(CH_2)_4—NH_2$

FIG. 17.10. Structure of *O*-lysylphosphatidylglycerol from *S. aureus*. The position of glycerol to which the lysyl residue is attached has not been established although it is shown attached to the 3′ position in this figure.

lysine-U-^{14}C into the lipid fraction required ATP and was sensitive to RNase. Addition of lysyl-U-^{14}C-sRNA to the system resulted in efficient incorporation of the isotope into lysylphosphatidylglycerol. Again, the incorporation of lysyl-sRNA was catalyzed by the particulate fraction of the enzyme extract.

References

Ailhaud, G. P. and Vagelos, P. R. (1966). *J. Biol. Chem.* **241**, 3866.

Alberts, A. W., Majerus, P. W., Talamo, B. and Vagelos, P. R. (1964). *Biochemistry* **3**, 1563.

Alberts, A. W., Majerus, P. W. and Vagelos, P. R. (1965) *Biochemistry* **4**, 2265.

Allison, M. J., Bryant, M. P., Katz, I. and Keeney, M. (1962) *J. Bacteriol.* **83**, 1084.

Bloch, K., Baronowsky, P., Goldfine, H., Lennarz, W. J., Light, R., Norris A. T. and Scheuerbrandt, G. (1961). *Federation Proc.* **20**, 921.

Bloomfield, D. K. and Bloch, K. (1960). *J. Biol. Chem.* **235**, 337.
Burger, M. M., Glaser, L. and Burton, R. M. (1963). *J. Biol. Chem.* **238**, 2295.
Chalk, K. J. I. and Kodicek, E. (1961). *Biochim. Biophys. Acta* **50**, 579.
Chung, A. E. and Law, J. H. (1964). *Biochemistry* **3**, 967.
Etemadi, A. H. and Lederer, E. (1965). *Bull. Soc. Chim. Biol.* **47**, 107.
Fulco, A. J. and Bloch, K. (1964). *J. Biol. Chem.* **239**, 993.
Fulco, A. J., Levy, R. and Bloch, K. (1964). *J. Biol. Chem.* **239**, 998.
Gastambide-Odier, M. and Lederer, E. (1960). *Biochem. Z.* **333**, 285.
Gastambide-Odier, M., Delaumeny, J. M. and Lederer, E. (1963). *Biochim. Biophys. Acta* **70**, 670.
Goldman, P. (1964). *J. Biol. Chem.* **239**, 3663.
Goldman, P., Alberts, A. W. and Vagelos, P. R. (1963). *J. Biol. Chem.* **238**, 1255.
Hill, D. L. and Ballou, C. E. (1966). *J. Biol. Chem.* **241**, 895.
Kaneda, T. (1963). *J. Biol. Chem.* **238**, 1229.
Kanfer, J. and Kennedy, E. P. (1964). *J. Biol. Chem.* **239**, 1720.
Kates, M. (1966). *Ann. Rev. Microbiol.* **20**, 13.
Kusunose, M., Kusunose, E., Kowa, Y. and Yamamura, Y. (1959). *J. Biochem. (Tokyo)* **46**, 525.
Lennarz, W. J. (1961). *Biochem. Biophys. Res. Commun.* **6**, 112.
Lennarz, W. J. and Talamo, B. (1966). *J. Biol. Chem.* **241**, 2707.
Lennarz, W. J., Light, R. J. and Bloch, K. (1962a). *Proc. Natl Acad. Sci. U.S.* **48**, 840.
Lennarz, W. J., Scheuerbrandt, G. and Bloch, K. (1962b). *J. Biol. Chem.* **237**, 664.
Lennarz, W. J., Nesbitt III, J. A. and Reiss, J. (1966). *Proc. Natl Acad. Sci. U.S.* **55**, 934.
Liu, T. Y. and Hofmann, K. (1962). *Biochemistry* **1**, 189.
Lynen, F. (1961). *Federation Proc.* **20**, 941.
Majerus, P. W., Alberts, A. W. and Vagelos, P. R. (1964). *Proc. Natl Acad. Sci. U.S.* **51**, 1231.
Majerus, P. W., Alberts, A. W. and Vagelos, P. R. (1965a). *J. Biol. Chem.* **240**. 618.
Majerus, P. W., Alberts, A. W. and Vagelos, P. R. (1965b). *Proc. Natl Acad. Sci. U.S.* **53**, 410.
Norris, A. T., Matsumura, S. and Bloch, K. (1964). *J. Biol. Chem.* **239**, 3653.
O'Leary, W. M. (1959a). *J. Bacteriol.* **77**, 367.
O'Leary, W. M. (1959b). *J. Bacteriol.* **78**, 709.
Pugh, E. L., Sauer, F., Moseley, W., Toomey, R. E. and Wakil, S. **J.** (1966). *J. Biol. Chem.* **241**, 2635.
Vagelos, P. R. (1964). *Ann. Rev. Biochem.* **33**, 139.
Wakil, S. J. (1958). *J. Am. Chem. Soc.* **80**, 6465.
Wegner, G. H. and Foster, E. M. (1963). *J. Bacteriol.* **85**, 53.
Zalkin, H., Law, J. H. and Goldfine, H. (1963). *J. Biol. Chem.* **238**, 1242.

18. BIOSYNTHESIS OF NUCLEIC ACIDS

I. Formation of Purine and Pyrimidine Nucleotides

A. Purine Nucleotides

1. De novo *formation of inosinic acid*

Much of the basic work on the *de novo* formation of purines was done with birds, which are good experimental animals since they excrete nitrogen as uric acid (see review of Buchanan and Hartman, 1959). Studies with isotopic carbon and nitrogen compounds revealed that carbon dioxide was a precursor of position 6 of uric acid, formate was a precursor of positions 2 and 8 and glycine was a precursor of positions 4 and 5 (Fig. 18.1). When the amino nitrogen of glycine was found to

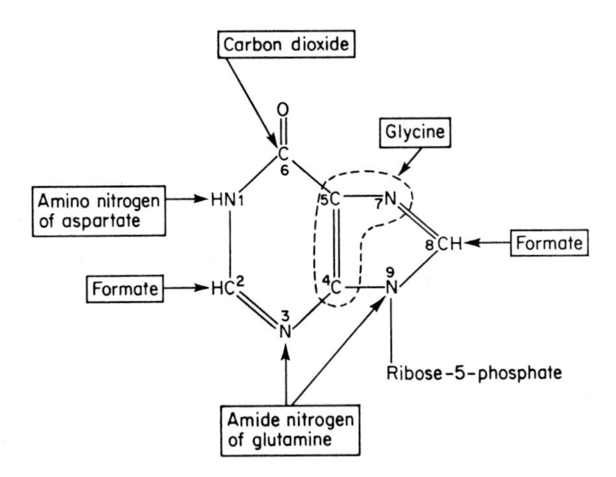

Fig. 18.1. Origin of carbon and nitrogen atoms of inosinic acid.

be a precursor of position 7, it was apparent that glycine was incorporated as a whole. The amide nitrogen of glutamine was the precursor of positions 3 and 9 while the amino nitrogen of aspartate was the precursor of position 1 of uric acid. The origin of the carbons of ribose and deoxyribose are discussed in Chapter 15.

Hartman and Buchanan (1958a) purified 5-phosphoribosylpyrophosphate amidotransferase (ribosylamine-5-phosphate: pyrophosphate phosphoribosyltransferase [glutamate-amidating]) from extracts of pigeon liver. This enzyme catalyzes the first reaction of purine biosynthesis, the formation of 5-phosphoribosylamine from 5-phosphoribosylpyrophosphate and glutamine (Fig. 18.2), which accounts for the incorporation of the nitrogen atom into position 9 of the purine ring. Buchanan and Hartman (1959) suggested that this reaction resulted in inversion of the configuration at carbon 1 of ribose since purine ribosides are of the β configuration while 5-phosphoribosylpyrophosphate has the α configuration. Nierlich and Magasanik (1965a) partially purified the enzyme from *A. aerogenes* and showed that low concentrations of purine ribonucleotides inhibited the action of the enzyme. This result was expected since it is now a common observation that the first enzyme of a biosynthetic pathway is subject to end-product inhibition. Hartman and Buchanan (1958b) also purified the second enzyme of the purine biosynthetic pathway, glycineamide ribonucleotide synthetase (ribosylamine-5-phosphate:glycine ligase [ADP]) and established that it catalyzed the condensation of glycine and 5-phosphoribosylamine to form glycinamide ribonucleotide. ATP was utilized in the reaction and served as a source of chemical energy for the formation of the amide linkage. Nierlich and Magasanik (1965b) also purified this enzyme from *A. aerogenes* and showed that it was not subject to inhibition by purine ribonucleotides.

The next enzyme of the purine biosynthetic pathway is phosphoribosylglycineamide formyltransferase (5′-phosphoribosyl-N-formylglycineamide:tetrahydrofolate 5,10-formyltransferase), which was first purified from chicken liver by Warren and Buchanan (1957). They established that the enzyme catalyzed the transfer of a formyl residue from either N^5,N^{10}-methenyltetrahydrofolate or N^{10}-formyltetrahydrofolate to glycineamide ribonucleotide to form formylglycineamide ribonucleotide (Fig. 18.2). In a later study, Hartman and Buchanan (1959) obtained a preparation of this enzyme free from cyclohydrolase, the enzyme which catalyzes interconversion of the two species of tetrahydrofolate, and showed that the actual formyl donor for this enzyme was N^5,N^{10}-methenyltetrahydrofolate.

Levenberg and Buchanan (1957b) prepared a protein fraction from pigeon liver which catalyzed the conversion of formylglycineamide ribonucleotide, glutamine and ATP to a compound which they identified as formylglycineamidine ribonucleotide (Fig. 18.2). The enzyme which catalyzes this reaction carries the Enzyme Commission designation of 5′-phosphoribosyl-formylglycineamide: L-glutamine amido-ligase (ADP). This reaction is responsible for the insertion of the nitrogen atom at position 3 of the purine ring. Levenberg and Buchanan (1957a) obtained another fraction from pigeon liver which catalyzed the conversion of formylglycineamide ribonucleotide, glutamine and ATP beyond the stage of formylglycineamidine ribonucleotide to another compound which they identified as 5-aminoimidazole ribonucleotide (Fig. 18.2). Levenberg and Buchanan (1957b) showed that the formation of the imidazole ring from formylglycineamidine ribonucleotide as the substrate required ATP. This enzyme was named 5′-phosphoribosyl-formylglycineamidine cyclo-ligase (ADP) by the International Commission on Enzymes.

Lukens and Buchanan (1959b) purified the carboxylase (5′-phosphoribosyl-5-amino-4-imidazolecarboxylate carboxy-lyase), which catalyzed the addition of carbon dioxide to 5-aminoimidazole ribonucleotide from chicken liver. They characterized the product as 5-amino-4-imidazolecarboxylate ribonucleotide. This reaction accounts for the origin of carbon 6 of the purine ring.

Lukens and Buchanan (1959a) obtained a soluble protein fraction from chicken liver which catalyzed the conversion of 5-aminoimidazole ribonucleotide, ATP, L-aspartate and bicarbonate to an intermediate which they identified as 5-amino-4-imidazole-*N*-succinocarboxamide ribonucleotide (Fig. 18.2). 5-Amino-4-imidazole carboxylate ribonucleotide was also converted to 5-amino-4-imidazole-*N*-succinocarboxamide ribonucleotide by a chicken liver enzyme fraction and in this case only ATP and L-aspartate were required (Lukens and Buchanan, 1959b). This study established the role of 5-amino-4-imidazolecarboxylate ribonucleotide as an intermediate in purine biosynthesis. The enzyme which catalyzed the addition of aspartate was given the Enzyme Commission name 5′-phosphoribosyl-4-carboxy-5-aminoimidazole-L-aspartate ligase (ADP). Miller *et al.* (1959) studied the enzyme which cleaved the dicarboxylic acid residue from 5-amino-4-imidazole-*N*-succinocarboxamide ribonucleotide in extracts of chicken liver and yeast and purified the enzyme 170-fold from baker's yeast. The products of the reaction were identified as 5-amino-4-imidazolecarboxamide ribonucleotide and fumaric acid.

Flaks *et al.* (1957) studied the final two enzymes involved in the formation of inosinic acid with protein fractions from chicken liver. One enzyme catalyzed the transfer of a formyl residue from a formyl derivative of tetrahydrofolate to 5-amino-4-imidazolecarboxamide ribonucleotide to produce 5-formamido-4-imidazolecarboxamide ribo-

$$P_2O_7^{3-}\text{—}C_5H_8O_4PO_3^{2-} \xrightarrow{\text{Glutamine}} NH_2\text{—}C_5H_8O_4PO_3^{2-} + \text{Glutamate} + PP_i$$

5-Phosphoribosyl-1-pyrophosphate → 5-Phosphoribosylamine

$$\xrightarrow[\text{ATP}]{\text{Glycine}} CH_2NH_2\text{—}C(=O)\text{—}NH\text{—}C_5H_8O_4PO_3^{2-} + \text{ADP} + P_i$$

Glycineamide ribonucleotide

$$\xrightarrow[\text{H}_4\text{-folate}]{\text{N}^5,\text{N}^{10}\text{ Methenyl-}} O{=}C(\text{—}CH_2NH\text{—}CHO)\text{—}NH\text{—}C_5H_8O_4PO_3^{2-} + H_4\text{-folate}$$

Formylglycineamide ribonucleotide

$$\xrightarrow[\text{ATP}]{\text{Glutamine}} H_2\overset{+}{N}{=}C(\text{—}CH_2NH\text{—}CHO)\text{—}NH\text{—}C_5H_8O_4PO_3^{2-} + \text{ADP} + \text{Glutamate} + P_i$$

Formylglycineamidine ribonucleotide

$$\xrightarrow{\text{ATP}} \text{(imidazole ring: HC—N=CH—N(}C_5H_8O_4PO_3^{2-}\text{)—C(}H_2N\text{)=)} + \text{ADP} + P_i$$

5-Aminoimidazole ribonucleotide

$$\xrightarrow{HCO_3^-} \text{(imidazole ring: }COO^-\text{—C—N=CH—N(}C_5H_8O_4PO_3^{2-}\text{)—C(}H_2N\text{)=)}$$

5-Amino-4-imidazolecarboxylate ribonucleotide

$$\xrightarrow[\text{ATP}]{\text{Aspartate}} \text{(imidazole ring: }COO^-\text{—CH(}CH_2\text{—}COO^-\text{)—NH—C(=O)—C—N=CH—N(}C_5H_8O_4PO_3^{2-}\text{)—C(}H_2N\text{)=)} + \text{ADP} + P_i$$

5-Amino-4-imidazole-*N*-succino-carboxamide ribonucleotide

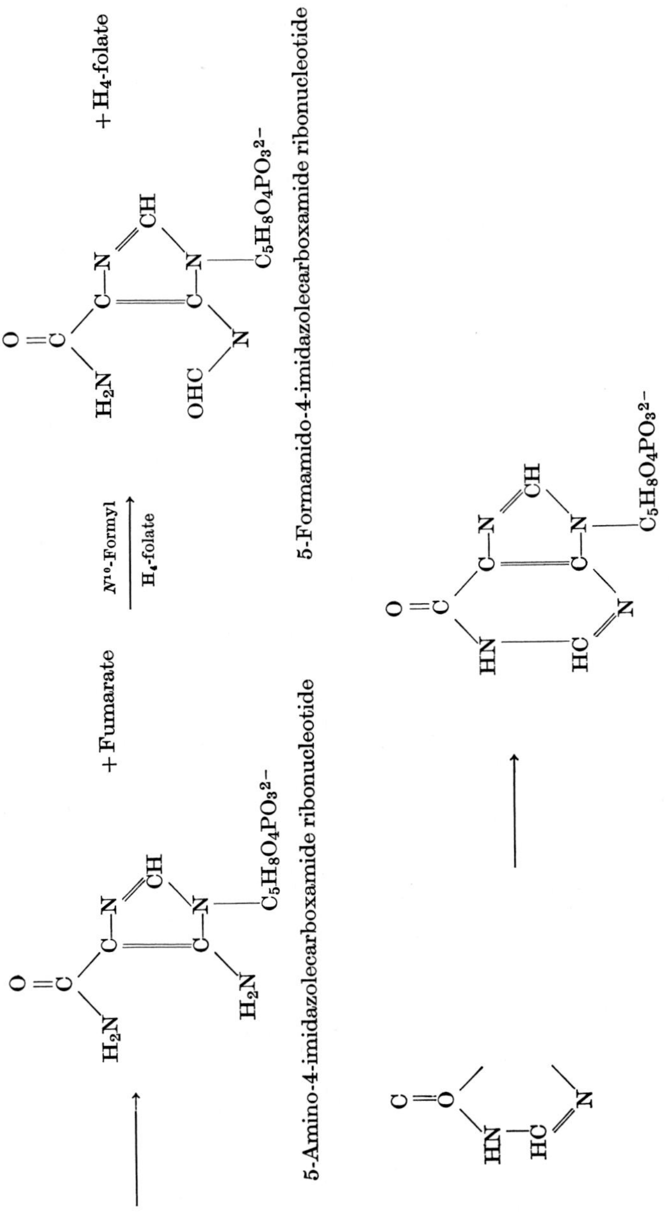

→ 5-Amino-4-imidazolecarboxamide ribonucleotide ($C_5H_8O_4PO_3^{2-}$) + Fumarate

$\xrightarrow[H_4\text{-folate}]{N^{10}\text{-Formyl}}$ 5-Formamido-4-imidazolecarboxamide ribonucleotide ($C_5H_8O_4PO_3^{2-}$) + H_4-folate

→ Inosinic acid ($C_5H_8O_4PO_3^{2-}$)

FIG. 18.2. Pathway for the biosynthesis of inosinic acid.

nucleotide. Hartman and Buchanan (1959) later showed that the actual formyl donor was N^{10}-formyltetrahydrofolate. This reaction accounted for the insertion of carbon atom 2 of the purine ring and the enzyme was named 5-amino-4-imidazolecarboxamide ribonucleotide transformylase (5′-phosphoribosyl-5-formamido-4-imidazolecarboxamide: tetrahydrofolate 10-formyltransferase). The enzyme preparations made by Flaks *et al.* (1957) also catalysed the closure of synthetic 5-formamido-4-imidazolecarboxamide ribonucleotide to inosinic acid and they named this enzyme inosinicase (IMP 1,2-hydrolase [decyclizing]).

2. *Conversion of inosinic acid to adenine and guanine ribonucleotides*

Lieberman (1956b) purified an enzyme from *E. coli* strain B, which catalyzed the formation of adenylosuccinate from inosinic acid, GTP and L-aspartate which he named adenylosuccinate synthetase (IMP: L-aspartate ligase [GDP]; Fig. 18.3). Unfractionated extracts of *E. coli* converted adenylosuccinate to adenylic acid. Carter and Cohen (1956) purified adenylosuccinase (adenylosuccinate AMP-lyase) enzyme which catalyzed the latter reaction from baker's yeast. They showed that adenylosuccinate was the substrate for the reaction and that the products were adenylic and fumaric acids (Fig. 18.3). This is probably the same enzyme which cleaves fumarate from 5-amino-4-imidazole-*N*-succinocarboxamide ribonucleotide, since the enzyme purified by Miller *et al.* (1959) from baker's yeast catalyzed both reactions. Also, Gots and Gollub (1957) isolated mutants of *E. coli* and *S. typhimurium* which required adenine and which lacked the ability to cleave fumarate from both 5-amino-4-imidazole-*N*-succinocarboxamide ribonucleotide and adenylosuccinate. The fact that the mutant lacked both enzyme activities also supports the idea that one enzyme catalyzes both reactions.

Lagerkvist (1958a) studied the route to guanine formation in pigeon liver extracts. He found that fortified extracts incubated with inosinic acid-5-^{14}C formed radioactive xanthylic and guanylic acids. The overall reaction required NAD, ATP and glutamine. Magasanik and his associates studied these reactions with purified enzymes prepared from *A. aerogenes*. One enzyme, inosinic acid dehydrogenase (IMP:NAD oxidoreductase), catalyzed the oxidation of inosinic acid with NAD as the electron acceptor (Magasanik *et al.*, 1957). Moyed and Magasanik (1957) purified the other enzyme, guanylic acid synthetase (xanthosine-5′-phosphate: ammonia ligase [AMP]) 300-fold and established that this enzyme catalyzed the amination of xanthylic acid with ammonia and ATP to form guanylic acid. The other products of the reaction were

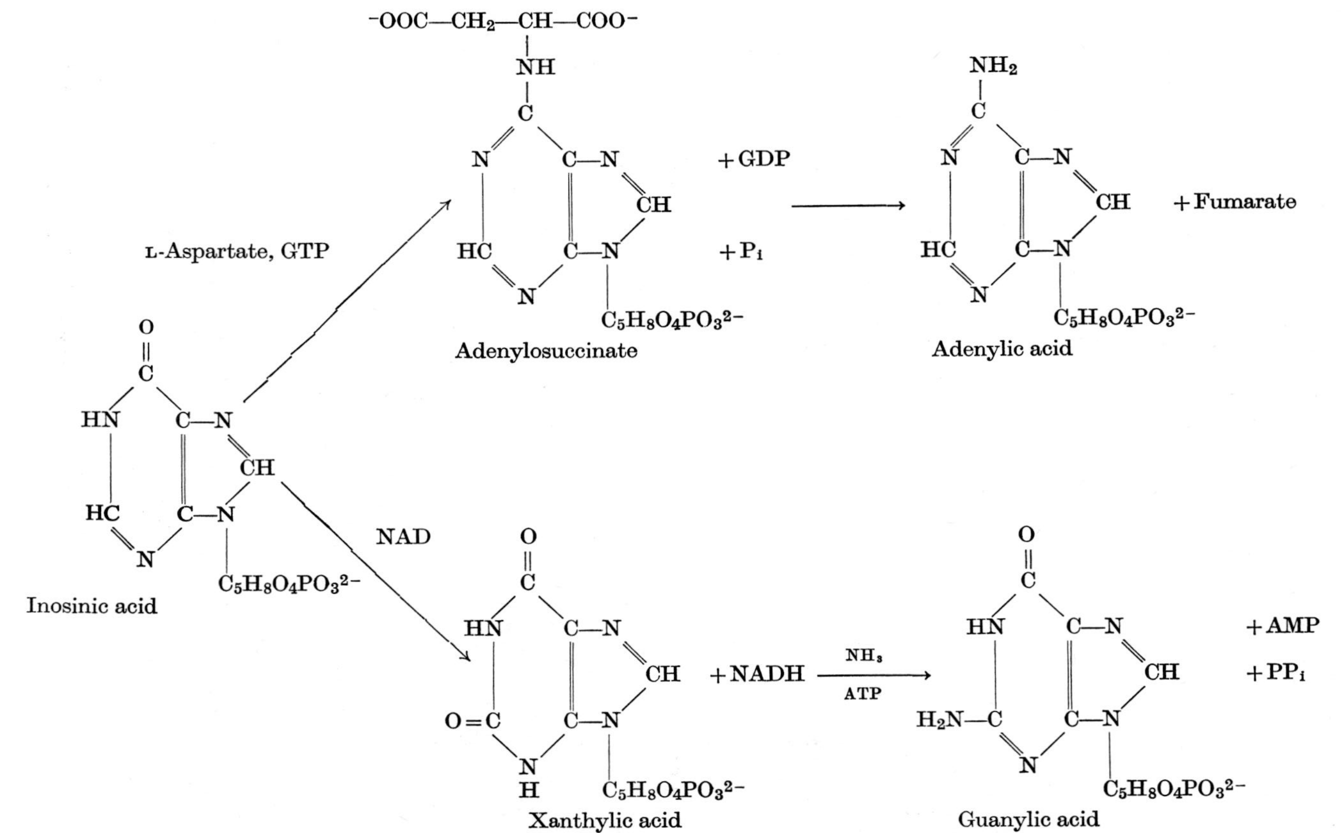

FIG. 18.3. Biosynthesis of adenylic and guanylic acids from inosinic acid in bacteria. In animals, glutamine is the amino donor for guanylic acid biosynthesis rather than ammonia.

AMP and pyrophosphate (Fig. 18.3). The corresponding enzyme in animals catalyzes a slightly different reaction, since it utilizes glutamine as the amino donor (Lagerkvist, 1958b).

3. *Reduction of purine ribonucleotides to deoxyribonucleotides*

Purine ribonucleoside diphosphates are reduced to the corresponding deoxyribonucleotides in *E. coli*, while the reduction takes place at the stage of nucleoside triphosphates in *L. leichmanii*. This process is discussed in Chapter 15, section III.C.1.

4. *Phosphorylation of purine ribonucleotides and deoxyribonucleotides*

ATP plays a central role in the phosphorylation of ribo- and deoxyribonucleotides, since it is the ultimate source of high energy phosphate for this process. Adenylate kinase (ATP:AMP phosphotransferase) has been detected in several species of bacteria (Oliver and Peel, 1956). The animal enzyme has been well characterized and catalyzes the following reaction:

$$\text{ATP} + \text{AMP} \rightarrow 2\text{ADP}$$

ADP is converted to ATP by oxidative and substrate level phosphorylation.

Hiraga and Sugino (1966) found that extracts of both normal and virus-infected *E. coli* contained at least five nucleoside monophosphate kinases; one enzyme acted on AMP and dAMP and another acted on GMP and dGMP. The pyrimidine kinases which were studied by these investigators are discussed in section I.B.2 of this chapter. Oeschger and Bessman (1966) purified the kinase for GMP approximately 7000-fold from *E. coli* and found that the enzyme catalyzed the following reaction:

$$\text{(d)ATP} + \text{(d)GMP} \rightarrow \text{(d)ADP} + \text{(d)GDP}$$

Pyrimidine nucleotides were inactive as substrates for this enzyme. Lehman *et al.* (1958a) prepared deoxyribonucleoside triphosphates for their studies of DNA biosynthesis with an enzyme fraction from *E. coli* which catalyzed the phosphorylation of dAMP, dGMP, dCMP and dTMP. It is likely that nucleoside diphosphates are converted to triphosphates by separate nucleoside diphosphate kinases in their enzyme preparations. Ratliff *et al.* (1964) crystallized a nucleoside diphosphate kinase (ATP:nucleoside diphosphate phosphotransferase) of broad specificity from yeast. This enzyme catalyzed the transfer of phosphate from ATP and dATP to virtually all biologically important ribo- and deoxyribonucleoside diphosphates.

5. *Formation of purine nucleotides from free bases and nucleosides*

Conversion of purine bases to the corresponding ribonucleotides. It is clear that free bases can be converted to purine nucleotides, since mutants blocked in the *de novo* pathway have been isolated which are able to grow with one of the four common purine bases, adenine, guanine, hypoxanthine and xanthine (Magasanik, 1962) (Fig. 18.4). The initial reaction in the utilization of purine bases is catalyzed by nucleotide pyrophosphorylases which catalyze the following general reaction:

$$\text{Purine} + \text{5-phosphoribosyl-1-pyrophosphate} \rightarrow \text{purine ribonucleoside-5}'\text{-phosphate} + \text{pyrophosphate.}$$

Carter (1959) purified IMP pyrophosphorylase (IMP:pyrophosphate phosphoribosyltransferase) from *E. coli* and reported that it catalyzed the formation of IMP and GMP from hypoxanthine and guanine respectively. This enzyme also catalyzed the conversion of 6-mercaptopurine to the corresponding ribonucleotide. Brockman *et al.* (1961) studied mutants of *S. faecalis* which had lost their sensitivity to purine analogs. Extracts of the parent strain catalyzed the formation of IMP, GMP, AMP and XMP from the corresponding bases and 5-phosphoribosylpyrophosphate (Fig. 18.4). Extracts of a mutant resistant to 6-mercaptopurine lacked IMP pyrophosphorylase. A mutant resistant to 2-fluoroadenine possessed a greatly reduced level of AMP pyrophosphorylase; another resistant to the action of 8-azaxanthine lacked XMP pyrophosphorylase. These observations led to the conclusion that there were three purine pyrophosphorylases in wild type *E. coli*, one for IMP and GMP, one for AMP and one for XMP.

Conversion of adenine to guanine nucleotides. Magasanik and Karibian (1960) showed that the adenine nucleus was converted to guanine ribonucleotides in *E. coli* by two pathways; in one pathway, the nitrogen atom at position 1 and the carbon at position 2 were lost and then replaced before guanine was formed. In the second pathway, adenine was converted to guanine without fission of the purine ring. Magasanik and Karibian suggested that the loss of positions 1 and 2 of adenine was because these two positions were incorporated into the imidazole ring in the biosynthesis of histidine (Fig. 16.7). The remainder of the adenine nucleotide, 5-amino-4-imidazolecarboxamide ribonucleotide, is an intermediate in purine biosynthesis (Fig. 18.2). This proposal was supported by the finding that addition of histidine to the growth medium

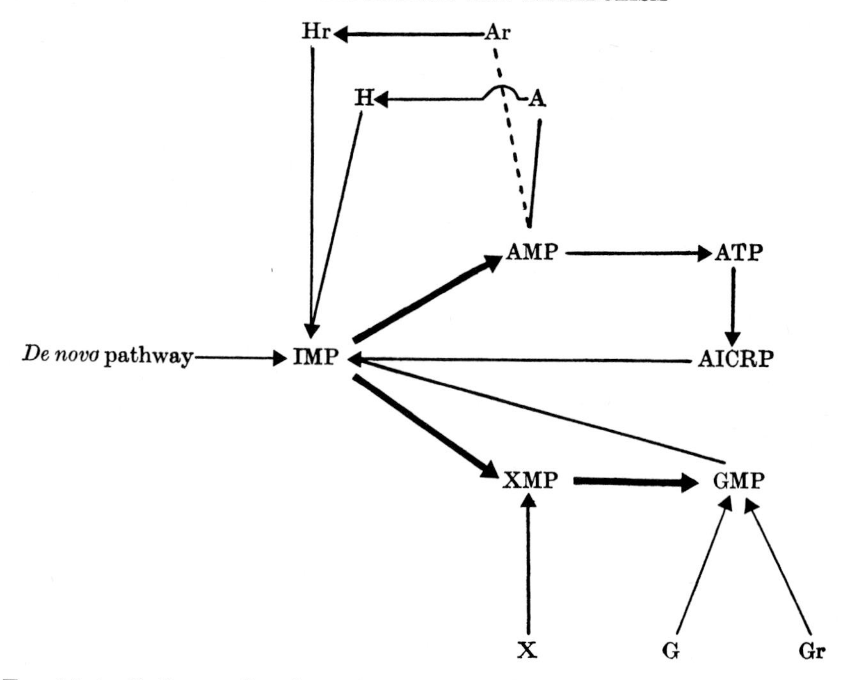

FIG. 18.4. Pathways for the utilization of purines and purine ribosides (fine arrows) and their relation to the *de novo* pathway of purine nucleotide synthesis (bold arrows). The abbreviations used are: A = adenine, Ar = adenosine, H = hypoxanthine, Hr = inosine, G = guanine, Gr = guanosine, AICRP = 5-amino-4-imidazolecarboxamide ribonucleotide. The dotted line from Ar to AMP indicates that this is a minor pathway of Ar utilization (Zimmerman and Magasanik, 1964).

resulted in formation of nucleic acid guanine only by the pathway in which the adenine nucleus remained intact. The pathway for the conversion of the intact adenine ring to guanine apparently proceeded through hypoxanthine (Fig. 18.4) since extracts of *E. coli* converted adenine to hypoxanthine (Zimmerman and Magasanik, 1964), probably as a result of the action of adenine deaminase (adenine aminohydrolase).

Conversion of guanine to adenine nucleotides. Mager and Magasanik (1960) deduced that there must be a pathway for the conversion of guanine to adenine nucleotides which did not involve reversal of the reactions leading from IMP to GMP in the *de novo* pathway of GMP biosynthesis. Mutants of several species of Enterobacteriaciae were known which could convert guanine to nucleic acid adenine, but not adenine to nucleic acid guanine. They discovered and isolated the key enzyme in the pathway from guanine to adenine nucleotides, GMP reductase (reduced-NADP:GMP oxidoreductase [deaminating]) and established that it catalyzed the reductive deamination of GMP to IMP:

$$GMP + NADPH \rightarrow IMP + NADP + NH_3$$

Thus, GMP is converted to IMP without the intermediate formation of XMP. IMP could then be converted to AMP by the *de novo* purine biosynthetic pathway.

Utilization of nucleosides. Zimmerman and Magasanik (1964) also obtained information regarding the pathways for the utilization of nucleosides using a mutant of *S. typhimurium* resistant to 6-mercaptopurine, and therefore assumed to possess a defective IMP pyrophosphorylase. This mutant was able to incorporate guanosine and inosine into nucleic acid purines, but not guanine and hypoxanthine. Most probably the nucleosides were converted to nucleotides by the action of a nucleoside kinase (Fig. 18.4). One curious point, however, is that these authors were unable to find any significant difference between the level of IMP pyrophosphorylase in the mutant and parent organism. However, it is possible that the enzyme may have been defective in some way and not functioning properly in the intact cell. Adenosine also was incorporated *in toto* since both the purine and ribose portions of the labeled nucleoside were incorporated to the same extent into nucleic acids of *S. typhimurium*. Utilization of adenosine may have occurred by way of inosine, since this was a major fate of adenosine in cell free extracts of the organism (Fig. 18.4).

B. Pyrimidine Nucleotides

1. De novo *formation of uridine-5′-phosphate*

Tissue slice preparations of animal organs capable of forming orotic acid incorporated ammonia into the nitrogen atom at position 1 of orotate, carbon dioxide into position 2 and L-aspartate into the remainder of the orotate molecule (see review of Crosbie, 1960). Wright *et al.* (1951) showed that pyrimidine biosynthesis in bacteria was similar to that in animals when they found that *L. bulgaricus* incorporated carbamoyl-L-aspartate, which was then an established intermediate in pyrimidine biosynthesis, into uracil and cytosine of nucleic acid.

The first reaction unique to the biosynthesis of pyrimidines is catalyzed by aspartate transcarbamylase (carbamoylphosphate:L-aspartate carbamoyltransferase). Reichard and Hanshoff (1956) purified the enzyme 100-fold from *E. coli* and established that it catalyzed the condensation of L-aspartate and carbamoylphosphate to form carba-

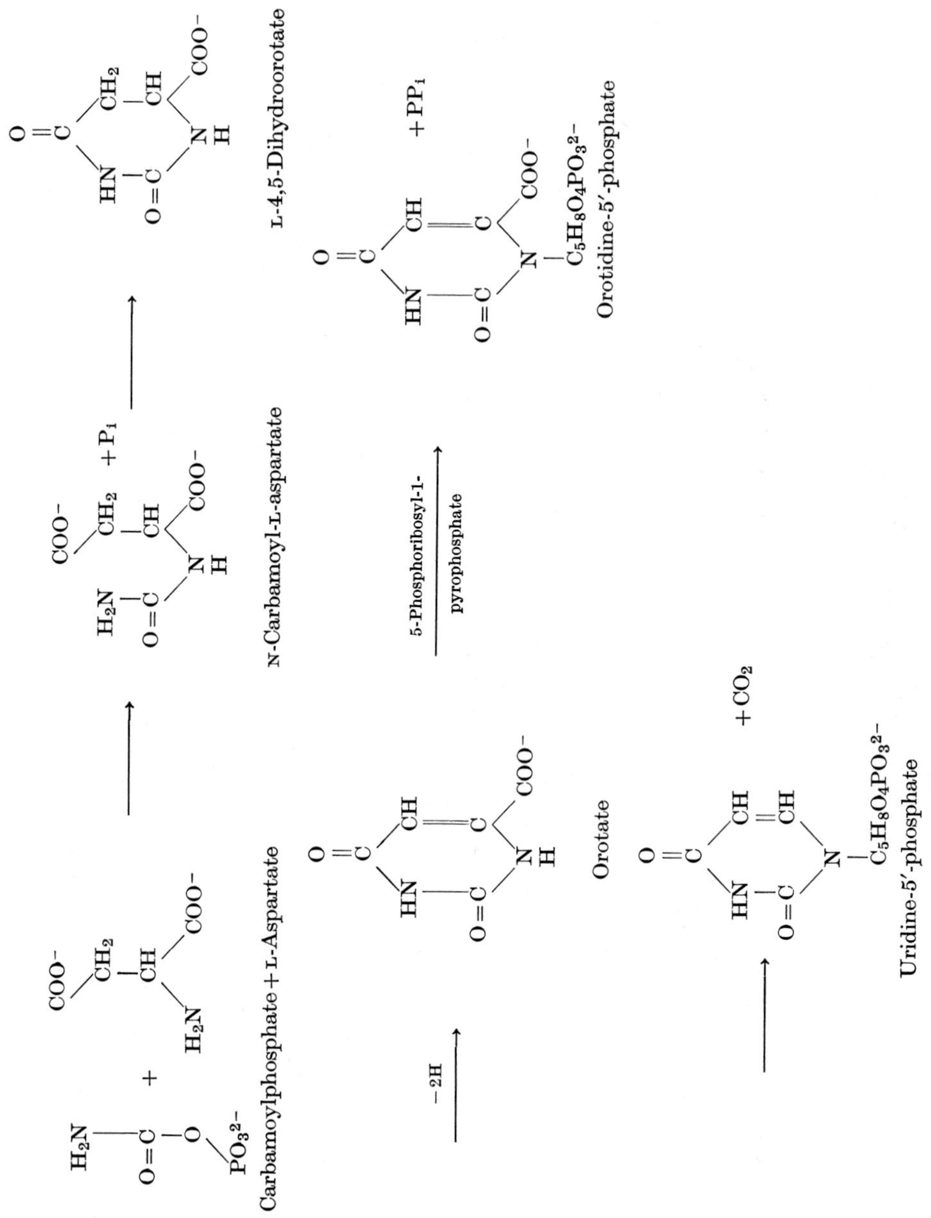

Carbamoylphosphate + L-Aspartate → N-Carbamoyl-L-aspartate $+ P_i$ → L-4,5-Dihydroorotate

$\xrightarrow{-2H}$ Orotate $\xrightarrow[\text{pyrophosphate}]{\text{5-Phosphoribosyl-1-}}$ Orotidine-5′-phosphate $+ PP_i$

→ Uridine-5′-phosphate $+ CO_2$

Fig. 18.5. Biosynthesis of uridine-5′-phosphate.

moyl-L-aspartate (Fig. 18.5). The enzyme was later obtained in a pure state from *E. coli* by Gerhart and Pardee (1962) who studied its regulation by CTP (see Chapter 16, section III.B).

The second enzyme of the pyrimidine biosynthetic pathway, dihydro-orotase (L-4,5-dihydro-orotate amidohydrolase), was discovered by Lieberman and Kornberg (1954), who were studying the degradation of pyrimidines in an organism which they isolated from an enrichment culture with orotate as the energy source. The isolate was given the name *Zymobacterium oroticum*. Lieberman and Kornberg established that this enzyme catalyzed the reversible formation of L-dihydro-orotate from carbamoyl-L-aspartate. Since Lieberman and Kornberg were actually working with a degradative pathway it was conceivable that dihydro-orotase was not involved in pyrimidine biosynthesis. However, Yates and Pardee (1956) showed that this enzyme was present in enzyme extracts prepared from *E. coli* grown in glycerol synthetic medium at a level high enough to account for the rate of pyrimidine biosynthesis in growing cells.

The third enzyme of the pyrimidine pathway is dihydro-orotate dehydrogenase (L-4,5-dihydro-orotate:oxygen oxidoreductase) which was also discovered by Lieberman and Kornberg (1953) in their study of orotic acid fermentation by *Z. oroticum*. Friedmann and Vennesland (1960) crystallized the enzyme from *Z. oroticum* and showed that it contained one mole of FMN, one of FAD and 2 moles of iron per mole enzyme. The flavin groups of the enzyme were reduced both by dihydro-orotate and NADH. Apparently the enzyme functions in the cell by accepting electrons from dihydro-orotate and then passing them to NAD.

Lieberman *et al.* (1955a) purified the last two enzymes involved in uridine-5′-phosphate formation and characterized the reactions. They established that orotidine-5′-phosphate was formed by the condensation of orotate and 5-phosphoribosyl-1-pyrophosphate by an enzyme which they named orotidylate pyrophosphorylase (orotidine-5′-phosphate:pyrophosphate phosphoribosyltransferase). This was the first report of a nucleotide pyrophosphorylase, and in the same volume of the *Journal of Biological Chemistry*, this research team also reported on the enzymic formation of 5-phosphoribosyl-1-pyrophosphate by the enzyme from pigeon liver (Kornberg *et al.*, 1955a) and on the formation of AMP, IMP and GMP by nucleotide pyrophosphorylases of yeast (Kornberg *et al.*, 1955b). Lieberman *et al.* (1955a) also purified orotidylate decarboxylase (orotidine-5′-phosphate carboxy-lyase) and established that it catalyzed the non-reversible decarboxylation of orotidylate to uridylate (Fig. 18.5).

2. *Conversion of uridine-5′-phosphate to cytidine and thymidine nucleotides*

The pathways for the conversion of UMP to UTP, CTP, dTTP and dCTP in *E. coli* are illustrated in Fig. 18.6. Lieberman *et al.* (1955b)

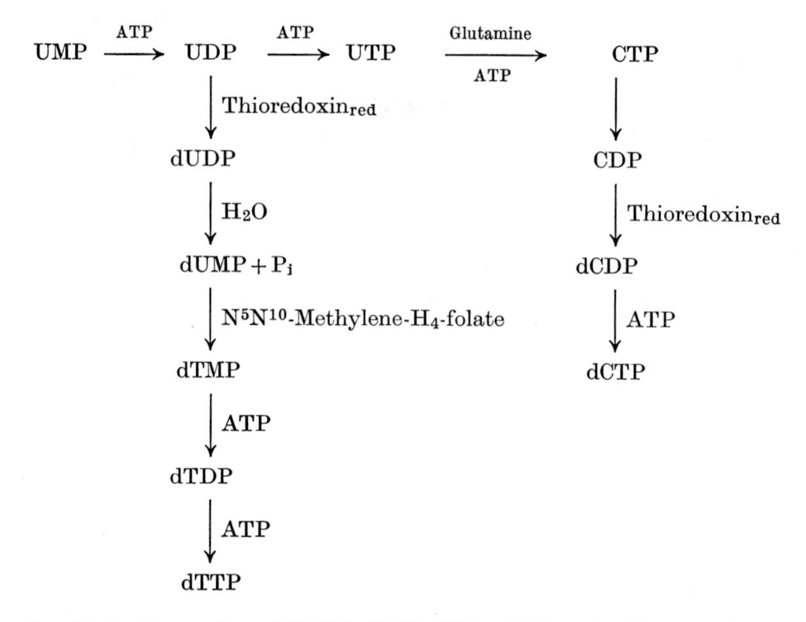

FIG. 18.6. Conversion of UMP to UTP, CTP, dTTP and dCTP in normal *E. coli*. In *L. leichmanii*, the reduction of ribonucleotides takes place at the stage of the triphosphates (Chapter 15, section II.C.1).

partially purified a nucleoside monophosphate kinase from yeast which catalyzed the formation of UDP from UMP and ATP. This enzyme also transferred phosphate from ATP to GMP. In their study of *E. coli* infected with MS2 virus, Hiraga and Sugino (1966) found a separate kinase for UMP in extracts of both normal and virus-infected cells. The nucleoside diphosphate kinase of yeast isolated by Ratcliff *et al.* (1964; section I.A.4). is also active on pyrimidine nucleoside diphosphates.

Lieberman (1956a) studied CTP formation from UTP with an enzyme from *E. coli* and found that ATP and ammonium ion were required. Long and Pardee (1967) extensively purified CTP synthetase (UTP:ammonia ligase ADP) and showed that glutamine would also serve as the nitrogen donor (Fig. 18.6). Since glutamine was effective at a lower concentration than ammonium ion, particularly in the presence of GTP, it appeared to be the physiological nitrogen donor. CTP acted as a competitive inhibitor of UTP at high substrate concentrations and thereby acted as a regulator of its own synthesis.

In *E. coli*, uridine and cytidine ribonucleoside diphosphates are reduced to deoxyribonucleotides, while in *L. leichmanii*, the triphosphates are reduced (Fig. 18.6). Since the first cytidine nucleotide is CTP, some mechanism must exist in *E. coli* for converting CTP to CDP prior to reduction. The nucleoside diphosphate kinase of yeast catalyzes the transfer of phosphate from CTP to ADP (Ratliff *et al.*, 1964) and this may be the case in *E. coli* as well. In addition, a certain amount of CMP would be formed in the biosynthesis of polysaccharides, such as the O-antigen, where CDP-sugar nucleotides are precursors. CMP could be converted to CDP by a nucleoside monophosphate kinase in *E. coli* active on both CMP and dCMP. Maley and Ochoa (1958) purified this enzyme 800-fold and established that a single enzyme catalyzed the transfer of phosphate from ATP to CMP and dCMP:

$$\text{ATP} + \text{(d)CMP} \rightarrow \text{(d)CDP} + \text{ADP}$$

Greenberg and Somerville (1962) reported that dUMP, the precursor of dTMP, was formed in *E. coli* by the action of a phosphatase which hydrolyzed dUTP and dUDP to dUMP. This enzyme is similar to another enzyme in *E. coli* which catalyzed a corresponding hydrolysis of dCTP and dCDP (Zimmerman and Kornberg, 1961). It is possible that one of the functions of dUTP phosphatase is to prevent the accumulation of dUTP and thereby to prevent its entrance into DNA.

The enzyme responsible for the formation of dTMP from dUMP is thymidylate synthetase, which catalyzes the following reaction (Wahba and Friedkin, 1962):

$$\text{dUMP} + N^5,N^{10}\text{-methylene-H}_4\text{-folate} \rightarrow \text{dTMP} + \text{H}_2\text{-folate}$$

In this reaction, tetrahydrofolate is oxidized to dihydrofolate. Blakley and McDougall (1962) prepared highly purified thymidylate synthetase from *S. faecalis*. The level of thymidylate synthetase increases approximately seven-fold in *E. coli* following infection by one of the T-even phages, apparently in response to the increased rate of DNA synthesis (Flaks and Cohen, 1959b).

E. coli contains enzymes capable of phosphorylating dCMP and dTMP to the triphosphates (Lehman *et al.*, 1958a). It is interesting that the enzyme preparation obtained by Lehman *et al.* did not act on dUMP, which may be a built-in safeguard to prevent the formation of dUTP which could be incorporated into DNA.

3. *Pyrimidine nucleotides formed in phage-infected bacteria*

Hydroxymethylcytosine replaces cytosine in the DNA of the T-even bacteriophages of *E. coli* (Table 3.7), and infection with these phages results in the induction of enzymes for the synthesis of deoxyhydroxymethylcytidylate (dHMP) from dCMP (Fig. 18.7). Zimmerman and

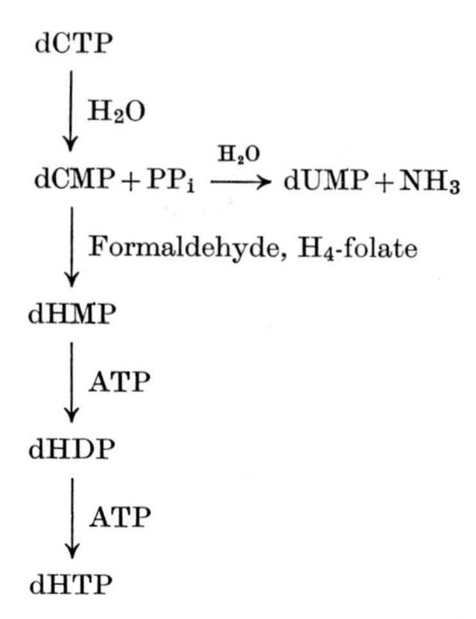

FIG. 18.7. Metabolism of dCTP in *E. coli* infected with a T-even bacteriophage.

Kornberg (1961) purified an enzyme from extracts of *E. coli* infected with T2r$^+$ which catalyzed the hydrolysis of dCTP and dCDP to dCMP:

$$\text{dCTP} + H_2O \rightarrow \text{dCMP} + PP_i$$

$$\text{dCDP} + H_2O \rightarrow \text{dCMP} + P_i$$

They were unable to detect the phosphatase in extracts of uninfected cells. This enzyme serves two purposes: it causes removal of dCTP, thereby preventing its entrance into the DNA of the infecting phage, and also provides a supply of dCMP, the substrate for dHMP synthesis.

Flaks and Cohen (1959a) found that extracts of *E. coli* infected with T-even phages contained an enzyme which formed dHMP from dCMP, formaldehyde and tetrahydrofolate. The enzyme was not present in extracts of uninfected cells or in cells infected with phages which did not contain hydroxymethylcytosine. Infection of *E. coli* with T2 resulted in the induction of a deoxynucleoside monophosphate kinase which phosphorylated dHMP with either ATP or dATP. This enzyme was purified by Bello and Bessman (1963) who showed that dTMP and dGMP were also substrates. Kornberg *et al.* (1959) found that dHMP was converted to dHTP with the ATP by extracts of *E. coli* infected with T2. Fleming and Bessman (1967) purified dCMP deaminase (deCMP aminohydrolase), an enzyme which was induced in *E. coli* following infection with T2. This enzyme catalyzed the conversion of dCMP to dUMP:

$$dCMP + H_2O \rightarrow dUMP + NH_3$$

The rate of catalysis of the enzyme was stimulated by dCTP and dHTP, but inhibited by dTTP, the product of dUMP metabolism. The function of the enzyme is to channel the flow of dCMP to dUMP when the intracellular concentration of dTTP is too low to meet the demands of phage DNA synthesis.

Bacteriophage ϕe, which parasitizes *B. subtilis*, contains 5-hydroxymethyluracil in its DNA in place of thymine. Infection of *B. subtilis* results in the induction of an enzyme which catalyzes the formation of 5-hydroxymethyldeoxyuridylate from dUMP, formaldehyde and tetrahydrofolate (Roscoe and Tucker, 1966).

4. *Formation of pyrimidine nucleotides from free bases and nucleosides*

Exogenous uracil, uridine, cytosine and cytidine are readily incorporated into nucleic acids of many species of bacteria, but under the usual conditions of growth no thymine and very little thymidine are incorporated (Bolton and Reynard, 1954; also see review of Magasanik, 1962). The pathways for the utilization of these compounds are shown in Fig. 18.8.

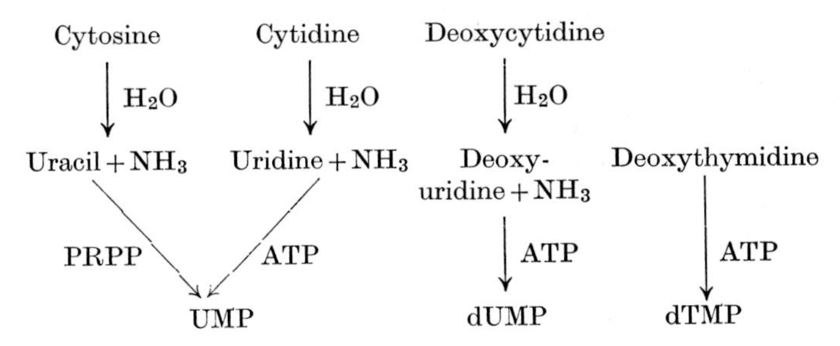

FIG. 18.8. Conversion of pyrimidines and pyrimidine nucleosides to nucleotides PRPP = 5-phosphoribosyl-1-pyrophosphate.

Crawford *et al.* (1957) studied uracil utilization in lactobacilli which required uracil or orotate for growth. They searched for and found UMP pyrophosphorylase (UMP: pyrophosphate phosphoribosyltransferase) in several species of lactobacilli and *E. coli* and purified the enzyme from *L. bifidus*. The reaction catalyzed by UMP pyrophosphorylase was:

Uracil + 5-phosphoribosyl-1-pyrophosphate → UMP + pyrophosphate

Brockman *et al.* (1960) isolated a strain of *E. coli* resistant to 5-fluorouracil and found that this organism was unable to incorporate uracil into nucleic acid. The mutant was found to lack UMP pyrophosphorylase, which confirms the role of this enzyme in the utilization of exogenous uracil.

Amos and Magasanik (1957) showed that both the uracil and ribose portions of uridine were converted to bacteriophage deoxythymidylic acid by *E. coli* infected with phage T1. Such a result would be obtained if uridine were converted directly to UMP by a kinase and then to dTMP by the reactions in Fig. 18.6. Uridine kinase (ATP:uridine 5′-phosphotransferase) of animal neoplastic tissue has been well characterized (Sköld, 1960), but little has been done with the bacterial enzyme. Purified deoxythymidine kinase of *E. coli* acts on deoxyuridine as well and converts it to dUMP (Okazaki and Kornberg, 1964b):

Deoxyuridine + ATP → dUMP + ADP

Cytosine, cytidine and deoxycytidine are utilized by *E. coli* by way of conversion to uracil or the corresponding uracil nucleoside (Fig. 18.8). Kream and Chargaff (1952) reported that cell-free extracts of *E. coli* caused the hydrolytic deamination of cytosine to uracil, and Wang *et al.* (1950) found that extracts of *E. coli* deaminated cytidine to uridine and deoxycytidine to deoxyuridine. Apparently, there have been no enzymes described which cause the direct conversion of cytosine, cytidine or deoxycytidine to CMP or dCMP.

Bolton and Reynard (1954) could not detect incorporation of thymine and deoxythymidine into the nucleic acids of *E. coli* as determined by the isotope competition test. However, Rachmeler *et al.* (1961) showed that radioactive deoxythymidine actually was incorporated into nucleic acids of *E. coli* as long as it remained intact but that an enzyme was induced by deoxythymidine which caused its decomposition to thymine. The enzyme was identified as thymidine phosphorylase (thymidine: orthophosphate deoxyribosyltransferase), which catalyzes the phosphorolysis of deoxythymidine and deoxyuridine (Razzell and Khorana, 1958):

Deoxythymidine + orthophosphate → thymine + deoxyribose-1-phosphate

Deoxyuridine + orthophosphate → uracil + deoxyribose-1-phosphate

The belief that thymidine phosphorylase was responsible for the prevention of thymidine incorporation into bacterial DNA was confirmed by the isolation of a mutant of *E. coli* able to incorporate thymidine into nucleic acid (Fangman and Novick, 1966) and found to lack thymidine phosphorylase. Deoxythymidine is converted to dTMP by deoxythymidine kinase (ATP:thymidine 5'-phosphotransferase) which was purified from *E. coli* by Okazaki and Kornberg (1964b):

$$\text{Deoxythymidine} + \text{ATP} \rightarrow \text{dTMP} + \text{ADP}$$

They found that this enzyme also catalyzed the phosphorylation of deoxyuridine and of analogs of uridine halogenated in the 5-position. Deoxycytidine, however, was not a substrate for the enzyme nor were purine deoxyribosides and purine and pyrimidine ribosides. Boyce and Setlow (1962) found that addition of deoxyadenosine to the growth medium of normal *E. coli* promoted the incorporation of thymine into nucleic acids. They proposed that in this case a transfer of the deoxyribosyl residue from deoxyadenosine to thymine occurred resulting in the formation of deoxythymidine.

II. Biosynthesis of DNA

A. Duplication of DNA

1. *DNA as the carrier of genetic information*

The identification of DNA as the active principle involved in inheritance came about as a result of the studies on transformation of bacteria from one genetic type to another. Griffith (1928) discovered this phenomenon when he found that mice injected with rough, avirulent strains of *D. pneumoniae* along with killed cells of a smooth virulent strain, frequently died of an infection with the smooth virulent strain. Griffith's discovery was verified and extended in the years following it (see review of Shaeffer, 1964). Avery *et al.* (1944) made one of the most outstanding contributions to the science of genetics when they isolated highly purified transforming principle from smooth type III pneumococci and brought about transformation of rough type II to smooth type III in the test tube. They identified DNA as the major constituent of purified transforming principle.

2. *Watson–Crick structure of DNA*

The next important advance was the proposal of Watson and Crick (1953a, b) for the structure of DNA. On the basis of X-ray diffraction studies of crystalline DNA (Wilkens *et al.*, 1953; Franklin and Gosling, 1953), they suggested that DNA was composed of two parallel chains of polynucleotides which twisted to form a helix. They proposed that the two chains were held together by hydrogen bonds between purines and pyrimidines of the two strands. At this time it was known that there was a distinct pattern to the proportions of purines and pyrimidines in DNA; adenine was equal to thymine and guanine was equal to cytosine (Chargaff, 1955; see also Table 3.7). Watson and Crick proposed that hydrogen bonds which held the two chains together were between the amino and keto substituents of the adenine–thymine and guanine–cytosine pairs (Fig. 3.14). Thus the two strands of DNA were complementary to each other but of opposite polarity.

3. *Chromosomal replication*

The beauty of the Watson and Crick proposal for the structure of DNA was that for the first time it suggested a molecular basis for the replication of DNA. Watson and Crick proposed that the two complementary strands untwined in some fashion and served as a template for the formation of complementary strands according to the rules of base pairing. Consequently, at the end of one generation, each helix of DNA should have one parental and one daughter strand (Fig. 18.9). At the end of two generations, one-half of the helices of the progeny should have one parental and one daughter strand, while the other half should have only daughter strands. Meselson and Stahl (1958) employed the technique of density gradient centrifugation to test the hypothesis. They found that DNA from *E. coli* grown in a medium with ^{15}N as the only nitrogen isotope could be separated by their technique from DNA of cells grown in a medium where ^{14}N was the sole nitrogen isotope. They grew *E. coli* in a synthetic medium with ^{15}N as the sole isotope of nitrogen so that almost all of the nitrogen of DNA was the ^{15}N isotope. The entire cell crop was then transferred to a medium where ^{14}N was the nitrogen isotope. DNA taken from cells after one generation in the latter medium migrated to a position half way between DNA-^{15}N and DNA-^{14}N; presumably this was the DNA with one ^{15}N (parental) strand and one ^{14}N (daughter) strand. DNA taken from cells after two generations separated into two bands in density gradient centrifugation, one at the position of DNA with both ^{15}N and ^{14}N and the other at the position of DNA with ^{14}N only. These results provided firm experimental support for Watson and Crick's proposal for the duplication of DNA.

Nagata (1963) and Yoshikawa and Sueoka (1963) obtained evidence for an oriented, linear duplication of the bacterial chromosome, which

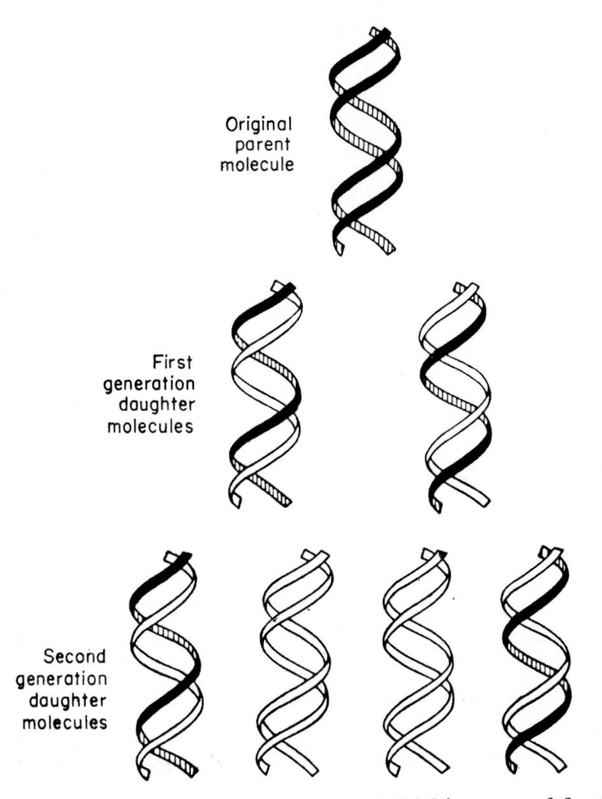

FIG. 18.9. Schematic illustration of Watson and Crick's proposal for the replication of DNA. At the end of one generation, each helix contains one strand of DNA derived from the parent and one strand derived from the daughter. In Meselson and Stahl's experiment, the ^{15}N strands would correspond to the black strands in this figure while the ^{14}N strands would correspond to the white strands. (Reproduced with permission from Meselson and Stahl, 1958).

would also be expected from the Watson and Crick proposal. Nagata used strains of *E. coli* which carried a prophage attached to a known position on the chromosome. When these organisms were grown in synchronous culture, the point in the cell cycle at which the number of prophage doubled, assumed to be the time at which that portion of the chromosome was duplicated, correlated closely with the position of the prophage on the chromosome. Yoshikawa and Sueoka (1963) used a somewhat different approach with ordinary batch cultures of *B. subtilis*. They showed mathematically that in an oriented duplication process, markers on the portion which was duplicated first would be encountered in the culture with a higher frequency than markers at the end of the chromosome duplicated last. They determined the frequency of markers in the culture and found that the frequency did indeed correlate closely with the position of the marker on the chromosome.

Cairns (1963a) obtained direct photographic evidence for the process of DNA duplication in *E. coli*. He grew the organism for two generations in a medium which contained thymidine-^{3}H, isolated intact chromosomes by a very gentle procedure and then spread the radioactive DNA on a photographic plate. After a sufficient period of time had elapsed, the film was exposed by the β-particles of tritium and yielded a picture of the isolated chromosome. Cairns found that most of the DNA was in the form of a circle with loops which were evidently the portion of the chromosome in the act of being duplicated (Fig. 18.10). He came to this conclusion since the loops contained more radioactivity, as judged by the number of silver grains in that portion of the picture. Cairns' experiments provided evidence that the replicating form of the chromosome was a circle, which poses some difficult mechanical problems. The unwinding of the DNA helix which must take place prior to replication of the strands and the subsequent rewinding would hopelessly snarl the helix if the circles were rigid. Cairns (1963b) proposed a mechanism to explain how duplication might occur (Fig. 18.11) without getting into this difficulty. He suggested that duplication always began at the same point on the chromosome, and that a "swivel" was built into the chromosome at this point to allow the strands to rotate. Consequently, the strands could unwind prior to duplication and complementary strands could be produced as the unwinding occurred.

B. Enzymic Formation of DNA

1. *DNA polymerase*

The first report of an enzyme which catalyzed the formation of DNA came from Kornberg's laboratory. Lehman *et al.* (1958a) purified an enzyme from *E. coli* which catalyzed the formation of a high molecular weight polynucleotide from dATP, dGTP, dCTP and dTTP. The enzyme, which they named DNA polymerase (deoxynucleoside

---

FIG. 18.10. Radioautograph of a chromosome of *E. coli* grown in a medium with tritiated thymidine. (Reproduced with permission from Cairns, 1963b). The segments XBY and CX are composed of two radioactive strands of DNA; the segments XAY and YC are composed of one radioactive and one non-radioactive strand of DNA. The scale at the bottom of the figure is 100μ. See Fig. 18.11 for an account of how this result would be obtained.

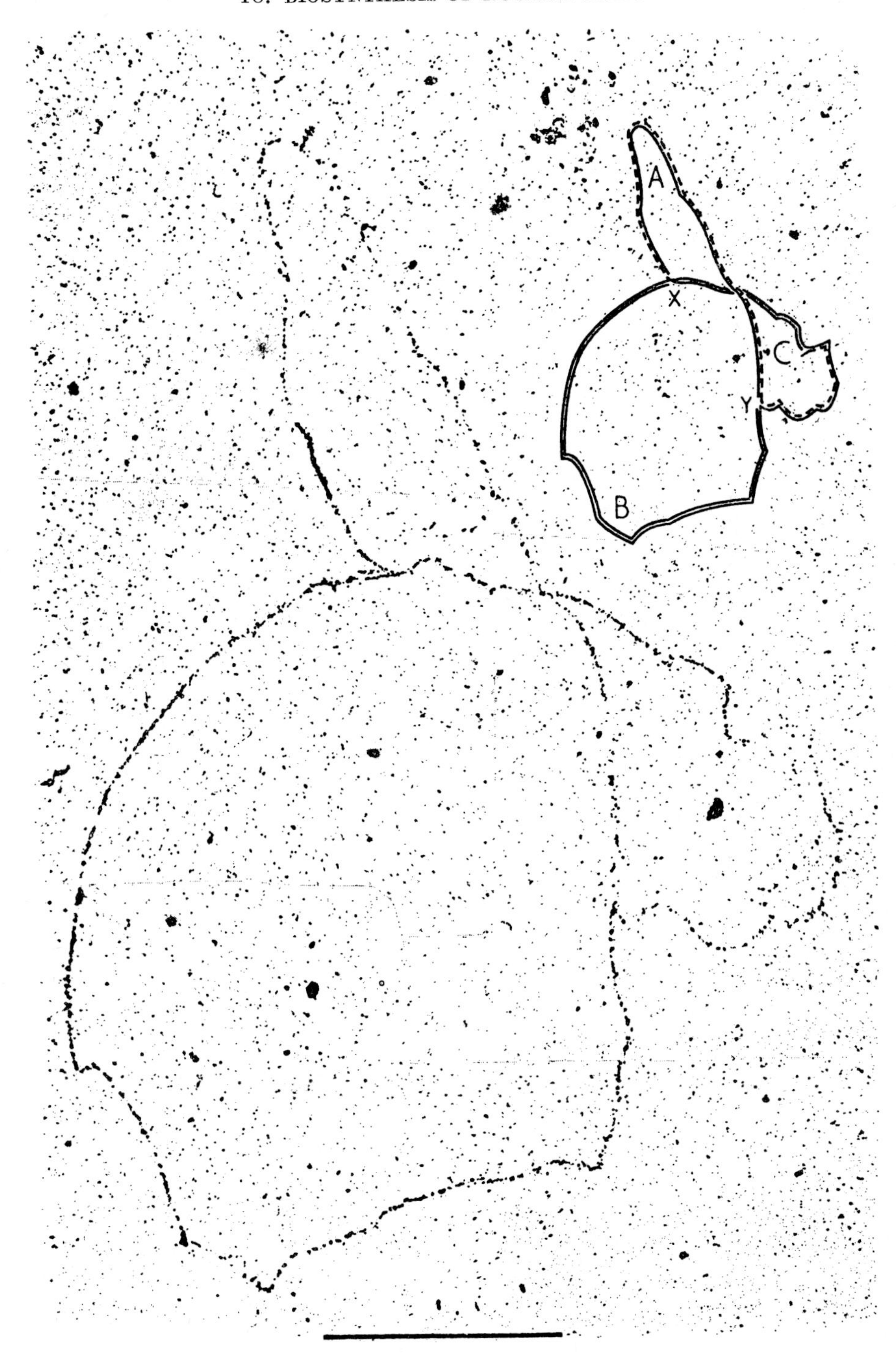
A
X
C
Y
B

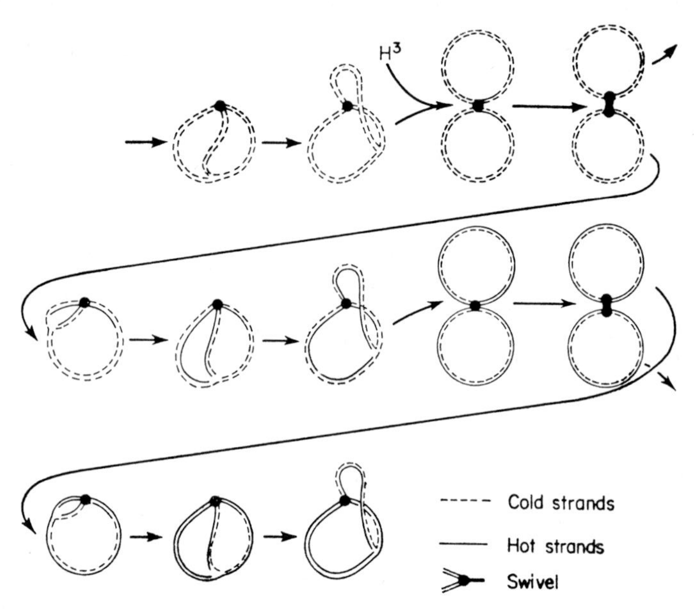

FIG. 18.11. Cairns' (1963b) proposal for the duplication of DNA. Tritiated thymidine was added at the point indicated to show how the chromosome in Fig. 18.10 became labeled; the final figure in this illustration corresponds to the chromosome in Fig. 18.10.

triphosphate:DNA deoxynucleotidyltransferase) required a DNA primer as well (Bessman *et al.*, 1958):

$$\text{n Deoxynucleoside triphosphate} + \text{DNA}_m \rightarrow \text{n pyrophosphate} + \text{DNA}_{n+m}$$

DNA polymerase was later obtained in a pure state by Richardson *et al.* (1964a) and was found to possess exonuclease II activity as well, which may be a property of the enzyme itself. DNA polymerase preparations have been obtained from *B. subtilis* (Okazaki and Kornberg, 1964a), *M. lysodeikticus* (Zimmerman, 1966), *E. coli* infected with T2 (Aposhian and Kornberg, 1962) and calf thymus (Bollum, 1960).

An exciting feature of the reaction was that the product contained the same molar proportions of bases as did the primer (Lehman *et al.*, 1958b). Kornberg and his associates set about characterizing the product carefully to see if the enzyme was indeed making a replica of the DNA primer. They developed the technique of "nearest neighbor" analysis (Josse *et al.*, 1961), a method for determining the frequency with which the 16 possible pairs of bases exist in DNA. The principle

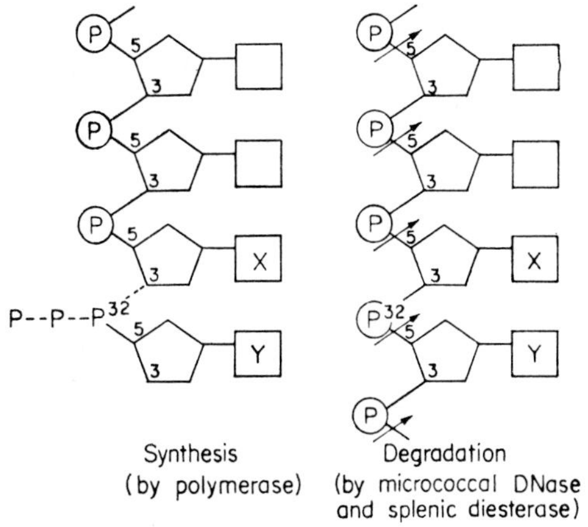

FIG. 18.12. Principle of the nearest neighbor analysis. (Reproduced with permission from Josse *et al.*, 1961).

of this method is shown in Fig. 18.12. In the reaction catalyzed by DNA polymerase, phosphate attached at position 5′ of deoxynucleoside triphosphate becomes attached to the 3′ position of the terminal nucleotide. Exhaustive treatment of the product with micrococcal deoxyribonuclease and spleen diesterase cleaves the polynucleotide to 3′ nucleotides, with the result that phosphate originally on deoxyribose attached to base Y is now attached to deoxyribose of base X, Y's nearest neighbor. The 3′ nucleotides were separated by electrophoresis, and the amount of ^{32}P in each nucleotide fraction was measured. The experiment was done four times, each time with a different ^{32}P labeled deoxynucleoside triphosphate. The data were then used to calculate the frequency of occurrence of the 16 possible combinations of pairs of bases, and this is shown in Table 18.1.

In order to interpret the data in Table 18.1, another feature of the Watson and Crick proposal for the structure of DNA must be considered, namely that the two strands have the opposite polarity (Fig. 18.13). If the strands have opposite polarity, then every TpA pair will be matched by another TpA; similarly the frequency of ApG will be equal to CpT, and GpA will be equal to TpC. In Table 18.1 roman numerals are used to designate the pairs which would be expected to have equal frequencies in DNA where the two strands have opposite polarity. The agreement with this prediction is excellent and thereby provides further strong support for the Watson and Crick structure. In a DNA helix where the strands have similar polarity, every TpA will be matched by an ApT pair which is clearly not the case. In Table 18.1 lower case

TABLE 18.1. Nearest neighbor frequencies of DNA synthesized by DNA polymerase with DNA from *M. phlei* as the primer. (Reproduced with permission from Josse *et al.*, 1961). The sums given at the bottom are the proportions of the bases in the enzymically synthesized DNA and agree well with the composition of the primer, namely thymine = 0·165, adenine = 0·162, cytosine = 0·335 and guanine = 0·338.

Reaction No.	Labeled triphosphate	Isolated 3′-deoxyribonucleotide			
		Tp	Ap	Cp	Gp
1	dATP[32]	a TpA 0·012	b ApA 0·024 I	c CpA 0·063 II	d GpA 0·065 III
2	dTTP[32]	b TpT 0·026 I	a ApT 0·031	d CpT 0·045 IV	c GpT 0·060 V
3	dGTP[32]	e TpG 0·063 II	f ApG 0·045 IV	g CpG 0·139	h GpG 0·090 VI
4	dCTP[32]	f TpC 0·061 III	e ApC 0·064 V	h CpC 0·090 VI	g GpC 0·122
	Sums	0·162	0·164	0·337	0·337

letters are used to designate pairs whose frequencies would be expected to be equal in a DNA helix with strands of similar polarity, and it is clear that the agreement with this model is very poor.

The final step in the purification of DNA polymerase by Richardson *et al.* (1964a) was chromatography on hydroxylapatite, which resulted in the separation of a minor protein component from the bulk of the DNA polymerase. Although the specific activity of DNA polymerase increased during purification when assayed with a synthetic DNA polymer, there was a loss in its activity when assayed with native DNA. Addition of the minor component to highly purified DNA polymerase restored its ability to replicate native DNA. The minor component was identified as a DNA phosphatase-exonuclease (exonuclease III) which catalyzed the hydrolytic removal of terminal 3′-phosphate

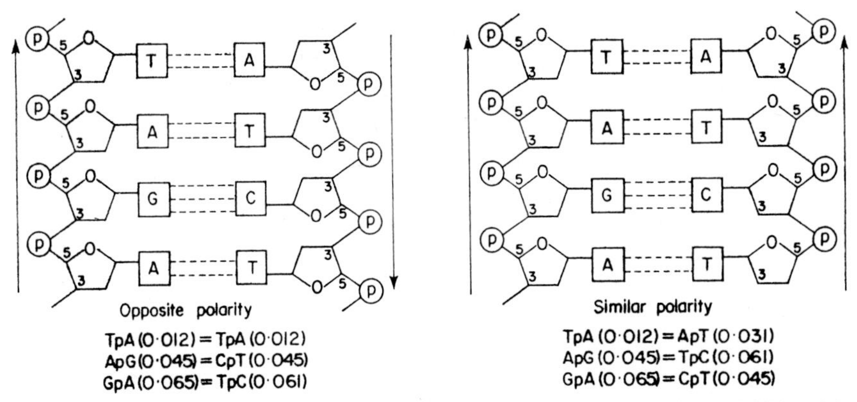

FIG. 18.13. Results of nearest neighbor analyses expected from DNA with strands of opposite and similar polarity. (Reproduced with permission from Josse *et al.*, 1961).

esters as well as the hydrolysis of double stranded DNA to produce 5′-mononucleotides (Richardson and Kornberg, 1964; Richardson *et al.*, 1964b). The presence of 3′-phosphate esters in DNA greatly inhibits the activity of DNA polymerase and apparently this linkage does exist in native DNA, although the means by which it is produced is unknown.

Goulian *et al.* (1967) used DNA polymerase to prepare biologically active DNA infective for spheroplasts of *E. coli*. They used single stranded DNA of phage $\phi \times 174$ (the (+) strand, Fig. 18.14) as a primer for DNA polymerase to produce double stranded circles of $\phi \times 174$ DNA. The reaction contained ^{32}P-deoxybromouridine-5′-triphosphate in place of dTTP and "joining enzyme", an enzyme which connects the 3′ and 5′ end of the polynucleotide to produce a circle (Fig. 18.14). This reaction mixture yielded two-stranded, circular DNA with ^{32}P-deoxybromouridylate located in the (−) strand. The circles were then treated with pancreatic deoxyribonuclease long enough to make a single break in approximately half the strands, and the mixture was heated and centrifuged by the density gradient technique to separate the various forms of DNA. The object of using deoxybromouridine-5′-triphosphate in the original mixture was to provide a means by which the complementary strands could be separated; this was possible because the strand containing bromouracil was more dense. The DNA of the reaction mixture was separated into three fractions, double-stranded circles, (−) strands and (+) strands. The fraction which contained the (−) strands contained both circular and linear forms and these were partially resolved by further centrifugation. Infectivity for

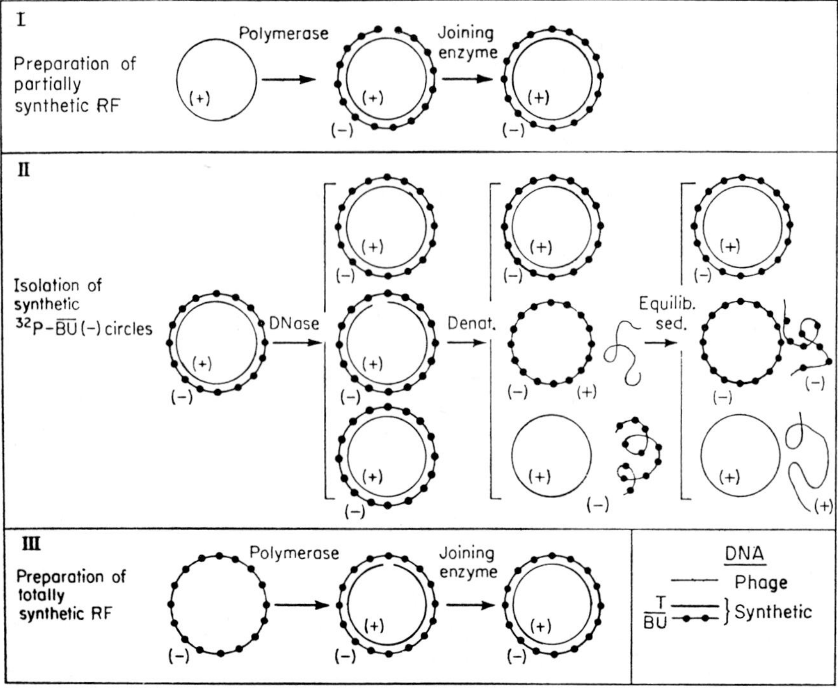

FIG. 18.14. The experimental plan used by Goulian *et al.* (1967) to prepare biologically active DNA with DNA polymerase and a primer of single-stranded, circular $\phi \times 174$ DNA. (Reproduced with permission from Goulian *et al.*, 1967). Abbreviations: RF = replicative form, $\overline{BU}$ = bromouracil.

spheroplasts was found to be a property of the circular strands. In addition, the synthetic (−) strands were used as templates to form (+) strands, now yielding fully synthetic, double-stranded $\phi \times 174$ DNA.

One important cell process is the repair of damage done to DNA by the organism's internal and external environment, since relatively minor breaks in DNA could be lethal to the cell. Richardson *et al.* (1964c) showed that DNA polymerase was able to carry out this function. They treated native, double-stranded DNA with exonuclease III, an enzyme which removed nucleoside-5′-monophosphates from the strand with the unesterified 3′-hydroxyl, leaving a large portion of the 5′-strand unpaired. When the partially digested DNA was treated with DNA polymerase, the missing portion was quickly repaired, yielding double-stranded DNA. When DNA from *B. subtilis* was treated in this fashion, much of the transforming activity which was lost by treatment with exonuclease III was restored.

2. *Methylation of DNA*

Methylation of the bases of DNA appears to occur mainly at the polynucleotide level. Gold and Hurwitz (1964a) purified an enzyme 400-fold from *E. coli* which catalyzed the transfer of a methyl group from S-adenosylmethionine to DNA according to the following equation:

$$\text{S-Adenosylmethionine} + \text{DNA} \rightarrow \text{methylated DNA} + \text{S-adenosylhomocysteine}$$

Homologous DNA, that is DNA from *E. coli*, was virtually inactive, presumably because it was already methylated, and consequently heterologous DNA was used as a substrate. Gold and Hurwitz (1964a) identified the methylated bases as 5-methylcytosine and 6-methylaminopurine, therefore a C–C bond was formed in one case and a C–N bond in the other. Gold and Hurwitz (1964b) found that native, double-stranded DNA was the best methyl acceptor. However, methylation of DNA caused no observable change in its ability to act as a primer for either DNA polymerase or RNA polymerase.

III. Biosynthesis of RNA

A. RNA Synthesis Primed by DNA

There is a good deal of evidence that protein synthesis is accompanied by formation of a specific type of RNA called messenger RNA (mRNA) which receives the code for protein synthesis from DNA (see review of Hurwitz and August, 1963). The process of making mRNA from a DNA template is referred to as transcription of the genetic code.

The enzyme which catalyzes the formation of mRNA, and possibly other types of RNA, is RNA polymerase (nucleosidetriphosphate: RNA nucleotidyltransferase). Furth *et al.* (1962) prepared highly purifield RNA polymerase from *E. coli* and established that the enzyme catalyzed the formation of a polyribonucleotide from ATP, GTP, CTP and UTP, that a primer of DNA was required, and that pyrophosphate was released in amounts equivalent to nucleosides incorporated into the product:

$$\begin{matrix} \text{ATP} \\ + \\ \text{GTP} \\ + \\ \text{CTP} \\ + \\ \text{UTP} \end{matrix} \xrightarrow{\text{DNA Primer}} \text{RNA} + \text{PP}_i$$

A considerable amount of work has been done to determine the role of the primer in RNA synthesis. Hurwitz *et al.* (1962) found that purified RNA polymerase from *E. coli* was able to use several types of DNA including single stranded, native and heat-denatured DNA, although the latter was somewhat less effective as a primer than native DNA. RNA polymerase preparations from *M. lysodeikticus* (Nakamoto and Weiss, 1962) and from *A. vinlandii* (Krakow and Ochoa, 1963) were able to utilize natural and synthetic polyribonucleotide primers as well as DNA, and it now appears that this is true of most preparations of RNA polymerase. The DNA primer does not appear to be altered in the course of RNA synthesis since the transforming activity of pneumococcal DNA, when used as a primer, was unaltered at the end of the reaction (Hurwitz *et al.*, 1962). Hurwitz and his associates established that both strands of DNA were copied under the conditions used to assay the enzyme in their laboratory. When double-stranded $\phi \times 174$ DNA was used as the primer, the molar proportions of bases in RNA were analogous to that of the primer indicating that both strands had been copied. Furthermore, no matter what was used as a DNA primer, the $A+U/G+C$ ratios of the RNA agreed very well with the $A+T/G+C$ ratios of the primer.

In the cell, however, only one of the two strands of DNA is copied by RNA polymerase. Marmur and Greenspan (1963) discovered this in experiments with the SP8 phage of *B. subtilis*. They managed to separate the two strands of SP8 DNA into a heavy (H) and a light (L) strand by heating to denature DNA, followed by density gradient centrifugation. The two strands were separable even though complementary because the H strand was rich in pyrimidines and slightly more dense than the L strand. They detected complementary RNA in extracts of phage infected *B. subtilis* by taking advantage of the fact that complementary strands of RNA and DNA will hybridize when properly heated and cooled and the hybrid strands can be isolated by density gradient centrifugation (Hall and Spiegelman, 1961). Marmur and Greenspan isolated radioactive RNA from *B. subtilis* infected with SP8 phage in a medium which contained ^{3}H-uridine and found that it combined only with the H strand, indicating that mRNA formed during phage synthesis was single-stranded and complementary to the H strand. Subsequent work has shown that only one of the DNA strands is copied in several *in vivo* systems (see review of Hayes, 1967).

Bremer *et al.* (1965) isolated and identified both the initial and growing end of the polyribonucleotide synthesized by RNA polymerase and thereby established the direction of chain growth (Fig. 18.15). The product was hydrolyzed with alkali and most of the bases were recovered

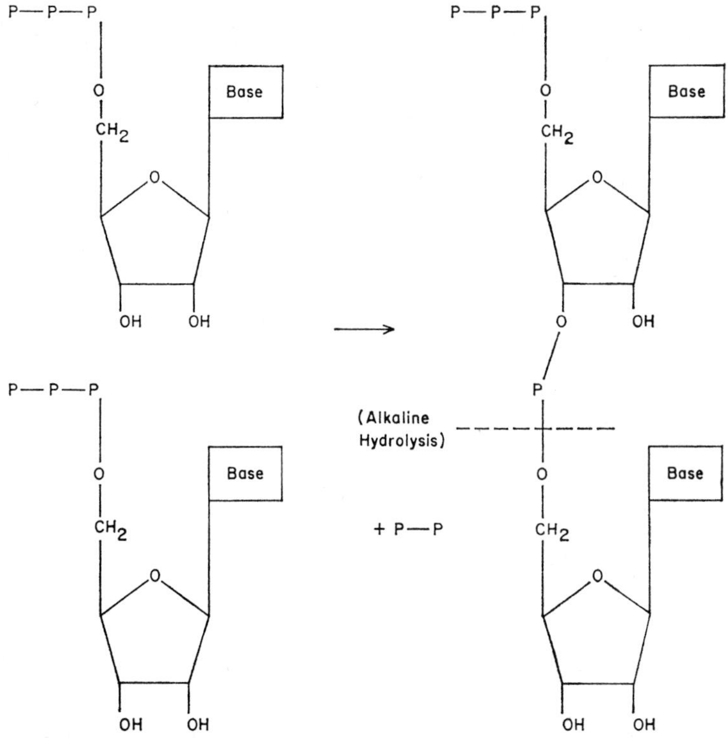

FIG. 18.15. Initial event in the formation of polyribonucleotide chain by RNA polymerase.

as nucleoside 2′- and 3′-phosphates, but two minor products were also recovered; one was identified as a nucleoside and the other was identified as a nucleoside tetraphosphate, probably nucleoside-3′-phosphate-5′-triphosphate. Such a result would be obtained if the chain grew as shown in Fig. 18.15, where the 3′-hydroxyl group of the initial nucleotide forms a phosphate ester linkage with the phosphate attached to the 5′ position of the incoming nucleotide. Alkaline hydrolysis of the product splits the phosphate bond at the position indicated in Fig. 18.15, forming a nucleoside tetraphosphate from the initial end of the chain and a nucleoside from the growing end of the chain, exactly as found by Bremer *et al.* (1965). Another interesting result was that the initial nucleotide in most cases was an adenine nucleotide.

B. RNA SYNTHESIS PRIMED BY RNA

Research on RNA-directed RNA synthesis was stimulated by the discovery that synthesis of RNA phages was accomplished without the formation of a DNA template (see review of Erikson and Franklin, 1966). The majority of RNA viruses which attack bacteria contain single-stranded RNA, the strand found in the virus itself being designated as the plus strand. Weissman *et al.* (1963) were the first to study a bacterial RNA polymerase primed by RNA when they purified the enzyme induced in *E. coli* by infection with the RNA phage MS2. The enzyme did not require an exogenous primer, however, the purified enzyme contained a large amount of RNA composed mainly of double-stranded MS2 RNA and hence contained its own primer. The product was also double-stranded RNA; however, only one strand was synthesized by the enzyme, and presumably the other strand came from the primer. The newly synthesized strand was a plus strand, that is, identical to that in the infecting virus.

Spiegelman and his associates studied two RNA polymerases which utilized RNA primers from *E. coli* infected with RNA viruses; one enzyme was formed following infection with MS2 virus (Haruna *et al.*, 1963) and the other was produced following infection with Qβ virus (Haruna and Spiegelman, 1965). Both enzymes, which Spiegelman named RNA replicases, required RNA as a primer, but what was more interesting, specifically required the RNA of the infecting virus which induced the enzyme.

Another important fact which was established in Spiegelman's laboratory was that the RNA produced was infective. Spiegelman *et al.* (1965) carried the reaction catalyzed by Qβ replicase through serial transfers with fresh enzyme and substrates. The enzyme continued to synthesize RNA, indicating that the product was now acting as a template for RNA synthesis. After enough serial transfers had been made so that less than one strand of the original Qβ RNA was present per tube, the RNA was tested for its ability to infect protoplasts of *E. coli*. The RNA was infectious and complete Qβ virus particles were produced. *This was the first instance of the production of a biologically active substance in the laboratory using a chemically defined, cell-free system.* Pace *et al.* (1967) found that two RNA structures were formed by Qβ replicase prior to the formation of the free plus strand; the first structure appeared to be a duplex composed of one plus and one minus strand, the second structure also appeared to be a duplex but with extra growing plus strands (Fig. 18.16). It was the latter structure which gave rise to mature Qβ RNA.

Bishop *et al.* (1967) studied the direction of synthesis of the nucleotide chain by Qβ replicase and found that the minus strand was synthesized in the 5′ to 3′ direction but that the plus strand was synthesized in the 3′ to 5′ direction (Fig. 18.16). In addition, they established that both

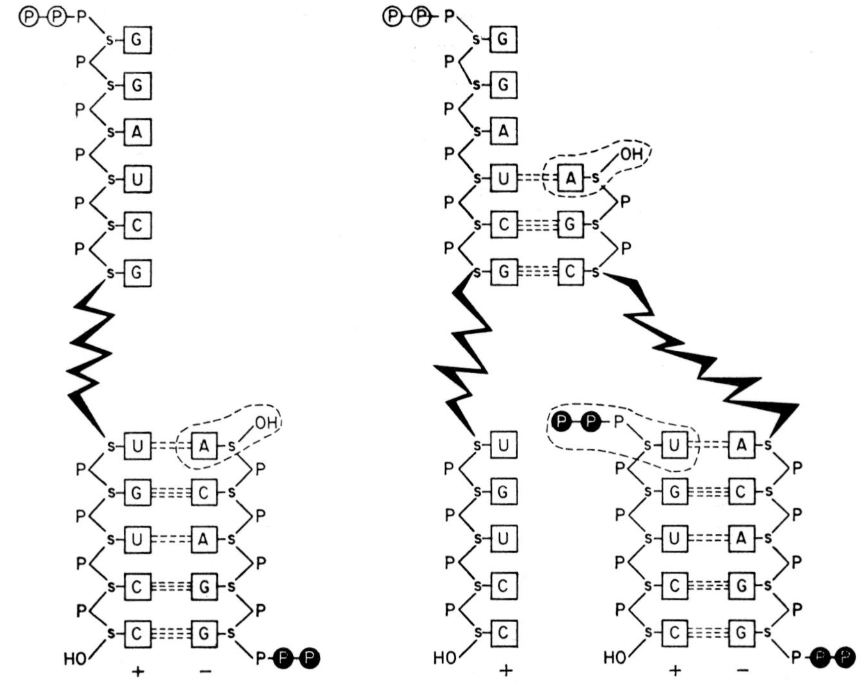

FIG. 18.16. Synthesis of plus and minus strands of Qβ RNA by Qβ replicase. (Reproduced with permission from Bishop *et al.*, 1967). The initial plus strand was provided as a primer and therefore did not contain label. The darkened circles represent radioactive phosphate.

the plus and minus strands had guanosine at the 5′ end and cytosine at the 3′ end of the chain. They primed their reaction with the plus strands of Qβ RNA and utilized nucleoside triphosphates labeled with ^{32}P in the β and γ phosphates to identify the base at the 5′ terminus. They found that guanosine was the only base which retained ^{32}P in duplexes and mature Qβ RNA, and hence must have been located at the 5′ terminus. By the rules of base pairing, this result established that cytosine was located at the 3′ terminus (Fig. 18.16). In those structures which contained growing plus strands, all four nucleoside triphosphates retained ^{32}P, which is evidence that these bases were at the growing end of the plus strand. If this is true, then the plus strand grows in the 3′ to 5′ direction.

C. METHYLATION OF RNA

Mandel and Borek (1963a) established that the methyl carbon of methionine was the source of methyl groups of methylated bases in

RNA of *E. coli*. They also showed that a methionine auxotroph of *E. coli*, which was starved for methionine, formed ribosomal RNA (rRNA) and soluble RNA (sRNA) deficient in methylated bases (Mandel and Borek, 1963b). Fleissner and Borek (1963) found that cell-free extracts of the mutant catalyzed transfer of the methyl group from methionine to methyl-deficient RNA isolated from the mutant during methionine starvation. This result established that methylation occurred at the stage of the polyribonucleotide rather than at the stage of the ribonucleotide.

Hurwitz and his associates (1964a) later purified several enzymes from *E. coli* which catalyzed the transfer of the methyl group of S-adenosylmethionine to sRNA according to the following equation:

$$\text{S-adenosylmethionine} + \text{sRNA} \rightarrow \text{S-adenosylhomocysteine} + \text{methyl-sRNA}$$

They separated six enzymes from *E. coli* and identified the methylated bases which were formed (Table 18.2). These enzymes were specific for

TABLE 18.2. RNA methylating enzymes isolated from *E. coli* by Hurwitz *et al.* (1964a).

Name	Methylated base isolated from sRNA
Guanine methylating enzyme I	1-Methylguanine
Guanine methylating enzyme II	1-Methylguanine
Guanine methylating enzyme III	7-Methylguanine
Adenine methylating enzyme	2-Methyladenine, 6-Methylaminopurine 6-Dimethylaminopurine
Cytosine methylating enzyme	5-Methylcytosine
Uracil methylating enzyme	Thymine

sRNA (Hurwitz *et al.*, 1964b). Hurwitz and his colleagues (1965) also purified enzymes from *E. coli* which catalyzed a similar methylation of rRNA.

D. ADDITION OF CCA TO sRNA

All types of sRNA contain the base sequence CCA at the 3′ hydroxyl end of the polyribonucleotide. Preiss *et al.* (1961) purified an enzyme from *E. coli* which catalyzed the incorporation of these bases into a specially prepared substrate which had some of the bases removed from the 3′ hydroxyl end of functional sRNA. The partially digested material now acted as a substrate for the purified enzyme which catalyzed the incorporation of two moles of CMP and one of AMP from CTP and ATP respectively; pyrophosphate was eliminated in the reaction.

References

Amos, H. and Magasanik, B. (1957). *J. Biol. Chem.* **229**, 653.

Aposhian, H. V. and Kornberg, A. (1962). *J. Biol. Chem.* **237**, 519.

Avery, O. T., MacLeod, C. M. and McCarty, M. (1944). *J. Exptl Med.* **79**, 137.

Bello, L. J. and Bessman, M. J. (1963). *J. Biol. Chem.* **238**, 1777.

Bessman, M. J., Lehman, I. R., Simms, E. S. and Kornberg, A. (1958). *J. Biol. Chem.* **233**, 171.

Bishop, D. H. L., Pace, N. R. and Spiegelman, S. (1967). *Proc. Natl Acad. Sci. U.S.* **58**, 1790.

Blakley, R. L. and McDougall, B. M. (1962). *J. Biol. Chem.* **237**, 812.

Bollum, F. J. (1960). *J. Biol. Chem.* **235**, 2399.

Bolton, E. T. and Reynard, A. M. (1954). *Biochim. Biophys. Acta*, **13**, 381.

Boyce, R. P. and Setlow, R. B. (1962). *Biochim. Biophys. Acta*, **61**, 618.

Bremer, H., Konrad, M. W., Gaines, L. and Stent, G. S. (1965). *J. Mol. Biol.* **13**, 540.

Brockman, R. W., Davis, J. M. and Stutts, P. (1960). *Biochim. Biophys. Acta*, **40**, 22.

Brockman, R. W., Debavadi, C. S., Stutts, P. and Hutchison, D. J. (1961). *J. Biol. Chem.* **236**, 1471.

Buchanan, J. M. and Hartman, S. C. (1959). *Advan. Enzymol.* **21**, 199.

Cairns, J. (1963a). *J. Mol. Biol.* **6**, 208.

Cairns, J. (1963b). *Cold Spring Harbor Symp. Quant. Biol.* **28**, 43.

Carter, C. E. (1959). *Biochem. Pharmacol.* **2**, 105.

Carter, C. E. and Cohen, L. H. (1956). *J. Biol. Chem.* **222**, 17.

Chargaff, E. (1955). *In* "The Nucleic Acids" (E. Chargaff and J. N. Davidson, eds), Academic Press, New York. Vol. I, p. 307.

Crawford, I., Kornberg, A. and Simms, E. S. (1957). *J. Biol. Chem.* **226**, 1093.

Crosbie, G. W. (1960). *In* "The Nucleic Acids" (E. Chargaff and J. N. Davidson, eds), Academic Press, New York. Vol. III, p. 323.

Erikson, R. L. and Franklin, R. M. (1966). *Bacteriol. Rev.* **30**, 267.

Fangman, W. L. and Novick, A. (1966). *J. Bacteriol.* **91**, 2390.

Flaks, J. G. and Cohen, S. S. (1959a). *J. Biol. Chem.* **234**, 1501.

Flaks, J. G. and Cohen, S. S. (1959b). *J. Biol. Chem.* **234**, 2981.

Flaks, J. G., Erwin, M. J. and Buchanan, J. M. (1957). *J. Biol. Chem.* **229**, 603.

Fleissner, E. and Borek, E. (1963). *Biochemistry*, **2**, 1093.

Fleming, W. H. and Bessman, M. J. (1967). *J. Biol. Chem.* **242**, 363.

Franklin, R. E. and Gosling, R. G. (1953). *Nature*, **171**, 740.

Friedmann, H. C. and Vennesland, B. (1960). *J. Biol. Chem.* **235**, 1526.

Furth, J. J., Hurwitz, J. and Anders, M. (1962). *J. Biol. Chem.* **237**, 2611.

Gerhart, J. C. and Pardee, A. B. (1962). *J. Biol. Chem.* **237**, 891.

Gold, M. and Hurwitz, J. (1964a). *J. Biol. Chem.* **239**, 3866.

Gold, M. and Hurwitz, J. (1964b). *J. Biol. Chem.* **239**, 3858.

Gots, J. S. and Gollub, E. G. (1957). *Proc. Natl Acad. Sci. U.S.* **43**, 826.

Goulian, M., Kornberg, A. and Sinsheimer, R. L. (1967). *Proc. Natl Acad. Sci. U.S.* **58**, 2321.

Greenberg, G. R. and Somerville, R. L. (1962). *Proc. Natl Acad. Sci. U.S.* **48**, 247.

Griffith, F. (1928). *J. Hyg.* **27**, 113.

Hall, B. D. and Spiegelman, S. (1961). *Proc. Natl Acad. Sci. U.S.* **47**, 137.

Hartman, S. C. and Buchanan, J. M. (1958a). *J. Biol. Chem.* **233**, 451.

Hartman, S. C. and Buchanan, J. M. (1958b). *J. Biol. Chem.* **233**, 456.
Hartman, S. C. and Buchanan, J. M. (1959). *J. Biol. Chem.* **234**, 1812.
Haruna, I. and Spiegelman, S. (1965). *Proc. Natl Acad. Sci. U.S.* **54**, 579.
Haruna, I., Nozu, K., Ohtaka, Y. and Spiegelman, S. (1963). *Proc. Natl Acad. Sci. U.S.* **50**, 905.
Hayes, D. (1967). *Ann. Rev. Microbiol.* **21**, 369.
Hiraga, S. and Sugino, Y. (1966). *Biochim. Biophys. Acta* **114**, 416.
Hurwitz, J. and August, J. T. (1963). *Progr. Nucleic Acid Res.* **1**, 59.
Hurwitz, J., Furth, J. J., Anders, M. and Evans, A. (1962). *J. Biol. Chem.* **237**, 3752.
Hurwitz, J., Gold, M. and Anders, M. (1964a). *J. Biol. Chem.* **239**, 3462.
Hurwitz, J., Gold, M. and Anders, M. (1964b), *J. Biol. Chem.* **239**, 3474.
Hurwitz, J., Anders, M., Gold, M. and Smith, I. (1965). *J. Biol. Chem.* **240**, 1256.
Josse, J., Kaiser, A. D. and Kornberg, A. (1961). *J. Biol. Chem.* **236**, 864.
Kornberg, A., Lieberman, I. and Simms, E. S. (1955a). *J. Biol. Chem.* **215**, 389.
Kornberg, A., Lieberman, I. and Simms, E. S. (1955b). *J. Biol. Chem.* **215**, 417.
Kornberg, A., Zimmerman, S. B., Kornberg, S. R. and Josse, J. (1959). *Proc. Natl Acad. Sci. U.S.* **45**, 772.
Krakow, J. S. and Ochoa, S. (1963). *Proc. Natl Acad. Sci. U.S.* **49**, 88.
Kream, J. and Chargaff, E. (1952). *J. Am. Chem. Soc.* **74**, 4274.
Lagerkvist, U. (1958a). *J. Biol. Chem.* **233**, 138.
Lagerkvist, U. (1958b). *J. Biol. Chem.* **233**, 143.
Lehman, I. R., Bessman, M. J., Simms, E. S. and Kornberg, A. (1958a). *J. Biol. Chem.* **233**, 163.
Lehman, I. R., Zimmerman, S. B., Adler, J., Bessman, M. J., Simms, E. S. and Kornberg, A. (1958b). *Proc. Natl Acad. Sci. U.S.* **44**, 1191.
Levenberg, B. and Buchanan, J. M. (1957a). *J. Biol. Chem.* **224**, 1005.
Levenberg, B. and Buchanan, J. M. (1957b). *J. Biol. Chem.* **224**, 1019.
Lieberman, I. (1956a). *J. Biol. Chem.* **222**, 765.
Lieberman, I. (1965b). *J. Biol. Chem.* **223**, 327.
Lieberman, I. and Kornberg, A. (1953). *Biochem. Biophys. Acta* **12**, 223.
Lieberman, I. and Kornberg, A. (1954). *J. Biol. Chem.* **207**, 911.
Lieberman, I., Kornberg, A. and Simms, E. S. (1955a). *J. Biol. Chem.* **215**, 403.
Lieberman, I., Kornberg, A. and Simms, E. S. (1955b). *J. Biol. Chem.* **215**, 429.
Long, C. W. and Pardee, A. B. (1967). *J. Biol. Chem.* **242**, 4715.
Lukens, L. N. and Buchanan, J. M. (1959a). *J. Biol. Chem.* **234**, 1791.
Lukens, L. N. and Buchanan, J. M. (1959b). *J. Biol. Chem.* **234**, 1799.
Magasanik, B. (1962). *In* "The Bacteria" (I. C. Gunsalus and R. Y. Stanier, eds), Academic Press, New York. Vol. III, p. 295.
Magasanik, B. and Karibian, D. (1960). *J. Biol. Chem.* **235**, 2672.
Magasanik, B., Moyed, H. S. and Gehring, L. B. (1957). *J. Biol. Chem.* **226**, 339.
Mager, J. and Magasanik, B. (1960). *J. Biol. Chem.* **235**, 1474.
Maley, F. and Ochoa, S. (1958). *J. Biol. Chem.* **233**, 1538.
Mandel, L. R. and Borek, E. (1963a). *Biochemistry* **2**, 555.
Mandel, L. R. and Borek, E. (1963b). *Biochemistry* **2**, 560.
Marmur, J. and Greenspan, C. M. (1963). *Science* **142**, 387.
Meselson, M. and Stahl, F. W. (1958). *Proc. Natl Acad. Sci. U.S.* **44**, 671.
Miller, R. W., Lukens, L. N. and Buchanan, J. M. (1959). *J. Biol. Chem.* **234**, 1806.
Moyed, H. S. and Magasanik, B. (1957). *J. Biol. Chem.* **226**, 351.
Nagata, T. (1963). *Proc. Natl Acad. Sci. U.S.* **49**, 551.

Nakamoto, T. and Weiss, S. B. (1962). *Proc. Natl Acad. Sci. U.S.* **48**, 880.
Nierlich, D. P. and Magasanik, B. (1965a). *J. Biol. Chem.* **240**, 358.
Nierlich, D. P. and Magasanik, B. (1965b). *J. Biol. Chem.* **240**, 366.
Oeschger, M. P. and Bessman, M. J. (1966). *J. Biol. Chem.* **241**, 5452.
Okazaki, R. and Kornberg, A. (1964a). *J. Biol. Chem.* **239**, 259.
Okazaki, R. and Kornberg, A. (1964b). *J. Biol. Chem.* **239**, 269.
Oliver, I. T. and Peel, J. L. (1956). *Biochim. Biophys. Acta* **20**, 390.
Pace, N. R., Bishop, D. H. L. and Spiegelman, S. (1967). *Proc. Natl Acad. Sci. U.S.* **58**, 711.
Preiss, J., Dieckmann, M. and Berg, P. (1961). *J. Biol. Chem.* **236**, 1748.
Rachmeler, M., Gerhart, J. and Rosner, J. (1961). *Biochim. Biophys. Acta* **49**, 222.
Ratliff, R. L., Weaver, R. H., Lardy, H. A. and Kuby, S. A. (1964). *J. Biol. Chem.* **239**, 301.
Razzell, W. E. and Khorana, H. G. (1958). *Biochim. Biophys. Acta* **28**, 562.
Reichard, P. and Hanshoff, G. (1956). *Acta Chem. Scand.* **10**, 548.
Richardson, C. C. and Kornberg, A. (1964). *J. Biol. Chem.* **239**, 242.
Richardson, C. C., Schildkraut, C. L., Aposhian, H. V. and Kornberg, A. (1964a). *J. Biol. Chem.* **239**, 222.
Richardson, C. C., Lehman, I. R. and Kornberg, A. (1964b). *J. Biol. Chem.* **239** 251.
Richardson, C. C., Inman, R. B. and Kornberg, A. (1964c). *J. Mol. Biol.* **9**, 46.
Roscoe, D. H. and Tucker, R. G. (1966). *Virology* **29**, 157.
Shaeffer, P. (1964). *In* "The Bacteria" (I.C. Gunsalus and R. Y. Stanier, eds), Academic Press, New York. Vol. V, p. 87.
Sköld, O. (1960). *J. Biol. Chem.* **235**, 3273.
Spiegelman, S., Haruna, I., Holland, I. B., Beaudreau, G. and Mills, D. (1965). *Proc. Natl Acad. Sci. U.S.* **54**, 919.
Wahba, A. J. and Friedkin, M. (1962). *J. Biol. Chem.* **237**, 3794.
Wang, T. P., Sable, H. Z. and Lampen, J. O. (1950). *J. Biol. Chem.* **184**, 17.
Warren, L. and Buchanan, J. M. (1957). *J. Biol. Chem.* **229**, 613.
Watson, J. D. and Crick, F. H. C. (1953a). *Nature* **171**, 737.
Watson, J. D. and Crick, F. H. C. (1953b). *Nature* **171**, 964.
Weissman, C., Simon, L. and Ochoa, S. (1963). *Proc. Natl Acad. Sci. U.S.* **49**, 407.
Wilkens, M. H. F., Stokes, A. R. and Wilson, H. R. (1953). *Nature* **171**, 738.
Wright, L. D., Miller, C. S., Skeggs, H. R., Huff, J. W., Weed, L. L. and Wilson D. W. (1951). *J. Am. Chem. Soc.* **73**, 1898.
Yates, R. A. and Pardee, A. B. (1956). *J. Biol. Chem.* **221**, 743.
Yoshikawa, H. and Sueoka, N. (1963). *Proc. Natl Acad. Sci. U.S.* **49**, 559.
Zimmerman, B. K. (1966). *J. Biol. Chem.* **241**, 2007.
Zimmerman, E. F. and Magasanik, B. (1964). *J. Biol. Chem.* **239**, 293.
Zimmerman, S. B. and Kornberg, A. (1961). *J. Biol. Chem.* **236**, 1480.

19. BIOSYNTHESIS OF PROTEINS

I. Amino Acid Activation

Several years ago Lipmann (1941) suggested that amino acyl phosphates were intermediates in the biosynthesis of proteins. It is now known that Lipmann's proposal was correct in principle although the details are more complex than originally proposed. Novelli's review (1967) is an excellent survey of the recent developments in the field of amino acid activation. Hoagland *et al.* (1956) first described activation of amino acids using a protein fraction which they obtained by acidification of a rat liver homogenate to pH 5. The pH 5 enzyme catalyzed the formation of carboxyl activated amino acids, which could be detected by conversion to the hydroxamic acid. The enzyme required L-amino acids and ATP and was inactive with D-amino acids. It was also possible to assay the reaction by measuring the exchange of pyrophosphate into ATP, and they proposed that an enzyme–amino acyl-AMP complex was formed. DeMoss and Novelli (1956) used the pyrophosphate exchange assay with cell-free extracts to show that several species of bacteria were capable of activating L-amino acids. In these early studies the activation of only eight amino acids was detected, but since then it has been possible to show that all amino acids which are found in proteins are activated.

The same enzyme which catalyzes the formation of the enzyme–amino acyl–AMP complex catalyzes the transfer of the activated amino acid to its specific sRNA, and consequently activating enzymes are referred to as amino acyl-sRNA synthetases. The overall reaction catalyzed by amino acyl-sRNA synthetases is:

$$\text{Enzyme} + \text{ATP} + \text{amino acid} \rightarrow \text{enzyme-amino acyl-AMP} + \text{PP}_i$$

$$\text{Enzyme-amino acyl-AMP} + \text{sRNA} \rightarrow \text{enzyme} + \text{amino acyl-sRNA} + \text{AMP}$$

Hoagland *et al.* (1958) showed that ^{14}C labeled amino acids incubated with their pH 5 fraction and ATP became attached to soluble RNA already present in the fraction. Zachau *et al.* (1958) studied the linkage between the amino acid and sRNA. They digested the amino acyl-sRNA complex with ribonuclease and isolated the amino acid-base conjugate. The base was identified as adenine and the amino acid was attached to either the 2′ or 3′ hydroxyl of ribose by an ester linkage, that is, attached to the terminus of sRNA which contained the 3′ hydroxyl. Berg *et al.* (1961) purified leucyl-, valyl-, isoleucyl- and methionyl-sRNA synthetases from *E. coli* and studied the reaction catalyzed by these enzymes. A specific sRNA was required for each amino acid and enzyme and it is now know that certain amino acyl-sRNA synthetases will function with more than one specific sRNA (Novelli, 1967). Norris and Berg (1964) determined that the enzyme-amino acyl-AMP complex formed by highly purified isoleucyl-sRNA synthetase, isoleucine and AMP was composed of one mole of each of these reactants. The same enzyme catalyzed the transfer of amino acid to its specific sRNA.

Several amino acyl-sRNA synthetases have been isolated from bacteria, many of them in nearly pure state (Table 19.1). In addition, Muench and Berg (1966) devised a method for the preparation of an enzyme fraction from *E. coli* which catalyzed the activation of all 20 natural amino acids.

TABLE 19.1. Amino Acyl-sRNA synthetases which have been purified from bacteria.

Amino Acid	Organism	Reference
Aspartate	*L. arabinosus*	Norton *et al.* (1963)
Asparagine	*L. arabinosus*	Hedgcoth *et al.* (1963)
Glutamate	*E. coli*	Lazzarini and Mehler, (1966)
Glutamine	*E. coli*	Lazzarini and Mehler (1966)
Isoleucine	*E. coli*	Baldwin and Berg (1966)
Leucine	*E. coli*	Berg *et al.* (1961)
Lysine	*E. coli*	Stern and Mehler (1965)
Methionine	*E. coli*	Berg *et al.* (1961), Heinrikson and Hartley (1967)
Phenylalanine	*E. coli*	Conway *et al.* (1962), Fangman *et al.* (1965), Stulberg (1967)
Proline	*E. coli*	Mehler and Jesensky (1966)
Tyrosine	*E. coli*	Calender and Berg (1966)
Tyrosine	*B. subtilis*	Calender and Berg (1966)

A good deal is known about the structure of sRNA since its relatively low molecular weight has made it more amenable to study than the larger nucleic acids. One of the best studied examples is alanine sRNA from yeast, which has a molecular weight of 26 600 and contains 77 bases (Holley *et al.*, 1965). In addition to the four conventional bases, alanine sRNA contains pseudouracil, dihydrouracil, thymine, dimethylguanine, hypoxanthine, 1-methylguanine and 1-methylhypoxanthine. All species of sRNA contain the base sequence pCpCpA at the terminus which contains the 3′ hydroxyl (Hecht *et al.*, 1959):

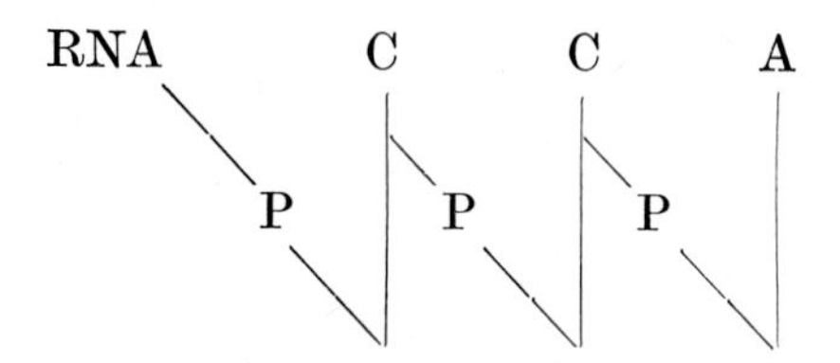

Holley and his associates (1965) worked out the complete base sequence for alanine sRNA from yeast (Fig. 19.1), the first ever to be reported for any kind of RNA. Only short segments of the molecule contained complementary sequences, and hence alanine sRNA cannot have a straightforward double stranded configuration. Holley and his associates

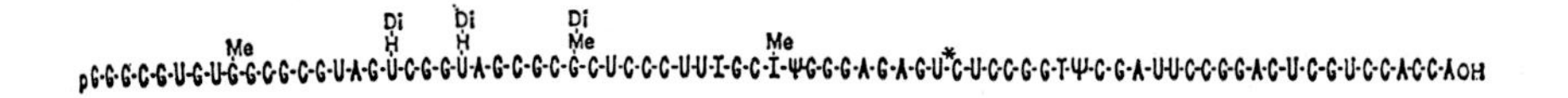

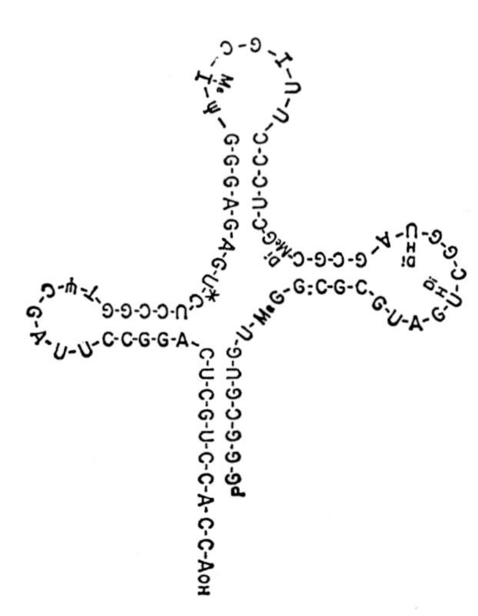

FIG. 19.1. Base sequence (above) and one possible conformation (below) of alanine sRNA from yeast. (Reproduced with permission from Holley *et al.*, 1965).

suggested several conformations which would be compatible with the primary structure, one of which is shown in Fig. 19.1. In addition to the binding site for the amino acid, sRNA also contains the codon recognition site, the anticodon. The codon for alanine is GCU (Table 19.2), and Holley *et al.* (1965) suggested that the CGG triplet in the east branch of the cloverleaf (Fig. 19.1) or the IGC triplet in the north branch were likely candidates for anticodons (hypoxanthine in the IGC triplet would correspond to guanine and would be read as GGC). Nirenberg *et al.* (1966) favored the IGC triplet on the basis of their studies on binding of amino acyl-sRNA to ribosomes in the presence of specific RNA triplets (section I.C., this chapter). If the codon and anticodon bind to each other in an antiparallel fashion (as in the case of DNA), then the base pairing would be:

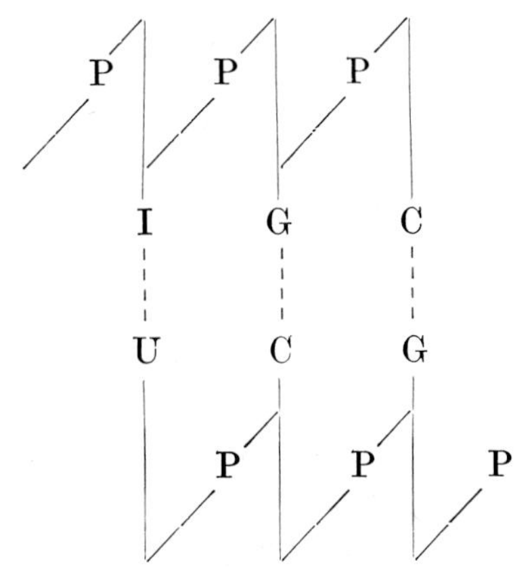

II. Transcription of the Genetic Code

The enzymic events of the transcription of the genetic code are described in Chapter 18, section III.A. A good deal of evidence indicates that the messenger which is transcribed corresponds to a length of chromosome which frequently includes the code for more than one enzyme, that is, the messenger is polycistronic (see review of Attardi, 1967). For example, Naono *et al.* (1965) took advantage of the diauxie phenomenon to isolate mRNA involved in the synthesis of proteins of the lactose operon. They inoculated *E. coli* into a medium with glucose and lactose, and when glucose was exhausted the organism adapted to growth on lactose by forming β-galactosidase and galactoside acetyltransferase (see section IV.A). RNA formed during the period of enzyme induction was extracted, isolated and identified as RNA complementary to DNA of the lactose region of the chromosome by hybridization with

TABLE 19.2. The genetic code (Crick, 1966; and Schweet and Heintz, 1966). The codon AUG also codes for formymethionine, the chain initiator.

1st ↓ 2nd →	U	C	A	G	↓ 3rd
U	Phe	Ser	Tyr	Cys	U
	Phe	Ser	Tyr	Cys	C
	Leu	Ser	Ochre	?	A
	Leu	Ser	Amber	Trp	G
C	Leu	Pro	His	Arg	U
	Leu	Pro	His	Arg	C
	Leu	Pro	Glun	Arg	A
	Leu	Pro	Glun	Arg	G
A	Ileu	Thr	Aspn	Ser	U
	Ileu	Thr	Aspn	Ser	C
	Ileu	Thr	Lys	Arg	A
	Met	Thr	Lys	Arg	G
G	Val	Ala	Asp	Gly	U
	Val	Ala	Asp	Gly	C
	Val	Ala	Glu	Gly	A
	Val	Ala	Glu	Gly	G

DNA from a strain of *S. marcescens* which contained the lactose operon of *E. coli*. The homologous RNA sedimented at approximately 26 S, which meant that it was large enough to account for the synthesis of all the proteins of the lactose operon.

III. Translation of the Genetic Code

A. Composition and Function of Ribosomes

Tissières *et al.* (1959) isolated ribosomes from cultures of *E. coli* in the exponential phase of growth, at which time ribosomes accounted for 30 % of the dry weight of the organism. These particles were composed of 63 % protein and 37 % RNA. They found that the magnesium concentration controlled the aggregation of ribosomes; by increasing the magnesium concentration, 30 S and 50 S ribosomes aggregated to form 70 S particles, and by increasing the concentration of magnesium even more, 100 S particles were formed which were aggregates of 70 S particles. The particles disaggregated when the magnesium concentration was decreased.

Protein synthesis takes place on polysomes, aggregates of 70 S ribosomes connected together with a strand of mRNA. Barondes and

Nirenberg (1962) showed that ribosomes of *E. coli* incubated with polyuridylic acid aggregated to form large complexes with sedimentation constants of 100 S to 130 S. Dresden and Hoagland (1965) developed a technique for isolating polysomes which were metabolically active from *E. coli* by weakening the cell wall with ethylenediaminetetraacetic acid and lysozyme and then opening the cells by treatment with a detergent. Polysomes were separated from other cell constituents by ultracentrifugation through a sucrose gradient. The fractions obtained by this treatment were tested for the ability to incorporate ^{14}C-leucine into proteins, and the polysome fraction accounted for approximately 80% of the label fixed by all fractions.

B. Binding of mRNA and sRNA to Ribosomes

1. *Binding of mRNA*

Okamoto and Takanami (1963) found that polyuridylic acid bound specifically to the 30 S ribosome but not to the 50 S ribosome. They prepared polyuridylic acid-^{32}P with polynucleotide phosphorylase (polyribonucleotide:orthophosphate nucleotidyltransferase), an enzyme which has been very useful in the study of nucleic acid and protein metabolism because the composition of the product can be controlled by varying the proportions of nucleoside diphosphates:

$$\text{n Nucleoside diphosphate} \rightarrow \text{polyribonucleotide} + \text{n P}_i$$

The enzyme does not require a primer although it can enlarge a polynucleotide which is added to the reaction mixture. Also, it does not require the presence of all four ribonucleotides, and in fact it will form a polyribonucleotide containing a single base if only one nucleoside diphosphate is added to the reaction mixture. When polyuridylic acid-^{32}P was mixed with 30 S ribosomes, polyuridylic acid and the ribosomes sedimented together in a sucrose gradient. The binding process apparently did not require a soluble enzyme but did require magnesium ion. The complex formed by ribosomes and polyuridylic acid sedimented much more rapidly than the 30 S ribosome alone, indicating that a ribosomal aggregate had been formed. Similar results were obtained when 70 S ribosomes were mixed with polyuridylic acid, but no complex was formed with 50 S ribosomes. When the polyuridylic acid-^{32}P enzyme complex was treated with ribonuclease, most of the ^{32}P was solubilized, but a finite amount, estimated to be about 25–30 base units in length, remained attached to the 70 S ribosome (Takanami and Zubay, 1964); presumably this is the length of the section of polynucleotide bound to the ribosome and protected from the action of ribonuclease. Takanami and Zubay (1964) calculated that a single polynucleotide chain of 25–30 residues would be approximately as long as the long dimension of the 30 S ribosome and suggested that mRNA was bound to the ribosome in this fashion. Moore (1966) also concluded that one mRNA chain per ribosome was bound by a study of the saturation of ribosomes with synthetic polynucleotides.

2. *Binding of amino acyl-sRNA to ribosomes*

Kaji and Kaji (1963) reported that the binding of amino acyl-sRNA to polysomes required the presence of a specific mRNA, that is, mRNA which contained the codon for that amino acid. Spyrides (1964) found that binding of amino acyl-sRNA required ammonium or potassium ions. Both groups of workers found that ^{14}C-phenylalanyl-sRNA (phenylalanine codon UUU, Table 19.2) was bound to ribosomes when polyuridylic acid was added to the incubation mixture, but not when polyadenylic acid was added. On the other hand, when polyadenylic acid was present, ^{14}C-lysyl-sRNA (lysine codon AAA, Table 19.2) was bound to the ribosome, but not when polyuridylic acid was present. Binding of amino acyl-sRNA occurred under conditions where polypeptide formation did not occur, hence binding must precede protein formation.

C. Nucleic Acid Code for Proteins

Crick's review (Crick, 1966) is a fascinating account of the events leading to the concept of a triplet, non-overlapping code for the synthesis of proteins. In 1961 Nirenberg and Matthaei published the results of their studies which dealt with the synthesis of protein (which was directed by a polyribonucleotide) in a cell-free system. Protein was formed when ribosomes, the supernatant fraction of a cell-free extract of *E. coli* obtained by centrifugation at 105 000 × g, ATP, a mixture of amino acids and high molecular weight RNA were incubated together. When polyuridylic acid was the RNA template, a polymer of phenylalanine was formed and the first codon assignment, UUU, was made to phenylalanine. Double-stranded RNA was not effective as a template. By the methods available at that time it was not possible to obtain synthetic RNA of definite polynucleotide sequence; nevertheless, by 1963 it was possible to assign some 50 triplets to 20 amino acids (Nirenberg *et al.*, 1963).

Nirenberg and Leder (1964) developed an assay technique which greatly facilitated work on the genetic code; they found that amino acyl-sRNA would bind to ribosomes in the presence of specific trinucleotides. The incubation mixture was filtered through a cellulose acetate membrane filter which retained the ribosomes and bound ^{14}C-amino acyl-sRNA, and the extent of the reaction could be estimated by measuring the number of counts fixed on the filter pad. By this time it was possible to make trinucleotides of known composition and sequence and now it became possible to determine the order of bases in the triplet. With the general use of this technique, it quickly became possible to determine the codes for all 20 amino acids plus glutamine and asparagine (Table 19.2; Nirenberg *et al.*, 1966; Matthaei *et al.*, 1966; Söll *et al.*, 1966). One fact which is apparent is that an amino acid can be coded for by more than one triplet, although usually the first two bases are important. This feature of the code is referred to as degeneracy. Apparently almost all 64 possible combinations of triplets of the four bases of RNA have a biological meaning. Two triplets indicated by *amber* and *ochre* in Table 19.2 do not code for an amino acid and apparently are used to signal termination of the peptide chain (section III.D.4). Most of the work leading to the code in Table 19.2 has been done with *E. coli*; however, the evidence indicates that the code applies to higher organisms as well and is therefore universal (Nirenberg *et al.*, 1966).

D. Peptide Bond Formation

1. *Direction of reading of mRNA*

Salas *et al.* (1965) established that the direction of reading of mRNA during protein synthesis was from the end which contained the 5′ phosphate to the end with the 3′ hydroxyl. They used an RNA template of polyadenylic acid with cytosine at the 3′ terminus. Several polypeptides were produced, and the major product contained three to five lysine residues. one of which was at the amino terminal end of the polypeptide. Since the codon for lysine, AAA, was at the 5′ end of the oligonucleotide and the *N*-terminal amino acid is the first one laid down (section III.D.3), this result indicated that the direction of reading of mRNA was from the 5′ end to the 3′ end. Asparagine, codon AAC, was found at the C-terminal end of the polypeptide, which further corroborated this interpretation. Thach *et al.* (1965) came to the same conclusion when they found that AAAUUU used as a template resulted in the formation of lysylphenylalanine.

2. *Chain initiation*

Waller and Harris (1961) obtained a clue to the initial reaction in protein synthesis by *E. coli* when they found that the distribution of *N*-terminal amino acids from ribosomal proteins of this organism was

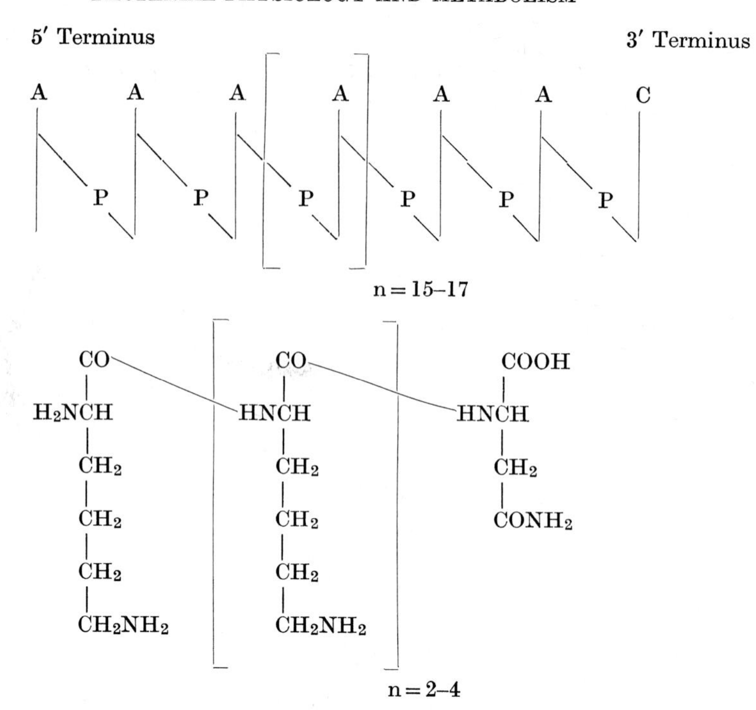

FIG. 19.2. Structure of the oligonucleotide which Salas *et al.* (1965) used as a template in a cell-free system to direct the formation of the lysyl asparagine polypeptide shown below. The oligonucleotide did not have phosphate at the 5′ terminus because this linkage was hydrolyzed by the method of preparation of the oligonucleotide.

non-random, since methionine and alanine were the two most frequently encountered *N*-terminal amino acids. The process is fairly well understood in the case of methionine, which is incorporated as the *N*-formyl derivative. Marcker and Sanger (1964) discovered the presence of *N*-formylmethionyl-sRNA in the methionyl-sRNA fraction of *E. coli*, and showed that formylation occurred after the attachment of methionine to sRNA. Marcker (1965) studied the formylation reaction with cell-free preparations of *E. coli* and N^{10} formyltetrahydrofolate as the formyl donor. He found two kinds of methionyl-sRNA in *E. coli*, now designated $sRNA_M$ and $sRNA_F$; methionine attached only to $sRNA_F$ could be formylated.

Clark and Marcker (1966) also found that when polyUAG was used as a template, methionyl-$sRNA_F$ and formylmethionyl-$sRNA_F$ were incorporated preferentially into the *N*-terminal position. When methionyl-$sRNA_M$ was used, methionine was incorporated internally.

Sundararajan and Thach (1966) studied the binding of formylmethionyl-sRNA to ribosomes in the presence of oligonucleotides of known composition, and came to the conclusion that AUG was the codon for formylmethionine. In addition, they established that the sequence AUG fixed the reading frame of the polynucleotide, a process they referred to as phasing. For example, ApApApUpG stimulated the binding of formylmethionyl-sRNA, but not lysyl-sRNA even though the lysine codon AAA is at the 5′ end of the polynucleotide. Experimental results such as these show that the mRNA is read like a role of movie film, one frame at a time. This feature of the genetic code is often referred to as its non-overlapping character. Hershey and Thach (1967) have also shown that GTP functions in the initiation of protein synthesis since it stimulates the formation of formylmethionylpuromycin. Puromycin is an antibiotic which acts as an analogue of the amino acyl end of amino acyl-sRNA and becomes covalently attached to the carboxyl end of the growing peptide chain by an amide linkage (Smith *et al.*, 1965).

A widely held conception of protein synthesis is that ribosomes have one binding site for peptidyl-sRNA and one binding site for amino acyl-sRNA (I and II, Fig. 19.3; Watson, 1964). It is difficult to obtain direct evidence for such a hypothesis, but there is a good deal of indirect evidence in support of this idea. Bretscher and Marcker (1966) obtained evidence that formylmethionyl-sRNA was bound preferentially to the peptidyl-sRNA site, which would be consistent with its role as a chain initiator. The evidence was that formylmethionyl-$sRNA_F$ reacted with puromycin to form formylmethionylpuromycin, but methionyl-$sRNA_M$ did not react.

3. *Chain elongation*

Peptide bond formation during protein synthesis proceeds from the *N*-terminal direction to the C-terminal direction. Bishop *et al.* (1960) first established this point using reticulocyte ribosomes incubated with ^{14}C-valine and then transferred to a cell-free system capable of synthesizing hemoglobin, but now incubated with non-radioactive valine. They found that the isotope was located mainly in the *N*-terminal portion of hemoglobin. The current concept of chain elongation during protein synthesis is shown in Fig. 19.3 (also see review of Watson, 1964). A central feature of this formulation is the postulation of two sites on the ribosome: site I, where peptidyl-sRNA is attached, and site II, where the incoming amino acyl-sRNA is attached, attracted to this point by hydrogen bonding between the anticodon of sRNA and codon of mRNA. Site I is known as the donor site, since the polypeptide

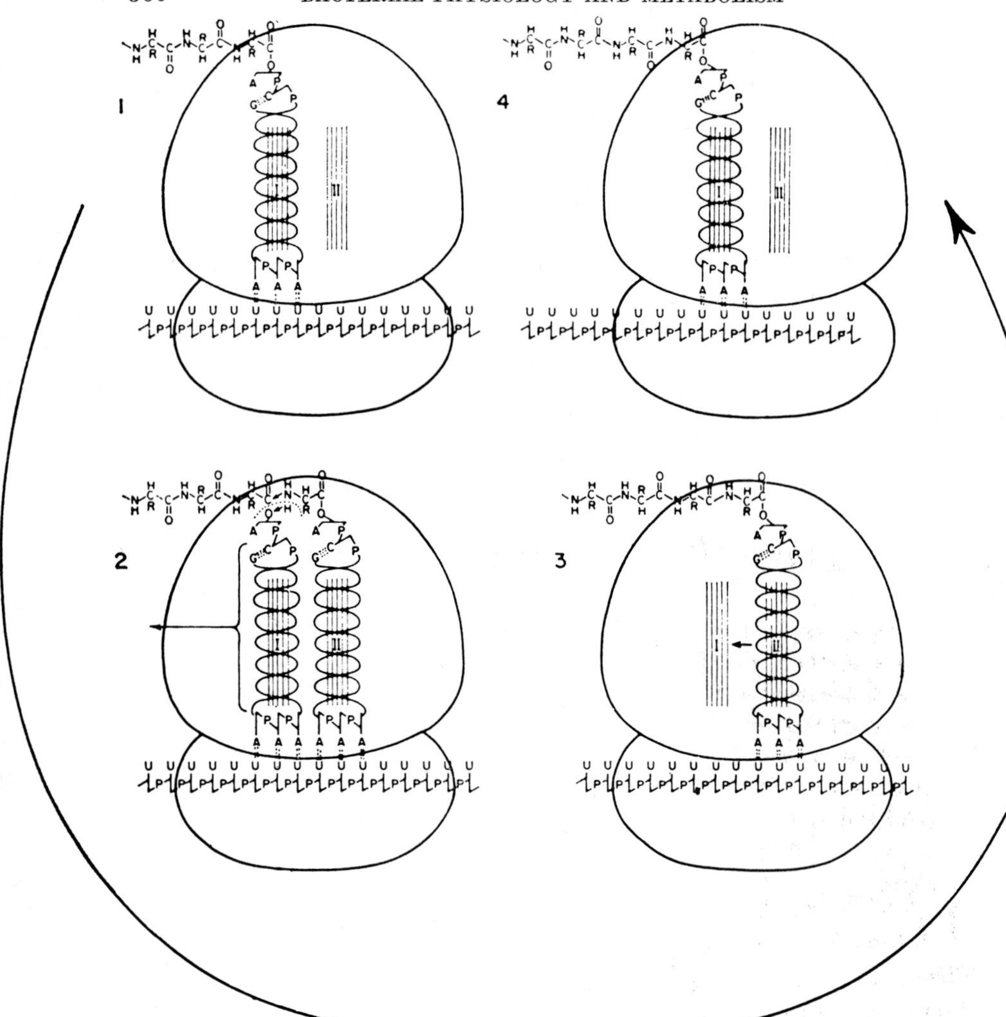

FIG. 19.3. Peptide condensation cycle illustrated with polyuridylic acid as the mRNA template. 1. Peptidyl-sRNA is located on site I, the donor site. 2. An amino acyl-sRNA binds to site II, the acceptor site, and attaches to mRNA by hydrogen bonding. The peptide bond is formed. 3. sRNA leaves site I. 4. Peptidyl-sRNA moves to site I; presumably this is the step which requires GTP. (Reproduced with permission from Nishizuka and Lipmann, 1966b.)

residue is donated to the free amino group of the acceptor amino acyl-sRNA. Following peptide bond formation, sRNA at site I is released from the ribosome and the elongated polypeptide chain now moves to site I; presumably mRNA moves as well.

There is a good deal of indirect evidence to support this concept. Allende *et al.* (1964) isolated two soluble factors from *E. coli*, now known as the T and G factors, which were required for peptide bond formation in a cell-free system. G factor was identified as an enzyme which hydrolyzed GTP, one molecule of GTP being split for every peptide bond formed (Nishizuka and Lipmann, 1966a). A logical place for the G factor to function is in the translocation of peptidyl-sRNA. However, the G factor is not involved in peptide bond formation itself. This is the activity of the T factor, the peptide synthetase. Lucas-Lenard and Lipmann (1966) resolved the T factor into two components, a stable one, T_s and an unstable one, T_u.

Yanofsky and his associates established the colinear relationship between the structure of the A protein of tryptophane synthetase and the segment of chromosome which carries the code for its primary structure (Fig. 19.4; see review of Yanofsky, 1964). A tremendous amount of work was required for this project; they collected a large number of mutants of *E. coli* which contained amino acid replacements in the A protein and determined the complete sequence of amino acids in the A protein of the normal enzyme and of the protein synthesized by the mutants. They also mapped the position of the mutation and found that the position of the amino acid replacement in the A protein correlated very closely with the map location of the mutation (Fig. 19.4). Sarabhai *et al.* (1964) also obtained evidence for the colinearity of the amino acid sequence of the head protein of T4 coliphage and the DNA of this phage (see following section).

4. *Chain termination*

A signal to terminate the synthesis of protein is required in order to prevent running-together of proteins whose code is located on adjacent cistrons. In addition, cleavage of sRNA from the completed peptide must occur at termination of protein synthesis.

A partial answer to the question of how chain termination occurs came from the study of amber mutants of coliphage T4. These mutants were unable to form a complete head protein in wild type *E. coli*, but could in certain mutants of *E. coli*, su^+ strains, which suppressed the mutation. Sarabhai *et al.* (1964) showed that infection of su^- by amber mutants resulted in formation of peptides of varying length, depending on which amber mutant was used. The map locations of amber mutations were known and the length of the peptide correlated with the map location of the mutation. This evidence provided additional support for the concept of colinearity of DNA and the protein synthesized, outlined in the preceding section. Stretton and Brenner (1965) studied

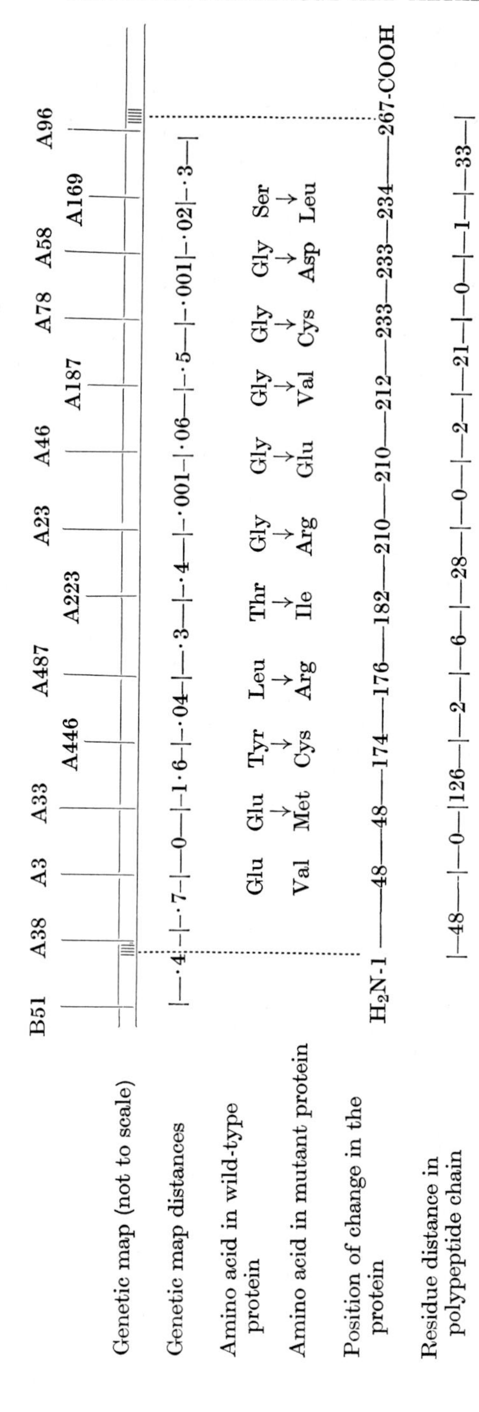

FIG. 19.4. Colinear relationship between genome of *E. coli* and structure of the A protein of tryptophane synthetase. (Reproduced with permission from Yanofsky *et al.*, 1967.)

one amber mutant H36, and isolated the head protein produced by infection of su+, and the peptide produced by infection of su−, comparing these with the amino acid sequence of the head protein of the parent phage. The head protein produced in the suppressor strain of *E. coli* contained a serine residue in place of glutamine present in the normal head protein (Fig. 19.5). The incomplete peptide formed by

T4D	Ala-Gly-(Val, Phe)-Asp-Phe-*GluN*-Asp-Pro-Ileu-Asp-Ileu-Arg-
H36 in *su*+	Ala-Gly-Val-Phe-Asp-Phe-*Ser*-Asp-Pro-Ileu-Asp-Ileu-Arg-
H36 in *su*−	Ala-Gly-Val-Phe-Asp-Phe

FIG. 19.5. Amino acid sequence of a segment of the head protein of phage T4D (top line), same segment of head protein of *amber* mutant H36 grown in a suppressor strain of *E. coli* (middle line), and corresponding segment of a peptide produced by H36 grown in a non-suppressor strain of *E. coli* (Stretton and Brenner, 1965).

H36 in su− *E. coli* terminated at the amino acid just prior to the replacement. The interpretation of these findings was that the amber mutation was a mutation to the chain termination signal. Brenner *et al.* (1965) deduced the code for termination in amber mutants and in another series of similar mutants named ochre mutants. The ochre mutation was very difficult to suppress, but ochre mutants could readily be converted to amber mutants. These investigators examined the effects of various mutagenic agents whose action is understood, such as 2-aminopurine and hydroxylamine, and came to the conclusion that UAG was the amber codon and UAA was the ochre codon. Because the ochre mutation is so difficult to suppress, it may be the natural termination codon.

IV. Regulation of Protein Synthesis

A. Transcriptional Control, Jacob and Monod's Operon Theory

In the preceding chapters there are numerous examples of enzyme induction and repression which make it clear that the formation of proteins is under very fine control which responds to the needs of the cell. Although the molecular basis for this control has not been completely established, it has been the subject of intensive research by several well-known groups of investigators. The paper by Jacob and Monod (1961) in which they proposed the operon concept has dominated thinking on the subject since its publication.

Jacob and Monod were impressed with the fact that exposure of *E. coli* to a gratuitous inducer, an analog of lactose which was non-metabolizable, resulted in the simultaneous induction of galactoside permease and β-galactosidase. Since that time it has also been established that galactoside acetyltransferase (acetyl-CoA:galactoside 6-O-acetyl-transferase) is formed as well. The latter enzyme catalyzes the acetylation of β-galactosides with acetyl-CoA:

$$\text{Acetyl-CoA} + \beta\text{-D-galactoside} \rightarrow \text{CoA} + \text{6-acetyl-}\beta\text{-D-galactoside}$$

Not only were the enzymes formed simultaneously, but their rate of synthesis was the same; this is known as coordinate induction. It was clear that the formation of these three enzymes was under control by the same regulator. In addition, it was also known that the structural genes for these enzymes were clustered, that is, located together on the chromosome of *E. coli* (Jacob and Monod, 1961; also see review of Clowes, 1964, for chromosome maps of bacteria). Jacob and Monod defined such a region of the chromosome as the operon, "the genetic unit of coordinate expression".

Jacob and Monod reasoned that the control of protein synthesis was exerted in a negative fashion, that is, a repressor substance was produced which prevented the expression of structural genes in the uninduced or repressed cell. When the inducer was added, it reacted with the repressor which inactivated the repressor and thereby allowed the synthesis of mRNA and protein. Jacob and Monod also realized that repression was a phenomenon fundamentally similar to induction, the difference being that the repressor became activated when a co-repressor was present and the activated complex prevented expression of the structural gene. Therefore, the co-repressor acted in repression in a fashion analogous to that of the inducer in enzyme induction.

These properties of the repressor are very much like those of allosteric enzymes where the effector causes a change in the conformation of the enzyme which may either activate or inactivate it (Chapter 16, section III.B). In addition, there is now good evidence that the repressor is a protein. Bourgeois *et al.* (1965) and Müller-Hill (1966) found that expression of *i* mutations in *E. coli*, mutations to constitutivity with respect to lactose fermentation, could be suppressed in su^+ *E. coli*. Considering the known action of suppressor strains on amber mutants, it seems likely that the chain termination codon was being read as an amino acid, allowing a complete repressor protein to be formed, but with an amino acid substitution. Gilbert and Müller-Hill (1966) actually purified a protein from *E. coli* which appears to be a product of the *i* gene. It was detected by its ability to bind isopropylthiogalactoside, a gratuitous inducer of the lactose operon. The protein was not detected in i^- mutants which produced an incomplete protein or in strains of *E. coli* where the *i* region was deleted from the chromosome.

There was a good deal of genetic evidence for postulating the kind of control outlined in the preceding paragraph. An *i* gene was known to be located on the chromosome near the *z*, *y* and *a* cluster for β-galactosidase, galactoside permease and galactoside acetyltransferase respectively. A mutation at the *i* locus resulted in an organism which was constitutive for lactose fermentation. Diploid *E. coli* which were heterozygous for the *i* gene were inducible, that is, the inducible condition was dominant. This result established that a substance was produced by the i^+ gene which was diffusable (or cytoplasmic in the terms of Jacob and Monod), since it acted not only on the operon on its own chromosome (*cis* effect) but also on the operon of the chromosome

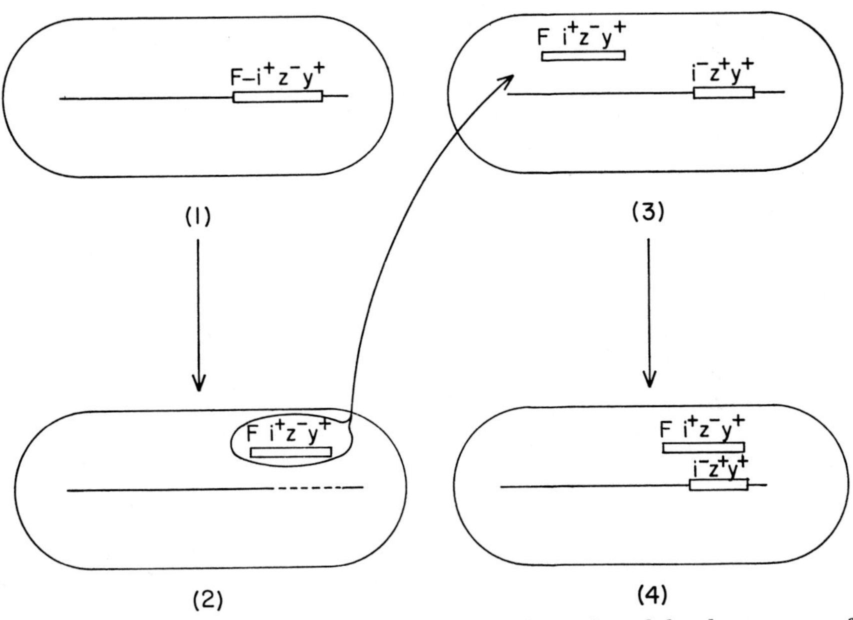

Fig. 19.6. Production of stable diploids for the *lac* region of the chromosome of *E. coli*. (1) the F particle of the Hfr strain shown here is located adjacent to the *lac* region. (2) The F′ strain is produced when the F particle detaches spontaneously carrying a segment of the chromosome with it and resulting in a chromosomal deletion (dotted line). The F particle now behaves as an episome and can be transferred to a recipient with a high frequency. (3) A strain of *E. coli* which contains an i^- gene and is constitutive for the ability to ferment lactose is infected with the F particle. (4) The F particle attaches to the chromosome of the recipient which now becomes inducible for lactose fermentation.

which contained the i^- gene (*trans* effect). The problem in this case was to produce stable diploids, which was accomplished in the following fashion (Fig. 19.6): strains of *E. coli* designated F′ strains were isolated where the F particle was free in the cytoplasm and contained a segment of the chromosome which included the *lac* operon. Therefore, in mating experiments in which the F particle was transferred, a small piece of the donor chromosome went along and was incorporated into the genome of the recipient, making it diploid for the *lac* region of the chromosome.

Jacob and Monod predicted that the target of the repressor would be located on the chromosome in the *lac* cluster and that it should be possible to isolate mutants of this locus with an altered affinity for the repressor. Jacob and Monod defined this site as the operator, *o*. If this concept were correct, diploids of the character o^+z^-/o^cz^+ (where o^c is the mutant which is unable to bind the repressor) should be constitutive for β-galactosidase, or operationally, the o^c condition should be dominant. They isolated mutants at the *o* locus and found that their predictions were indeed fulfilled. They also discovered that the *o* locus affected only the operon on its own (*cis*) chromosome, which again would be expected since the *o* gene did not produce a diffusible product as did the *i* gene. The *cis* effect was demonstrated in the following manner: certain z^- mutants were known to produce an altered protein with no enzymic activity but which did cross-react with antibody against the normal enzyme. Diploids of the type o^+z^+/Fo^cz^- produced the z^- protein constitutively, but produced normal β-galactosidase, the product of the z^+ gene, only in the presence of the inducer period. Therefore, the operator controlled only the genes on its own chromosome.

The Jacob and Monod theory can be characterized as proposing that the primary site of regulation of protein synthesis occurs at the stage of transcription. Attardi *et al.* (1963) obtained very strong evidence in favor of transcriptional control when they found that the amount of mRNA complementary to the *lac* operon increased markedly during enzyme induction. The theory fits very well for the *lac* operon, and closely, although not perfectly, for several other systems (see review of Vogel and Vogel, 1967; Brenner, 1965). In some cases the fit is not so good; for example, structural genes for four of the eight enzymes involved in the biosynthesis of arginine in *E. coli* are clustered, the rest are scattered through the chromosome. The four clustered genes are not coordinately repressed and derepressed although they are subject to regulation by the same R gene which is located outside of the cluster (see review of Vogel and Vogel, 1967).

B. TRANSLATIONAL CONTROL

There are also indications that control mechanisms may act at the level of translation. One of the most interesting cases was described by Roth and Ames (1966) in a study of the histidine biosynthetic pathway in *S. typhimurium*. The structural genes for the enzymes of the histidine pathway are clustered; however, Roth *et al.* (1966) isolated four classes of regulatory mutants, that is organisms which were derepressed with respect to the enzymes of the histidine pathway. The loci for three of the regulatory genes were located outside the cluster and the fourth was located inside the cluster. Roth and Ames (1966) characterized one isolate with a mutation in the *his*S region, located outside the cluster, and found that the organism possessed a defective histidyl-sRNA synthetase. Roth and Ames suggested that histidyl-sRNA might regulate protein synthesis at the level of translation. For example, if the codon for histidyl-sRNA preceded AUG, the codon for formylmethionine, histidyl-sRNA, might act to prevent the initiation of protein synthesis. In histidine starvation the concentration of uncharged sRNA specific for histidine would be increased and it would have the opposite effect, namely of allowing protein synthesis to occur. Neidhardt (1966) has also reported that a strain of *E. coli* with a defective valyl-sRNA synthetase is derepressed for certain enzymes of the valine-isoleucine pathway.

Cline and Bock (1966) proposed a model for regulation of protein synthesis where the primary site of regulation is at the level of translation. They noted that there were cases where a single mutation affected both repression and feedback inhibition of an enzyme and they proposed that there was a connection between these two phenomena in control of protein synthesis. Somerville and Yanofsky (1965) reported one such example of a mutation which affected both repression and feedback inhibition. The cluster of four structural genes on the *E. coli* chromosome which code for the enzymes which synthesize tryptophane from anthranilic acid behave like an operon. Somerville and Yanofsky isolated a series of mutants which could not be repressed by growth in the presence of usually effective concentrations of tryptophane and which produced an altered anthranilate synthetase not inhibited by tryptophane. Cline and Bock also noted that enzymes which are subject to allosteric effectors frequently, if not always, contain a regulatory and a catalytic subunit (Chapter 16, section III.B). Cline and Bock suggested that folding of the nascent enzyme occurred as protein synthesis proceeded and that regulatory subunits attached to the growing enzyme resulting in a conformational change in the polypeptide.

There is reason to believe that folding does occur during protein synthesis, since cases are known of nascent enzymes still attached to ribosomes and possessing catalytic activity (Kihara *et al.*, 1961; Zipser, 1963). In repression such a change in conformation could prevent the further synthesis of polypeptide by a number of possible mechanisms; for example, it might push the C-terminal end of the peptide away from the incoming amino acyl-sRNA, thus preventing peptide bond formation. In enzyme induction, the effector would cause a change which would align the nascent peptidyl-sRNA in the proper configuration on the ribosome, allowing protein synthesis to proceed.

At the present time it is still not clear whether control of transcription or translation is the more important regulatory process in protein synthesis. However, in such a complex process as protein synthesis, it seems reasonable that control mechanisms exist at both levels and that transcriptional control may be more important in certain cases while translational control is more important in the synthesis of other proteins.

References

Allende, J. E., Monro, R. and Lipmann, F. (1964). *Proc. Natl Acad. Sci. U.S.* **51**, 1211.

Attardi, G. (1967). *Ann. Rev. Microbiol.* **21**, 383.

Attardi, F., Naono, S., Rouviere, J., Jacob, F. and Gros, F. (1963). *Cold Spring Harbor Symp. Quant. Biol.* **28**, 363.

Baldwin, A. N. and Berg, P. (1966). *J. Biol. Chem.* **241**, 831.

Barondes, S. H. and Nirenberg, M. W. (1962). *Science* **138**, 813.

Berg, P., Bergmann, F. H., Ofengand, E. J. and Dieckmann, M. (1961). *J. Biol. Chem.* **236**, 1726.

Bishop, J., Leahy, J. and Schweet, R. (1960). *Proc. Natl Acad. Sci. U.S.* **46**, 1030.

Bourgeois, S., Cohn, M. and Orgel, L. E. (1965). *J. Mol. Biol.* **14**, 300.

Brenner, S. (1965). *Brit. Med. Bull.* **21**, 244.

Brenner, S., Stretton, A. O. W. and Kaplan, S. (1965). *Nature* **206**, 994.

Bretscher, M. S. and Marcker, K. A. (1966). *Nature* **211**, 380.

Calender, R. and Berg, P. (1966). *Biochemistry* **5**, 1681.

Clark, B. F. C. and Marcker, K. A. (1966). *J. Mol. Biol.* **17**, 394.

Cline, A. L. and Bock, R. M. (1966). *Cold Spring Harbor Symp. Quant. Biol.* **31**, 321.

Clowes, R. C. (1964). *In* "The Bacteria" (I. C. Gunsalus and R. Y. Stanier, eds), Academic Press, New York. Vol. v, p. 253.

Conway, T. W., Lansford, E. M. and Shive, W. (1962). *J. Biol. Chem.* **237**, 2850.

Crick, F. H. C. (1966). *Cold Spring Harbor Quant. Biol.* **31**, 3.

DeMoss, J. A. and Novelli, G. D. (1956). *Biochem. Biophys. Acta* **22**, 49.

Dresden, M. and Hoagland, M. B. (1965). *Science* **149**, 647.

Fangman, W. L., Nass, G. and Neidhardt, F. C. (1965). *J. Mol. Biol.* **13**, 202.

Gilbert, W. and Müller-Hill, B. (1966). *Proc. Natl Acad. Sci. U.S.* **56**, 1891.

Hecht, L. I., Stephenson, M. L. and Zamecnik, P. C. (1959). *Proc. Natl Acad. Sci. U.S.* **45**, 505.
Hedgcoth, C., Ravel, J. and Shive, W. (1963). *Biochem. Biophys. Res. Commun.* **13**, 495.
Heinrikson, R. L. and Hartley, B. S. (1967). *Biochem. J.* **105**, 17.
Hershey, J. W. B. and Thach, R. E. (1967). *Proc. Natl Acad. Sci. U.S.* **57**, 759.
Hoagland, M. B., Keller, E. B. and Zamecnik, P. C. (1956). *J. Biol. Chem.* **218**, 345.
Hoagland, M. B., Stephenson, M. L., Scott, J. F., Hecht, L. I. and Zamecnik, P. C. (1958). *J. Biol. Chem.* **231**, 241.
Holley, R. W., Apgar, J., Everett, G. A., Madison, J. T., Marquisee, M., Merrill, S. H., Penswick, J. R. and Zamir, A. (1965). *Science* **147**, 1462.
Jacob, F. and Monod, J. (1961). *J. Mol. Biol.* **3**, 318.
Kaji, A. and Kaji, H. (1963). *Biochem. Biophys. Res. Commun.* **13**, 186.
Kihara, H. K., Hu, A. S. L. and Halvorson, H. O. (1961). *Proc. Natl Acad. Sci. U.S.* **47**, 489.
Lazzarini, R. A. and Mehler, A. H. (1966). *In* "Procedures in Nucleic Acid Research" (G. L. Cantoni and D. R. Davies, eds), Harper and Row, New York and London. P. 409.
Lipmann, F. (1941). *Advan. Enzymol.* **1**, 99.
Lucas-Lenard, J. and Lipmann, F. (1966). *Proc. Natl Acad. Sci. U.S.* **55**, 1562.
Marcker, K. (1965). *J. Mol. Biol.* **14**, 63.
Marcker, K. and Sanger, F. (1964). *J. Mol. Biol.* **8**, 835.
Matthaei, J. H., Voight, H. P., Heller, G., Neth, R., Schöch, G., Kübler, H., Amelunxen, F., Sander, G. and Parmeggiani, A. (1966). *Cold Spring Harbor Symp. Quant. Biol.* **31**, 25.
Mehler, A. H. and Jesensky, C. (1966). *In* "Procedures in Nucleic Acid Research" G. L. Cantoni and D. R. Davies, eds), Harper and Row, New York and London. P. 420.
Moore, P. B. (1966). *J. Mol. Biol.* **18**, 8.
Muench, K. H. and Berg, P. (1966). *In* "Procedures in Nucleic Acid Research" (G. L. Cantoni and D. R. Davies, eds), Harper and Row, New York and London. P. 375.
Müller-Hill, B. (1966). *J. Mol. Biol.* **15**, 374.
Naono, S., Rouviere, J. and Gros, F. (1965). *Biochem. Biophys. Res. Commun.* **18**, 664.
Neidhardt, F. C. (1966). *Bacteriol. Rev.* **30**, 701.
Nirenberg, M. W. and Leder, P. (1964). *Science* **145**, 1399.
Nirenberg, M. W. and Matthaei, J. H. (1961). *Proc. Natl Acad. Sci. U.S.* **47**, 1588.
Nirenberg, M. W., Jones, O. W., Leder, P., Clark, B. F. C., Sly, W. S. and Pestka, S. (1963). *Cold Spring Harbor Symp. Quant. Biol.* **28**, 549.
Nirenberg, M. W., Caskey, T., Marshall, R., Brimacombe, R., Kellogg, D., Doctor, B., Hatfield, D., Levin, J., Rottman, F., Pestka, S., Wilcox, M. and Anderson, F. (1966). *Cold Spring Harbor Symp. Quant. Biol.* **31**, 11.
Nishizuka, Y. and Lipmann, F. (1966a). *Proc. Natl Acad. Sci. U.S.* **55**, 212.
Nishizuka, Y. and Lipmann, F. (1966b). *Arch. Biochem. Biophys.* **116**, 344.
Norris, A. T. and Berg, P. (1964). *Proc. Natl Acad. Sci. U.S.* **52**, 330.
Norton, S. J., Ravel, J. M., Lee, C. and Shive, W. (1963). *J. Biol. Chem.* **238**, 269.
Novelli, G. D. (1967). *Ann. Rev. Biochem.* **36**, 449.
Okamoto, T. and Takanami, M. (1963). *Biochim. Biophys. Acta* **68**, 325.

Roth, J. R. and Ames, B. N. (1966). *J. Mol. Biol.* **22**, 325.
Roth, J. R., Anton, D. N. and Hartman, P. E. (1966). *J. Mol. Biol.* **22**, 305.
Salas, M., Smith, M. A., Stanley, W. M. (Jr.), Wahba, A. J. and Ochoa, S. (1965). *J. Biol. Chem.* **240**, 3988.
Sarabhai, A. S., Stretton, A. O. W., Brenner, S. and Bolle, A. (1964). *Nature* **201**, 13.
Schweet, R. and Heintz, R. (1966). *Ann. Rev. Biochem.* **35**, 723.
Smith, J. D., Traut, R. R., Blackburn, H. M. and Monro, R. E. (1965). *J. Mol. Biol.* **13**, 617.
Söll, D., Cherayil, J., Jones, D. S., Faulkner, R. D., Hampel, A., Bock, R. M. and Khorana, H. G. (1966). *Cold Spring Harbor Symp. Quant. Biol.* **31**, 51.
Somerville, R. L. and Yanofsky, C. (1965). *J. Mol. Biol.* **11**, 747.
Spyrides, G. J. (1964). *Proc. Natl Acad. Sci. U.S.* **51**, 1220.
Stern, R. and Mehler, A. H. (1965). *Biochem. Z.* **342**, 400.
Stretton, A. O. W. and Brenner, S. (1965). *J. Mol. Biol.* **12**, 456.
Stulberg, M. P. (1967). *J. Biol. Chem.* **242**, 1060.
Sundararajan, T. A. and Thach, R. E. (1966). *J. Mol. Biol.* **19**, 74.
Takanami, M. and Zubay, G. (1964). *Proc. Natl Acad. Sci. U.S.* **51**, 834.
Thach, R. E., Cecere, M. A., Sundararajan, T. A. and Doty, P. (1965). *Proc. Natl Acad. Sci. U.S.* **54**, 1167.
Tissiéres, A., Watson, J. D., Schlessinger, D. and Hollingworth, B. R. (1959). *J. Mol. Biol.* **1**, 221.
Vogel, H. J. and Vogel, R. H. (1967). *Ann. Rev. Biochem.* **36**, 519.
Waller, J.-P. and Harris, J. I. (1961). *Proc. Natl Acad. Sci. U.S.* **47**, 18.
Watson, J. D. (1964). *Bull. Soc. Chim. Biol.* **46**, 1399.
Yanofsky, C. (1964). *In* "The Bacteria" (I. C. Gunsalus and R. Y. Stanier, eds), Academic Press, New York. Vol. v, p. 373.
Yanofsky, C., Drapeau, G. R., Guest, J. R. and Carlton, B. C. (1967). *Proc. Natl Acad. Sci. U.S.* **57**, 296.
Zachau, H. G., Acs, G. and Lipmann, F. (1958). *Proc. Natl Acad. Sci. U.S.* **44**, 885.
Zipser, D. (1963). *J. Mol. Biol.* **7**, 739.

AUTHOR INDEX

Numbers in italic indicate the page on which references are listed at the end of each chapter.

C

H

I

M

N

S

SUBJECT INDEX

D

E

F

G

H

I

J

K

Q

R

T

U

V

W

X